HOTEL DREAMS

Studies in Industry and Society
Philip B. Scranton, Series Editor

Published with the assistance
of the Hagley Museum and Library

HOTEL DREAMS

LUXURY, TECHNOLOGY, AND URBAN AMBITION IN AMERICA, 1829–1929

Molly W. Berger

Johns Hopkins University Press
Baltimore

This book has been brought to publication
with the generous assistance of the
Hagley Museum and Library.

Johns Hopkins Paperback edition, 2016

2 4 6 8 9 7 5 3 1

Johns Hopkins University Press
2715 North Charles Street
Baltimore, Maryland 21218-4363
www.press.jhu.edu

The Library of Congress has cataloged the hardcover edition of this book as follows:

Berger, Molly W., 1948–
Hotel dreams : luxury, technology, and urban ambition in America, 1829–1929 /
Molly W. Berger
p.   cm. — (Studies in industry and society)
Includes bibliographical references and index.
ISBN-13: 978-0-8018-9987-4 (hardcover : alk. paper)
ISBN-10: 0-8018-9987-7 (hardcover : alk. paper)
1. Hotels—United States—History—19th century. 2. Luxuries—
United States—History—19th century. I. Title.
TX909.B37 2011
917.306090'34—dc22   2010025988

A catalog record for this book is available from the British Library.

ISBN-13: 978-1-4214-1992-3
ISBN-10: 1-4214-1992-0

*Special discounts are available for bulk purchases of this book. For more information,
please contact Special Sales at 410-516-6936 or specialsales@press.jhu.edu.*

Johns Hopkins University Press uses environmentally friendly book materials,
including recycled text paper that is composed of at least 30 percent post-
consumer waste, whenever possible. All of our book papers are acid-free, and
our jackets and covers are printed on paper with recycled content.

CONTENTS

HOTEL DREAMS

INTRODUCTION

I N THE EARLY 1980s, after moving into a new house in suburban Cleveland, we used our last remaining dollars to take our four young children on a trip east. In Philadelphia, we stayed at a large chain hotel, which for us at the time was a bit of a step up. My younger daughter, all of ten years old, kept asking me if this was a "fancy" hotel. It was a perfectly respectable hotel, utilitarian and somewhat upscale but, in truth, not fancy. I did not know how to respond. Wanting neither to lie nor to dampen her sense of adventure, I answered ambiguously that this was a very nice hotel, that, indeed, we were lucky to be able to stay here.

On our last day, we took the children for a glimpse of New York City. We warned them of the unusual things we might see—this was definitely not the Midwest—and, sure enough, once we parked the car, an encounter with a mélange of six-foot-tall dancing vegetables validated every prediction. But the drive from Philadelphia had been long, and our most pressing concern was to find a bathroom. We trooped into the lobby of the Essex House, one of the tall sentries that guard the southern boundary of Central Park. My daughter, dwarfed by the enormous ceilings and overwhelmed by the luxuriously dark wood, blazing chandeliers, and towering flower arrangements, sighed deeply and declared knowingly, "Now *this* is a fancy hotel."

My daughter's reaction was one based on cultural knowledge. She could not quite define what a luxury hotel was, but even at the age of ten, she knew one when she saw one. The lobby's design and décor set out to convey a class-bound ethos of extravagance, beauty, and exclusiveness that expressed whom the hotel meant to attract and serve as its clientele. Yet there we were, *les touristes américains,* bedecked in the quintessential T-shirts, shorts, and sneakers, dragging kids and baby, trying to act nonchalantly, as though we belonged. This was an American palace. What did it take to belong? Our American birthright supported our presumption that we could claim equal access to any public space. This particular public space, however, asked for a certain level of economic achievement or, alternately, a convincing performance that—through dress or manners—would convey the affect of that achievement. Fortunately, we negotiated the performance successfully. This anecdote nonetheless highlights the dissonance between egalitarianism and elitism and between reality and pretense that fuels a nearly universal curiosity about "fancy" hotels and gives weight to their history.

Our family experience struck at the essential core of American identity. Even though buttressed by an ideology of equality, we would not have been surprised to be told we did not belong there, a message that would have been conveyed, ironically, by working-class security guards or desk clerks.

This book is about fancy hotels—specifically, large, urban, commercial luxury hotels. After working on this project, I can confidently say that *everyone* has a hotel story. The historical record confirms that these extravagant buildings have captivated Americans since they first appeared in American cities in the late 1820s. The role of the commercial luxury palace, which occupied a complicated position as a private enterprise with public access, has always been contested in this democratic country, and its history is rife with arguments both justifying and criticizing its development. The conflict between elite sovereignty and egalitarian entitlement that manifests itself in my family's anecdote is one that lies both at the heart of our nation's history and at the center of controversies that often dogged the development and popularity of urban luxury hotels. This underlying tension fed the public's fascination with luxury hotels, making these buildings an important urban venue for sorting out competing political, economic, social, and cultural ideas.

Hotel Dreams explores the development of the American luxury hotel, how it functioned culturally in American cities and how, as a key urban institution, it spread across the nation carrying social and cultural meanings, particularly about technology and progress. I came to this project after reading the chapter "Palaces of the Public" in Daniel Boorstin's *The National Experience,* the second of his sweeping three-volume history, *The Americans.*[1] As a student of the history of technology, I was intrigued by Boorstin's characterization of these buildings as showcases for emerging technologies, and my early research and writing followed along those lines. As I began my research, the elaborate descriptions of increasingly innovative technologies such as plumbing, heating, and lighting systems jumped off the pages of newspapers and magazines. A bank of indoor water closets generated as much excitement in the late 1820s as electric lighting and telephones did toward the end of the nineteenth century. As technology and industry transformed the nation during that century, hotel developers tuned in to the excitement of the age, incorporating every new invention into their buildings, often years or even decades before they became readily affordable to the general public. One primary hallmark of a luxury hotel became its technology, creating a definition of modern luxury that promised a level of consumer comfort and ease that only these new conveniences could provide. This is not to say that functionality and efficiency were enough to stir the public's interest. But the coupling of extravagant traditional luxury appointments and sensual aristocratic décor with the energy of American knowhow helped to create an institution that developers and civic leaders parlayed into a story of urban and national progress.

What I call "technological luxury" provides one of the concepts that drive this study. On the one hand, it emerges in linear fashion from straightforward ideas about progress. Over the course of one hundred years, the serial integration of successive technologies resulted in buildings of increasing complexity that measurably changed the comfort of guests for the better. During the nineteenth century, when once-separate ideas about progress and technology were beginning to merge into a unified concept that equated them as two sides of the same coin, urban leaders believed that a first-class hotel endowed with technological improvements said much about a city's ability to keep pace in national and world economies. Yet these innovations, by their very nature, contributed to ever-changing and escalating definitions of luxury and luxury's role in a democracy. For example, in 1829, travelers rejoiced over the aforementioned single communal bank of water closets. By 1929, expectations about taste, cleanliness, and privacy dictated nothing less than a fully equipped *en suite* bathroom for each room with fixtures of the newest and most fashionable design.[2] Technological luxury fed an insatiable nationwide competition to build the largest and most expensive hotel in the world and not only created measurable standards for competition among hotel entrepreneurs— it was far easier, after all, to tally toilets, lengths of pipe, and dollars than to adjudicate matters of taste—but also drew hotel projects into national debates about progress, the growing power of capitalists, modernity, urbanity, and issues of race, class, and domesticity.

This history of America's luxury hotels begins in the 1820s with the elite stewardship of large public projects draped in the cloak of democratic ideology. It ends one hundred years later with the development of projects built on Fordist principles that offered "first-class" accommodations at a price middle America could afford. These one-hundred-plus years coincide with what historians are beginning to call "the long nineteenth century," a conceptualization that privileges the chronology of historical processes over traditional political demarcations—such as the Age of Jackson or the Civil War and Reconstruction—or dated boundaries such as the numerical beginning and end of the century. The rise of the luxury hotel is embedded in two aligned national developments, the rise of American capitalism and industrialization. Both of these processes accelerated dramatically beginning in the second and third decades of the nineteenth century and catapulted through the next one hundred years toward the cataclysm brought on by the onset of the Great Depression in 1929. While this characterization takes a broad brush to complex histories, it nonetheless recognizes historical continuities that can be disrupted by pausing the story at the turn of the twentieth century. As an integral part of the evolution of both capitalism and industrialization, luxury hotel development followed the ebb and flow of economic cycles and drew heavily on technological innovations and industrial maturation. The concept of the long nineteenth century places the rise of the luxury hotel into its appropriate temporal period.[3]

The luxury hotel, emerging as it did in the late 1820s, was a product of nation building and, as such, served to authenticate and shape critical attitudes about American politics, economic development, and culture. These buildings occupied an extraordinary place in antebellum American cities. Not only did they physically dominate the urban landscape, the structures also set the standard for future urban development that would extend geographically across the nation and chronologically over the next hundred years. By the mid-nineteenth century, they served as models for new construction in many parts of the world. The luxury hotel was one of the most important institutions through which American urban elites forged a highly unified identity that was both local and specific to place and, at the same time, representative of national ideals. This identity drew on capitalist and liberal democratic principles and projected a dynamic vision of a modern, urban, and progressive society. Not merely or even primarily utilitarian, luxury hotels served as important symbolic structures. They reinforced and guided a particular vision of American society for the community, the nation, and the world that celebrated commerce, technological creativity, and the promise of wide-open access to a rich American life. As institutional stages for the exchange of information among those engaged in defining the nation's work, luxury hotels contributed to a burgeoning civil society that existed at the intersection of political and economic networks. They supported the growth of a new public sphere, different from the one Jürgen

Habermas identifies in the seventeenth and eighteenth centuries, which operated in deliberate disregard for status. Instead, American urban luxury hotels intentionally conferred a contrived and highly prized social standing upon all those who visited and lived within.[4]

Luxury hotels first appeared in the larger cities on America's eastern seaboard, not in Europe, as many people assume. They were, as one historian declared in 1930, an American invention.[5] The buildings were fantastically large and complex and something entirely new to travelers from both sides of the Atlantic. During the long nineteenth century, luxury hotels were novel, inspired, and highly visible expressions of a city's ranking in the world. As such, they were—as corporate skyscrapers would become in the early twentieth century—symbols of a city's vibrant commercial health. In the early part of the nineteenth century, however, luxury hotels suggested much more than a narrow commercial representation; they signified the triumph of a new political system, the growth of America as an industrialized nation, the development of a modern economic system, and the maturation of a refined American culture. As large and small cities alike became immersed in global markets, these hotels served as a recognizable way to trumpet local initiative as well as to assess development against national standards.

In the 1820s, these extraordinary buildings began to imprint themselves on the national consciousness. On June 18, 1827, an editorial in the *National Intelligencer* dubbed them "palaces of the public," a phrase that continues to intrigue historians, even as it did contemporaries, for the self-satisfied images of egalitarian access and entitlement to aristocratic pleasures that it inspired. Luxury hotels were the only American buildings that approached European palaces in size or décor. European architectural traditions drew primarily on the three symbols of power and wealth: churches, castles, and palaces.[6] The American luxury hotel of the 1820s and 1830s was very nearly an Americanized derivative of the three. It stood in secular reverence for the emerging economic power of capitalism, provided living arrangements and protected urban space for an upper-class elite, and adapted the symbol of royal political power to the democracy. Large modern urban luxury hotels served as displays of nationalist supremacy, and none too subtly, either, as foreign observers regularly noted.

This book also looks at the important ways that luxury hotels contributed to the long nineteenth century's process of urbanization. The hotel itself came to be widely known as a "city within a city," thus providing an opportunity to investigate what it meant to have a large technologically crafted structure represent the developing city outside its walls. Hotels provided a range of services that replicated the city.

These included accommodations, meals, and meeting space for commercial and political groups and individuals, especially in the decades before the telegraph and the telephone enabled remote and instant communication. Merchants occupied a high percentage of the hotel's prime public space, providing conveniences such as ticketing and postal services as well as access to retail shopping. In addition to transient guests, permanent residents—who included bachelors, single women, married couples, and couples with children—engaged rooms on a long-term basis. Hotels, too, were influential commercial customers, purchasing fuel, foodstuffs, furniture, and other supplies in enormous quantities and employing large kitchen, housekeeping, maintenance, and managerial staff. Indeed, portions of the household staff lived in basement or attic dormitories, more accurately representing the dynamics of the general society.

Given the mix of people, functions, and spatial specialization, the metaphor of "a city within a city" gathered considerable strength over the course of the century. Here were the travelers, busily meeting with friends or completing business transactions. Upstairs and straddling long corridors in a seemingly endless stream, private bedrooms and suites provided personal domestic shelter. The lobby often served as an exchange where local merchants met to transact business, including such activities as slave auctions in the antebellum South. Sex-segregated parlors enabled men and women to receive guests, socialize before or after an evening at the theater, or seek respite during arduous downtown shopping trips. If the hotel was indeed a city unto itself, it was one tightly controlled by a system of management and its architecture, both of which determined and regimented the movements and actions of the people within. The hotel, then, presented a class-defined, idealized version of life outside its walls. Yet, even the stoutest of walls could not segregate groups of people neatly. Looking at the hotel in its entirety—forgoing traditional divisions between the back of the house and front of the house— reveals a more realistic, pluralistic, and contested version of the city, raising interesting questions about the ways in which entrepreneurs and urban elites believed they could construct an idealized urban environment.

Few historians consider buildings as actors in the historical process. Hotels are examples of what a leading architectural-cultural historian calls the "urban cultural landscape," a conceptualization that combines the material world with the cultural universe and imaginative visions that created it.[7] The historical significance of hotels rests not simply in their material culture but in the way in which that materiality—the massiveness, the mechanics, the artifacts that filled the buildings—was so thoroughly integrated with and inseparable from the social and cultural world that gave it life. The meaning that we might infer from either the material or the social world is quite diminished without considering the other. For example, it would be impossible to comprehend the building's design without an understanding of shifting gender, class, and race relations. The social and cultural context—which changed considerably over the course of the long nineteenth century—not only gives meaning to the design but also highlights the many contradictions and conflicts that resulted from combining ideologically oppositional domestic and commercial functions.

It is at the intersection of the commercial and the domestic that much of the critical commentary about hotels emerged in the nineteenth century. As a historian looking back on the sources, I was able to discern fairly unambiguous attitudes about these buildings and their mostly elite clientele in the 1820s and 1830s. As the century advanced, the buildings themselves became larger, requiring a less restrictive customer base to stay profitable, and they did so at the same time that the middle class coalesced into a recognizable entity. Many historians have ably demonstrated that much of this new middle-class identity was forged through the acquisition of goods and the presentation of self. Luxury hotels were nothing if not exaggerated representations of the aggrandized glory of capitalist production, and so reactions bounded back and forth between celebrations of American capitalist potential on the one hand and scathing denunciations of consumption and entitlement on the other. No social class was immune from either commentary. The proliferation of an active mass-market press brought these controversies to a wide public and fueled debate.

Not surprisingly, ideas about class and entitlement tie into the long nineteenth century's growing consumer culture. Looking at consumer culture through the lens of the luxury hotel adds several new perspectives to our understanding of it. First, it demonstrates that Americans drew on the practices of eighteenth-century European consumers and participated in mass consumption experiences as early as the 1820s, rather than later in the nineteenth century, as studies grounded in the department store have shown. Indeed, luxury hotels set a precedent for department store décor and amenities. The luxurious appointments, ladies' parlors, and eclectic designs based on fantasy and exotic worlds were fully developed in luxury hotels before their appropriation by post–Civil War department stores. Second, consumers at luxury hotels purchased an experience rather than a product that could be taken home at the end of the day. One aspect of the hotel critique railed against the notion of tenancy as opposed to ownership, of "renting" a luxury environment and calling it one's own without actually having the means to acquire it.

While studying hotels contributes to the literature that focuses on experience, such as that which looks at theater, carnivals, museums, and other forms of participatory entertainment, hotel life adds a further dimension to performance whereby guests inserted themselves into a theatrical set and played at roles that may or may not have been consonant with their "real" lives.

Finally, as the counterpoint to mass consumption, our understanding of mass production is expanded if we include large luxury hotels in the mix of manufactories. Considering the production and sale of intangibles such as restaurant meals, hot showers, and a good night's sleep in mass quantities enlarges the parameters of mass production to include service and experience. By the twentieth century's second decade, hotels with rooms numbering one thousand and up became obligatory in America's largest cities. Businesses such as these operated on principles of economy of scale, efficiency, and mass production in order to remain profitable and competitive. By this time, men such as E. M. Statler and E. J. Stevens had perfected a system by which they could provide thousands of clean functional hotel rooms and many thousands of meals to their customers each day. No longer subjected to debates about entitlement, these men built lavishly decorated and efficient hotels that were unambiguously directed at a traveling middle-class market composed of salesmen, tourists, and conventioneers.

This history proceeds chronologically along two complementary paths. I tell the story of hotel development through case studies of individual hotels that move forward through time and geographically across the United States. Focusing on a building's "biography" enables us not only to "know" the building descriptively but, more importantly, to delve into the personalities, conflicts, and context that help us understand the culture that produced it. As one anthropologist proposes, biographies of "things" reveal cultural meanings, in much the same way that biographies of people do. We ask, more or less, the same questions: about status, where the thing comes from, who made it, what might be an ideal career for a particular thing, what happens as it ages, how its use changes over time, and what happens when it reaches the end of its usefulness. The answers to these questions "make salient what might otherwise remain obscure."[8] In the case of the luxury hotel, the building's biography tells several parallel stories. One is of the political and economic world that generated the project. Another follows the hotel's social organization and its instantiation of and influence on changing American social practices. And yet a third story is of technological development and the way in which technology becomes endowed with specific cultural meanings.

As with so many of our civic institutions, this one gets its start in Boston.

Because an important part of the story is the national diffusion of a unitary ideal expressed in local language, the story continues by looking at hotels in Philadelphia, San Francisco, New York, and Chicago. Each of these individual hotels aspired to be the largest in the world. The public debate sparked by their design, construction, and openings exposed expectations, ideologies, and conflicts in ways that changed over time. The case studies alternate with interpretive essays about hotel society and culture that focus on class, gender, race, material culture, technological progress, and capitalist development. Together, these two forms of analysis connect ideas about material culture to broader issues in American history.

This book focuses on urban commercial luxury hotels. There are many other kinds of hotels. These include resort, railroad, humbler "one-hundred room" hotels common to small cities and towns, residential, and single-room occupancy hotels, with many variations among them, including motels, boarding houses, and inns.[9] I choose to focus on the urban commercial luxury hotel primarily because I believe that all other hotel forms derive from it. Understanding the development of the luxury hotel serves as a platform for delving into other hotel forms, and other historians have taken that route. Most books that historicize other kinds of hotels begin—as they should—with discussions of the early luxury hotels.[10]

In his 1907 book, *The American Scene,* Henry James reflected on his visit to the Waldorf-Astoria, the scale and exuberance of which mesmerized him. James's commentary reads both as cynical and sincere. He saw the Waldorf-Astoria as exuding the "essence of the loud New York story" in a way that made the hotel, he asserted, a synonym for American civilization, reflected in both the building itself and the people within. "One is verily tempted to ask," he wrote, "if the hotel-spirit may not just *be* the American spirit most seeking and most finding itself." James reveled in the great "promiscuity" of hotel life. Whereas in Europe, a building such as the Waldorf-Astoria would be only for "certain" people, in the United States the great breadth of folks teeming within its walls rendered the place "a new thing under the sun." The hotel served as a backdrop to the American drama. American society had found in this "prodigious public setting, so exactly what it wanted," a commercial, sumptuous, public building that admitted anyone who could pay the "tariff." James compared the hotel—as others had before him—to a temple built to both an idea and an ideal. "Here," he stated, "was a world whose relation to its form and medium was practically imperturbable." Reading this passage, one has the sense of being gently mocked, yet James concluded by harkening once more to the moment's perfection and—rather than chance it being spoiled by "some false note"—never returned to the Waldorf-Astoria again.[11]

This book seeks to make explicit the connection between the form and society of the American luxury hotel, a relationship teeming with meaning and history. The luxury hotel is rich ground for historians trying to understand American life and culture because it is an institution that is similar to many others, such as homes, restaurants, meeting facilities, business exchanges, social clubs, and even factories. Yet, as James observed, it is unlike any other institution because it is all these things at once. As a key institution embedded in the growth of capitalism and the process of industrialization, the urban luxury hotel is a significant venue for interrogating the way technology functions culturally in American society. This integration of social, cultural, economic, and technological themes expands our understanding of urban life in the long nineteenth century.

CHAPTER ONE

....

THE EMERGENCE
OF THE AMERICAN FIRST-CLASS HOTEL,
1820S

....

"All at Hand, and All of the Best"

O N APRIL 15, 1848, Philip Hone, diarist and one of New York City's "best" citizens, gave voice to what he often called the "go-ahead" age. Describing what he considered "the magical performance of the lightning post," Hone revealed his enchantment with the heretofore unimaginable changes wrought by technology, in this case recounting the speed at which news of the markets had traveled fourteen hundred miles by telegraph from New York City to Minneapolis in a matter of hours. Hone underscored his sense of wonder by noting, "I was once nine days on my voyage from New York to Albany."[1] Living during a time of incredible and often incomprehensible change, Hone marveled at modern technology's ability to annihilate traditional parameters of time and space. Indeed, his diary is replete with careful notations on the speed at which people, goods, and information traversed both the North American continent and the Atlantic Ocean.

Hone's adult life corresponded with what one historian termed the "birth of the modern," the fifteen years between 1815 and 1830 when "the matrix of the modern world was largely formed."[2] Not coincidentally, the modern luxury hotel emerged during this same period as a significant urban institution. These grand new hotel buildings filled the need for accommodations for the increasing number of travelers made possible by developments in communication and transportation systems and the expansion of an international capitalist economic system. As a product of this development and a center of human activity, modern luxury hotels came to dominate the urban landscape, taking their place alongside the government buildings that served as signposts of a maturing nation. It was a place where the world of strangers and the world of goods intersected, where ideas about commerce, national ideology, and culture expressed themselves in buildings designed

Barroom dancing. John Lewis Krimmel, c. 1820. This watercolor depicts a dance scene inside an American tavern barroom and illustrates the "promiscuous" mixing of classes and people that Isaac Weld so deplored. A slippered African-American fiddler provides the music, while gentlemen observe the dancing couple. On the fringes of the room, others engage in amorous adventures. A bountiful supply of liquor fuels the activities. *Courtesy, Library of Congress.*

concerts, meetings, and political events. These might offer liveried servants and well-decorated private bedchambers. Elite taverns also provided reliable access to foreign newspapers. This appealed to upper-class travelers, fortifying their knowledge of foreign and domestic issues and heightening their ability to engage in political and economic discussions.[7]

Toward the end of the eighteenth century, the word *hotel* came into use in America to describe the elite group of taverns and inns that catered to upper-class patrons.[8] The word was an adaptation from the French, for whom it described a nobleman's city residence, the word *palais* being reserved for houses of kings, princes, and princesses, and *maison* referring to homes of the bourgeoisie.[9] After the French Revolution, Americans often used the word *hotel* interchangeably with *tavern* and *inn* when referring to urban establishments, but proprietors employed the word *hotel* to indicate that their buildings offered more refined accommodations.[10] Still, hotelkeepers used the word somewhat arbitrarily. Establishments might be listed in the city directory one year as a tavern and the next as a hotel. Historians often credit New York's 1794 City Hotel, located at 123 Broadway, as being the first establishment to have been purposefully built as a hotel.[11] Hone refers to the City Hotel regularly in his diary with fondness, at least until its upper stories burned in 1833. Nonetheless, it first appeared in the city directory as the City Tavern, illustrating the plasticity of the two terms at the time.[12] An 1800 trade directory that compiled "all the occupations and trades ... practiced in the city" listed 158 tavern keepers and no hotelkeepers. The directory entry for the City Hotel was the "Tontine City Tavern," an equally common name for the hotel that referred to its method of financing.[13] At the same time, a quick look at New York City newspapers reveals numerous references to events at various "hotels."[14]

The City Hotel marked an evolutionary step toward the development of the modern hotel. Contemporaries considered the building immense, with its five stories and 137 rooms; however, the architecture and furnishings assumed a plain and unpretentious air, in keeping with the less ornate early American style favored in the early years of the new republic. The building's exterior looked like an oversized square house. By 1811, the hotel had expanded to include a public barroom and coffee room with windows looking out on the street, which allowed the hotel to serve as a theater of social class for hotel patrons and passersby alike. The hotel's ballroom functioned as a society venue for dancing schools, balls, musical performances, and banquets. The manager cleverly advertised that the hotel's location in the fashionable district offered "irresistible attractions to gentlemen of taste."[15] Even with the initial ambiguity associated with its name, the City Hotel set itself apart from taverns and inns in several important ways. Its specially built structure distinguished it from taverns and smaller converted houses and created a category of buildings that could be more clearly defined as hotels. Its financing, in the form of a tontine association, whereby the last surviving investor inherited the property, was an early corporate structure that lifted the hotel into the realm of capitalist enterprise.[16] The manager's conscious efforts to court a fashionable clientele indicated a demand for exclusive accommodations that provided—in contrast to the city's numerous inns and taverns—a superior level of service and comfort. While still only an intimation of what was to come, New York's City Hotel established a precedent for urban hotel development.

In addition to the City Hotel, there were five other large hotel projects conceived

City Hotel (New York). The City Hotel was also known variously as the City Tavern and the Tontine City Tavern. This engraving shows both Trinity and Grace churches near the hotel's prime location on Broadway. People in the heavily draped windows peer out at the street below. *Warshaw Collection of Business Americana—Hotels, Archives Center, National Museum of American History, Smithsonian Institution.*

between 1793 and 1808, although only two were actually constructed, of which only one was used as a hotel. In Washington, D.C., a lottery supported the building of the Union Public Hotel, whose owner's financial blunders resulted in its being purchased by the new federal government for use by the postal service. Weld mentioned the Union Public Hotel building in his diary but observed that in 1796 the structure was roofed in but remained unfinished. It was "anything but beautiful," he wrote dismissively.[17] Additional projects planned for Boston, Washington, D.C., and Newport, Rhode Island, also failed to move beyond bids for financing, despite the sponsorship of their up-and-coming architects, Charles Bulfinch in the case of Boston and Benjamin Henry Latrobe in Washington, D.C. The fifth project, the 1808 Boston Exchange Coffee House and Hotel, was an enormous multipurpose building that ultimately burned because its great height exceeded the reach of firefighting abilities.[18] These schemes were premature in their anticipation of the nation's burgeoning development. American cities had not yet grown to the point that they could support the scale of enterprise promoted by the hotels' developers. Nonetheless, entrepreneurs and architects clearly envisioned these buildings as grand enterprises meant both to overpower existing urban buildings and to accommodate the needs of an emerging capitalist economy that depended, in part, on the mobility of its participants.

By the mid-1820s, a change seems to have occurred, particularly in the commercial and political centers on the nation's eastern seaboard. Improvements in transportation included the development of steamboats, railroads, canals, engineered roads, and packet ships to and from Europe, enabling more and more people to travel. The Boston newspaper *The American Traveller* commenced publication in 1825 and observed in its inaugural issue that an "astonishing revolution has been wrought in navigation and is extended to all modes of travelling."[19] In 1827, the *National Intelligencer* declared that a new kind of hotel had emerged to meet the needs of all these travelers. As examples, the editorial described the recent renovation and expansion of two Washington hotels, the Mansion Hotel and the Indian Queen, and the opening of the newly constructed National Hotel, which transformed six row houses into one large complex. These business ventures opened to accommodate the seasonal requirements of politicians attending legislative sessions. "The splendor of modern hotels," the editorial concluded, "has obtained for them the appellation of 'the palaces of the public' and really the elegance of some of them here, and elsewhere, almost justifies the phrase." The writer juxtaposed these hotels with "genteel and excellent boarding houses" as buildings of superlative size, "elegance, good keeping and comfort."[20]

The same *National Intelligencer* article reprinted a letter from the *Baltimore Gazette* comparing the new National favorably to Baltimore's City Hotel, the country's most renowned hostelry at the time. In the early nineteenth century, Baltimore was one of the nation's leading shipping and manufacturing centers, and Barnum's, located in the heart of the city near Monument Square, served its commercial and traveling public. The hotel created a stir when it opened in September 1826, owing in large part to the experience and expertise of owner and manager David Barnum, who had already acquired an excellent reputation as a hotelkeeper based on his previous work at Boston's Exchange Coffee House and at Baltimore's Indian Queen. One local newspaper called the City Hotel "a noble experiment," a characterization that referred not only to the novelty of the building's great size but also to the degree of financial risk Barnum assumed by opening such a building. Moreover, the reference to it being a "noble" experiment implied that Barnum's undertaking encompassed some civic consciousness, that Barnum conceived the project as contributing to the public good.

The hotel's four stories stood above a ground level that housed service businesses

Calvert Street looking north from Baltimore Street. The cover of the sheet music for "The Very Last Polka" features Barnum's City Hotel toward the end of the street on the left, at the corner of Calvert and Fayette, near Battle Monument. A crowd of partyers leaves the hotel at dawn after a ball, their carriages taking them home after a night of merriment. *Cator Print 184, Courtesy, Enoch Pratt Free Library, Maryland's State Library Resource Center, Baltimore, Maryland.*

such as bathing rooms, a barber shop, a post office, and law and ticket offices. The interior contained 172 "ample and splendid" bedrooms, private family suites, an enormous 86-by-30-foot dining room, plus another room of equal size to accommodate "public dinner parties" and balls. Barnum's also provided a reading room furnished with the nation's leading newspapers, in itself a clear indication of the proprietor's sense that the hotel and its guests were at the center of the young republic's commercial and political development.[21] Gaslight chandeliers illuminated all the public rooms. Baltimore's Gas-light Company—the nation's first company organized to manufacture gas—supplied the needed gas, making the hotel one of the first public buildings to incorporate industrial lighting. Comments from travelers concerning food, accommodations, and service were universally adulatory. All complimented "King David" on his civility and good nature; some mentioned the comfortable bedroom arrangements, the plentiful attentive servants, the hearty table, and the popularity of the place.[22]

Following the custom of the times, the barroom doubled as the business office and offered a place for both men of the city and male travelers to meet, socialize, and conduct business. A private, limited stock offering financed the $130,000 required for the construction and renovations, but the hotel remained privately held rather than incorporated under a state charter.[23] The hotel's central location made it convenient for travelers and those conducting business in the city. Its accommodations enabled large numbers of people to congregate. For example, the hotel hosted the first national nominating convention of a political party in December 1831, when the National Republicans nominated Henry Clay for president.[24] The concentration of services on the ground floor was the first step toward the goal to satisfy every traveler's anticipated needs, eventually leading to the hotel's designation as a "city within a city."

From the beginning, Barnum's placed itself in a class of its own. By the time the *National Intelligencer* anointed this class of hotels as palaces, standards set by Barnum's determined entry into this elite classification. A letter from "A New Yorker" reprinted in the *Baltimore Gazette and Daily Advertiser* praised Barnum's as the pinnacle of refinement and good taste and expressed the "wish of one who hopes soon to see New York furnished with a public house of equal if not superior splendour."[25] Hotels like Barnum's and Washington's National offered a loosely defined level of superior service that had not yet replicated itself in the nation's largest cities. Barnum's won a considerable amount of national attention, yet its place in history remains obscured by the more widespread and enduring attention lavished on Boston's Tremont House, which followed three years later. Nonetheless, Baltimore's 1820s hotels serve as an instructive point of departure for understanding the difference between a "first-class" hotel and the modern luxury hotels that soon followed.

A First-Class Hotel

Because of Barnum's and other competing hotels, Baltimore is a good place to begin to understand what characteristics led contemporaries to define a hotel as "first-class," the phrase commonly used in the early nineteenth century to describe a luxury establishment. In November 1826, an advertisement for the recently re-modeled Indian Queen Hotel and Baltimore House appeared in the *Baltimore Gazette and Daily Advertiser*. Despite the building's provenance as an eighteenth-century mansion house, the Indian Queen competed directly with Barnum's for business.[26] The hotel's manager explicitly pointed out that he was delivering what travelers expected from a first-class hotel in the late 1820s. Indeed, Mr. G. Beltz-hoover succeeded Barnum as manager of the Indian Queen. Thus, Beltzhoover's advertisement sought the same elite clientele and could only attract them by com-municating that his hotel offered the same level of service as the "new establish-ment." Beltzhoover changed the Indian Queen's name to "Baltimore House," hop-ing to upgrade his business's image as one of the finer hotels.

It was common practice for newspapers to serve as boosters for their advertis-ers, so it is not surprising that an editorial appeared in the *Baltimore Gazette and Daily Advertiser* on the same day as the advertisement in question. Yet this editorial and ad appeared just five days after a similarly detailed ad placed by Barnum in the same paper.[27] Juxtaposed in this way, the advertisements suggest a small drama in the relationship between the newspaper and its advertisers as well as between the hotelmen themselves. The editorial recommended Mr. Beltzhoover for his "spir-ited and very satisfactory manner" and praised the hotel's recent remodeling and redecorating, ensuring "to the Guests everything of comfort, convenience and el-egance which can reasonably be desired in a Hotel." Referring to Barnum's and the widespread attention it had been receiving as the "subject of deserved panegyric," the *Gazette* reminded the public that Baltimore's older hotels deserved as much support and commendation. In addition, the *Gazette* reassured its readers that Mr. Beltzhoover's claims were "not exaggerated."[28]

Beltzhoover's hotel fronted one hundred feet on Market Street and 223 feet on Hanover Street. For hotels—like many things in America—bigger meant better. First-class hotels established themselves as such first of all by great size. Beltzhoover described his public rooms as being "splendidly fitted up." The bedchambers had been furnished with "new, neat and suitable furniture." The Baltimore House of-fered "elegantly furnished" family suites with a separate, private entrance on Mar-ket Street. Interior details, expensive materials, and elegant fabrics created a setting that represented a growing American cultivation of refinement in terms of wealth and taste that set the hotel's patrons apart from the coarse and vulgar lower orders of the democracy.[29] First-class hotels, then, required beautiful interiors that, in ad-dition to offering comfort, substantiated one's station in life. The separate entrance protected women and children from having to make their way through crowds of men lingering at the main entrance or walk through the boisterous barroom in order to reach the public dining room or parlors or their own suites. Hotels with family suites welcomed and accommodated the civilizing influence of women.

Beltzhoover next emphasized that his house satisfied a traveler's basic needs: a "clean house and chambers, attentive servants, and all that is embraced in a good Table." A first-class hotel professed to guarantee cleanliness in both its public and private rooms. Travelers commented on the relative cleanliness of hotels. For example, in 1833, James Stuart complained about Barnum's, saying that, "The arrangements of the house, considering its great size, appear to me very good; but the frequent use of tobacco renders it impossible to keep a house of public entertainment so clean as it should be."[30] Responsive, faultless personal attention marked the service of a first-class hotel, as did plentiful food and wine of superior quality. Room rates typically included the cost of four daily meals: breakfast, dinner, tea, and supper. The potential for disastrous dining was everpresent, for, as one traveler commented, "If the devil sends cooks to any part of the world, it is to the United States."[31] Perhaps the *National Intelligencer* summarized the idea of first-class best when describing Washington D.C.'s National: "The style of accommodation cannot be excelled—everything from the table to the bar, to the minutest article which luxury or convenience may require, are all at hand, and all of the best."[32]

First-class hotels were situated in a central, convenient location. The Baltimore House housed the stage office to the Western, Southern, York, and Philadelphia stagecoach lines, adding to the convenience of travelers. Next door, trustworthy hostlers provided stabling facilities and carriages. The hotel was near the business district, wharves, and steamboat landings. Beltzhoover enthused that his hotel sat in the midst of "the enlivening, gay crowds which throng and pass in review the great thoroughfare of the city." A first-class hotel became part of the energy and pageantry of the best and most active part of the city, while at the same time providing security from discomfiting urban dangers and congestion. Beltzhoover promised a "trusty watchman" to guard the house interior at night from potential intruders and as, if not more, importantly, the danger of fire. At all hours, the barkeep stood available to assist travelers to and from the steamboats. Lastly, Beltzhoover pledged that his proficient hotelkeeping would "keep the establishment equal to any in the Union in all respects." Attentive, expert, and experienced management insured a first-class house.

The first-class hotel of the mid-1820s distinguished itself, then, by its prime location, impressive size, cleanliness, attentive service, private bedchambers and family suites, excellent and plentiful food, elaborate tasteful decoration, abundance of beautiful public rooms, city services such as ticket offices, barber shops, bathing rooms, and, above all, skillful administration on the part of an experienced reputable hotelkeeper. All of these needed to be generously provided, comprehensive, and "unsurpassed in excellence."[33] Additionally, hotels encouraged urban sociability by providing public parlors for personal entertainment and by hosting large commemorative events at which city officials welcomed renowned travelers such as Lafayette or venerated public servants such as Secretary of the Treasury Richard Rush.[34] The guests themselves often lent their cachet to the hotel. Authors of widely read contemporary North American travel literature generally patronized the same hotels. They not only regaled readers about hotel amenities (and shortfalls) but also more or less established an itinerary and travel guide for an American grand tour. Moreover, a prominent visitor could always be tracked at the city's best hotel. Thus, the patronage of wealthy and distinguished customers insured the continuing business of similarly classed people.

The Context of Change

The features that defined a first-class hotel were both qualitative and highly subjective. In travel accounts, good service often translated into abundant ice water or the ability to dine in one's room at "off" hours without provoking resentment on the part of the hotelkeeper. Fine fabrics and comfortable surroundings were ephemeral kinds of evaluative measures that fell prey to the capricious nature of changing fashions and the reality of wear and tear. Definitions of comfort differed from traveler to traveler according to individual and cultural expectations, especially for foreign travelers. As western nations entered fully into the modern age, hotels took on the characteristics of the changing world around them. Like the rationality that typified modern development, boosters began to enumerate the qualities of a modern hotel objectively so that dollars invested, the number of baths and water closets, and feet of frontage became the measures of luxury. The modern urban hotel became a site through which ideas characteristic of an evolving modernity—luxury, egalitarianism, progress, technological and economic development, consumerism, management, and efficiency—unfolded, allowing the hotel to represent many significant and varied aspects of modern American society and culture.

Citizens of the late eighteenth and nineteenth centuries were acutely aware of living in momentous times. With the organization of life changing around them, and laden with memories and knowledge of a "premodern" life, observers found themselves in two worlds, a condition both disconcerting and exhilarating. Modernity, as Baudelaire suggested, embodied this conflict between a traditional immutable world and the unceasing destructiveness of modern development, the tension between continuity and change.[35] In 1829, the *American Traveller* reprinted an unattributed excerpt from Thomas Carlyle's seminal *Edinburgh Review* essay of that same year, "Signs of the Times." Using as its headline Carlyle's famous phrase, "The Mechanical Age," the paper chose to highlight Carlyle's description of the ways in which new processes and machines had replaced and denigrated traditional ways of life. "On every hand," Carlyle wrote, "the living artisan is driven from his workshop, to make room for a speedier, inanimate one. . . . Even the horse is stript of his harness, and finds a fleet fire-horse yoked in his stead. Nay, we have an artist that hatches chickens by steam—the very brood-hen is to be superseded!"[36]

Everything in the world was succumbing to a technocratic idea of progress and invention. Mountains had been moved to make way for the railroads; steamships had made the ocean a "smooth highway." Even the chicken was made to conform to a mechanical way of life. Carlyle brayed, "We war with rude nature; and, by our resistless engines, come off always victorious, and loaded with spoils." This progressive vision of the mechanical age embraced ideas about the scientific and technological domination of nature that promised with inevitable certainty a new freedom from the traditional scourges of hunger, disease, and natural calamity.[37] "How much better fed, clothed, lodged, and in all outward respects, accommodated, men now are, or might be," Carlyle observed.

Yet Carlyle also acknowledged the potential cost of modern progress in the growing distance between rich and poor, the loss of personal autonomy and its effect on "modes of thought and feeling." He perceived "a mighty change in our whole manner of existence." Not limited to methods of production or the compression of time and space that was the hallmark of new modes of transportation and communication, the new mechanical age manifested itself as well in "politics, arts, religion, morals." Men had lost confidence in their individual strengths and in "natural force of any kind." The danger lay in their growing "mechanical in head and in heart, as well as in hand." These alternating doubts and enthusiasms, the new faith in linear progress, the drama of technological development, the alteration of "old relations," and the fear and excitement of a new age brought a heightened sense of drama to urban development and life. While modern life promised adventure, power, growth, and transformation, it also carried the threat of destroying an entire social structure, knowledge base, and material world.[38]

Modern life embraced the idea of progress and offered an optimistic worldview that believed that new conditions of production, transportation, communication, and consumption would result in the ultimate perfectibility of civilization. Classical or traditional theories of history had asserted that civilizations lived through life cycles analogous to that of living beings, characterized by periods of formation, youth, mature vigor, and then, ultimately and inevitably, decay and ruin brought on by the infections of pride and luxury. Americans in the new republic seriously undertook to prolong the youthful stage of their new nation by actively living a life consistent with republican values that promoted the public good over self-interest. Republicans, though, were of differing minds. While some enjoyed the creature comforts of life, many others warned against luxurious living and conspicuous consumption because they feared that, as in the case of ancient Rome, slavishness to luxury would bring an end to the success of their political experiment. For example, in 1832 Lydia Maria Child admonished young American housewives against the lure of traveling and public amusements in her housekeeping manual *The American Frugal Housewife*. "A luxurious and idle *republic*! Look at the phrase!— The words were never made to be married together; everybody sees it would be death to one of them," she warned.[39]

When Child wrote these words, luxury hotels had already established themselves as important symbolic structures in American cities. Clearly, contested views of the world existed side by side. Even while classical republican theory feared a cyclical demise wrought by decadence and vice, modern republicans heeded a contradictory belief in progress that gave as much weight to material prosperity as to intellectual, moral, and spiritual progress.[40] The American Revolution served as evidence of progress through its rejection of Europe's ancient model and its recognition and protection of a new political, commercial, and social order.[41] These conflicting ideas represented the growing tension between republican values and those associated with a commercialized economy and market society. By the 1820s, the belief in progress had begun to assume for many a technocratic vision that emphasized liberal individualism, commercial prosperity, and utilitarianism.

This culture of change also included the rehabilitation of the very concept of luxury, transforming it from a potentially negative vice to a positive force in American and western culture. In the early republic, rejection of luxury demonstrated a profound moral commitment to the public good rather than simply the choice of an ennobling, austere, private lifestyle. The change in luxury's status was tied to the growth of commerce. The expanding commercial network, with its manufacturing and trading centers and new inventions and production methods, represented a dramatic shift away from land-based power toward the cities and their commercial activity.[42] Influenced by theorists such as Adam Smith and philosopher David Hume, advocates of commerce argued that the production of luxury goods was linked to the working class's rising material expectations and would fuel the expansion of national wealth.[43] The War of 1812 energized the nation's foreign policy and resulted in a noticeable shift whereby classical republican theory gave way to politics dominated by market capitalism, validating, among other things, the centrality of commerce and the indispensable role of luxury in generating a civilized society and expanding economy.[44] The modern luxury hotel appeared in the late 1820s as the perfect manifestation of these new and evolving ideas about commerce, luxury, and progress and combined them with a characteristically American enthusiasm for technological innovation. Together, these cultural shifts helped produce a new concept of technological luxury that the hotel adopted to inaugurate a century-long process of redefining "all at hand, and all of the best."

The Modern Luxury Hotel

At the end of the 1820s, then, out of this crucible of change, a new genre of modern luxury hotels took shape. While incorporating the characteristics of earlier first-class hotels, newer hotels established modern standards for management, design, and the integration of technology that fully represented the economic, social, and cultural climate from which they had emerged. These hotels continued to address the needs of both travelers and city residents in ways that privileged comfort and convenience. Owners developed mechanisms that not only enabled them to expand and succeed but also signaled the hotel's immersion in the worlds of commerce and consumerism. Innovations that set these new modern hotels apart from their predecessors included the adoption of commercial architecture, the movement toward financing projects through modern corporate structures, the separation of management from ownership, and the integration of technological systems into the building design. This last development not only fueled an ongoing competition that effectively limited the useful lifespan of buildings but also established technology as fundamental to definitions of luxury and comfort.

Lewis Mumford, in a series of 1952 lectures, reflected on the symbolic role of architecture and its unique ability to combine symbolism and function, art and technics. In discussing the indivisibility of art and practicality, Mumford noted, "Even in the plainest esthetic choices of materials, or of proportions, the builder reveals what manner of man he is and what sort of community he is serving."[45] The significant shift in the luxury hotel's architectural design demonstrated the building's growing importance in the American city of the nineteenth century. Baltimore's City Hotel, like New York's City Hotel thirty years before, retained the look of a mansion house, albeit an enormously large one. Indeed, one of the more popular names for large city hotels was "Mansion House." In Stuart's recollection of his visit to Barnum's hotel, he described the evening meal with "Mr. Barnum, the landlord, a very portly figure, sitting at the top of his table, and doing the honours in the same manner that a private gentleman would do in his own house in Britain."[46] Barnum conducted his house and himself almost as though he were entertaining privately. In a similar setting, Frances Trollope described a hotel in Memphis, sometime between 1828 and 1832, at which she preferred to dine alone in her room but was informed that "the lady of the house would consider the proposal as a personal affront."[47] To Mumford's point, the architecture of the house spoke to the way the proprietors conducted their hotels, supervising their guests personally in the genteel setting of a mansion house. As in private mansions, a brick exterior symbolized wealth and substance; its size and tasteful symmetry communicated the refined material possessions and lifestyle inside.

The new hotels were different. These adopted Greek Revival architecture, a style that, in the early republic, represented a political economy that married concepts of democracy and commerce and cloaked America's public buildings—its banks, markets, and custom houses—in an aura of sturdy traditionalism. Modern hotels from 1829 onward appropriated commercial architecture and did so on a scale that dominated the urban landscape. Promoters of important city hotels employed professional architects whose names were identified with significant buildings and modern building techniques. In addition to providing written plans for their clients, these architects contracted the work to local craftsmen or specialists from other cities in order to obtain the very best materials and craftsmanship. Working as agents of corporations, architects both engaged local artisans to showcase local production and sought outside sources to validate and promote their sophisticated taste and extensive knowledge. In either case, the architect supervised the construction of what became increasingly larger, more complex, and more technical buildings.

The increasing size and cost of the new hotels dictated that they be financed through a modern corporation. With its 172 bedrooms, Barnum's was a very large hotel, possibly the largest on the eastern seaboard.[48] Even at a similar size, the newer hotels used more expensive building materials (granite rather than brick), more elaborate decorations (stained glass rotunda ceilings), and incorporated more sophisticated technology into the building (indoor plumbing, water closets, remote signaling devices). The cost of the buildings soared. Unlike joint-stock ventures or private partnerships, a corporation benefited from limited liability, thus protecting investors from larger losses. Capping the amount of risk encouraged investment and facilitated the mobilization of capital.[49] This was especially important in the case of an uncertain and unproven project such as a major hotel that early nineteenth-century projectors regarded first as a public improvement and second as a profit-generating business.

The separation of management from ownership logically followed the turn to incorporation. Larger city taverns, contrary to folkloric images, were often leased to tavern keepers, who then conducted the establishment on behalf of the building's owner.[50] This arrangement carried forward so that, in the late 1820s, when hotels came to be owned by corporations, investors continued and strengthened this existing management structure. Modern hotel companies leased their buildings to a hotel manager who was then responsible for outfitting the hotel with furniture, draperies, dishes, linens, and so forth. The manager owned the hotel furnishings. Thus, the manager assumed a considerable financial stake in the hotel's success, but one that he could sell or take with him if he chose to move on. Hotelkeepers, as in other professions, trained apprentices informally at first and often developed family hotelkeeping dynasties. As hotels became larger and more complicated structures, specialized management positions soon resulted in a formalized industry supported by regional and national organizations, journals, and eventually university-based training programs. In this way, hotels followed the pattern of other modern business enterprises, characterized by hierarchies of middle and top management employees.[51]

Finally, technological systems became the hallmark of a modern hotel. As an example, Barnum's lighted its rooms with chandeliers fueled by gas. Newer hotels incorporated technological innovations as they developed throughout the long nineteenth century. These included indoor plumbing, steam heat, electric call bell systems, and patent locks as well as path-breaking design elements meant to segment operations and control the flow of goods and people. Most importantly, these buildings incorporated the idea of modern technology into their definition of luxury. This concept of technological luxury defined hotels as modern, progressive, and luxurious and lent substance to their symbolic role through the merger of art and technology. Technological innovation drove the size, cost, and development of new hotels for one hundred years. Hotel ephemera such as stationery, envelopes, advertising brochures, promotional booklets, and postcards promoted the call bells, steam heat, elevators, lighting, and plumbing that characterized modern creature comforts.[52] Moreover, the technological systems fueled the rapid obsolescence of these buildings as entrepreneurs pushed the boundaries of technological applications with each successive enterprise. Technological luxury captured modern society's essence through its progressive, dynamic nature and the inevitable complete destruction and devaluation of its own past.[53] The very words signify the modern relationship that binds production and consumption with capitalism and bourgeois gentility.

Modern luxury hotels actively worked to project a progressive image for the men who built them.[54] The architecture, corporate financing, management techniques, and technology represented an enthusiastic and optimistic engagement with a market and industrial worldview. Hotel promoters viewed the modern luxury hotel as an essential part of their city's competitive commercial edge within a global marketplace. Because it straddled and combined the public and private sectors, the hotel needed to respond both to the commercial market and to the private ideals of refinement that resulted from the accumulation of wealth and cultural knowledge. The modern hotel was both the mechanism and the product of a market society. A large, elegant commercial hotel acknowledged the commercial success of the men who built it while attracting the patronage of like-minded men and their women. Inside, the elaborate and elegant interior underscored the benefits of a commercial society, contributed to the comfort of its guests, and confirmed the role of luxury as a symbol of a progressive, cultured civilization and an engine of economic health.

....

CHAPTER TWO

THE TREMONT HOUSE, BOSTON, 1829

....

"A Style Entirely New"

Two twenty-foot-tall Doric granite columns anchor opposite corners of Institute Park in Worcester, Massachusetts, just across the street from the American Antiquarian Society. The columns are a mystery to most Worcester residents today: the columns' monumental proportions are at odds with the small urban park, and the strange juxtaposition between the two underscores the columns' seeming purposelessness.[1] Close imitations of those of the Doric Portico in Athens, these columns are two of the four that graced the main entrance of Boston's Tremont House, the nation's first modern luxury hotel.[2] Their simple elegance and the history and memories they evoked so impressed Stephen Salisbury III, a Worcester philanthropist, that he rescued them at the time of the hotel's 1895 demolition and had them moved the forty miles from Boston to the park that had once been a part of his family farm. Now, removed from their original context, they serve as monuments to a bygone Boston institution that had shaped the growth of that city during the long nineteenth century.

The published records of travelers to Boston in the 1830s and 1840s give testimony to the idea that the Tremont House was something new and different, not just in the United States, but in England and on the Continent as well. Wealthy European and English visitors wrote about their impressions of the new world in scores of published volumes. As the nation's foremost hostelry, the Tremont House not only housed travelers but also was itself a focus of inspired commentary.[3] As the Honorable Charles Augustus Murray, a young member of the British aristocracy, wrote in the mid-1830s, "On arriving, I drove (as every traveler *must* do, *bon gré, mal gré*) to that first and most complete of hostelry monopolies, the Tremont House, which is certainly one of the largest and best-conducted establishments in the world."[4] Yet Murray shared sentiments of a different nature with a Scottish traveler and war veteran, Captain Thomas Hamilton, both of whom found the reality of the simple Greek architecture disappointing. They emphasized this in both instances, however, because their expectations had been aroused by the hotel's far-reaching reputation as "one of the proudest achievements of American genius" and "worthy of all the spare admiration at [a stranger's] disposal."[5] Like many other travelers, Murray and Hamilton found the Tremont noteworthy as a distinctive product of American creativity and invention.

The Tremont House, because of its reputation and its place in history, serves well as a starting point for the history of American luxury hotels. The sense of novelty that surrounded the hotel and pervaded the commentary of its time demonstrates that the Tremont House differed substantially from its predecessors. Even though Barnum's Baltimore hotel enjoyed an enviable reputation equal to that of the Tremont's for comfort and service, the Tremont House was something of a tourist attraction. Its imposing Greek Revival architecture and size contributed to this, but of greater significance was the civic importance bestowed on the hotel by Boston's business, political, and cultural leadership.

A consortium of Boston businessmen built the Tremont in response to a perceived need for appropriate accommodations for the many wealthy—especially southern and foreign—business travelers to that city, then the nation's center for banking, shipping, and textiles.[6] In addition to housing travelers, the Tremont served as a social and political center. Bostonians held large dinners and club functions there and transacted business in the bustling barroom. William Havard Eliot, a member of one of Boston's most prominent families, shepherded the financing and construction of the building, with nearly the entire upper class of Boston contributing financial support. In this way, the project replicated typical elite behavior in other arenas such as health care and education, for which the city's elite assumed the stewardship of charitable and civic projects.

The Tremont's story is set at a time when the structure of urban life was undergoing profound change. Older patterns of social and economic organization based on artisanal production were giving way to the pressures of a developing market society, dramatic urban growth, the new mechanical age, and the growth of financial power deriving from industrialization. In Boston, merchants who had made fortunes in an increasingly tenuous China trade found themselves looking for new avenues of investment.[7] The hotel represented a collaboration of interests between the entrenched merchant class—the "men of the mast"—and the new and powerful textile manufacturers for whom innovation and risk-taking was the stuff of the

The Tremont House, Tremont Street, Boston. Note the iron fence to the left that encloses the Granary Burying Ground that so bothered certain Bostonians as well as the men loitering under the front portico that bothered others. *Warshaw Collection of Business Americana—Hotels, Archives Center, National Museum of American History, Smithsonian Institution.*

new age. Despite the growing influence of industrial processes and organization that threatened more traditional modes of production, there appeared to be a mutually respectful working relationship between the financiers and the tradesmen who built the hotel. The building embodied and symbolized the dynamics of the decade, incorporating both traditional and modern perspectives on development, urban life, and elite culture, paying tribute to an older way of life even as it boldly ushered in the modern age.

Specific political, economic, and cultural influences combined to determine the form of the modern luxury hotel as defined by the Tremont House. The changing nature of capitalism, which fueled the rehabilitation of luxury, the growing courtship of refinement by the upper classes, and a budding consumer culture, provided the context for its development. The result was an extravagant and capacious building that represented old and new wealth and encouraged the patronage of the established elite as well as newcomers to the upper and upper-middle classes. The Boston Whigs' displacement from national political power resulted in a new system through which the Boston elite consolidated their cultural power in private and civic institutions. The Tremont House fit this pattern of combining corporate, financial, and civic goals. The corporate shareholders who executed the hotel project viewed it as an enterprise conceived for the public good on its behalf. As self-appointed cultural mentors for the nation, they extended their influence to the urban landscape through the plans and construction of a nationally recognized first-class hotel.[8]

The Tremont tied concepts of civic importance to architectural design, function, and technological innovation, and its legacy influenced an entire century of hotel construction. The design, by architect Isaiah Rogers, created a standard that others borrowed and elaborated upon in all parts of the country throughout the antebellum years. Coinciding with the acceleration of America's industrial revolution, Rogers's inclusion of technological systems in the building's design marked the beginning of a technological competition within the hotel industry. As a result of his work on the Tremont House, Rogers earned numerous commissions for hotels from Maine to Alabama, from New York City to Nashville. In addition, towns and cities across the country appropriated the Tremont's name so as to invoke the modern urban sophistication symbolized by the Boston hotel.[9] As a project of prominent Bostonian businessmen, the Tremont became a monument to themselves and their city as well as a prototype for hotel development throughout the growing nation.

The Case for a New Hotel

The Tremont House was the pet project of William Havard Eliot, a lawyer and member of a leading Boston merchant family.[10] In the 1820s, Boston was a national financial center. This was true especially for banking and insurance, most particularly for the import of teas and beautiful textiles from China, and as the venue for exchange between southern cotton planters and Massachusetts textile manufacturers. As a result, the city attracted a great many travelers of substantial means. The *American Traveller* took note of the "vast increase of traveling and the multiplication of the means of conveyance."[11] Road improvements in particular facilitated faster travel, enabling more people to be on the move as the cost of travel fell. Good roads, fine stagecoaches, canals, and steamboats all contributed to the increasingly frenetic mobility characteristic of the young nation.

In 1818, fire destroyed the Exchange Coffee House, Boston's largest and most comfortable hotel. Andrew Dexter Jr., a man of dubious repute, built the hotel in

1806-07, financing it under speculative circumstances based on the circulation of bank bills that, as one nineteenth-century historian noted, had no actual capital behind them and were worth nothing.[12] In June 1809, the building was sold at a public auction, and subsequently, Harrison Gray Otis, the "nineteenth-century king of Boston society," organized a new company of investors.[13] David Barnum, who eventually built and managed Baltimore's City Hotel, had managed the Exchange Coffee House, which, under his administration, developed a fine reputation for elegance and service.[14] An 1817 history of Boston described the hotel as a "mammoth affair of seven stories" costing $500,000, and being "far in advance of the wants of its day."[15] Dexter intended that the first floor be used as a business exchange; however, the merchants, out of longstanding habit, preferred conducting their business outside on the street. While known primarily as a hotel, the building served multiple purposes, housing eleven printing offices, several Masonic lodges, business offices, and society rooms, as well as the more traditional rooms associated with hotels, such as a ballroom, an observatory, billiard rooms, bar, several lounges, and a large dining room.

When the hotel burned (the fire started in the attic and thus was always beyond the reach of fire fighters), Boston remained without a hotel "corresponding to the expectations and wants of the numerous visiters [sic] of the Metropolis of New England."[16] Dexter's scheme left Bostonians bitter, leading one wag to comment that the hotel "was conceived in sin, brought forth in iniquity, but it is now purified by fire."[17] The Panic of 1819, the nation's first experience with the modern economy's boom and bust cycles, ensued shortly thereafter, creating a sobering economic climate that, in combination with memories of the failed enterprise, dampened local enthusiasm for investment in a new hotel. In Eliot's words, "The unfortunate experience of those who had previously engaged in similar enterprises was sufficient to deter any individual from undertaking a work of so much labor, cost, and hazard."[18] Bostonians, aware of failed projects in other cities, remained cautious.

Ongoing competition for trade dollars with other states generated intense discussion in the Commonwealth on internal improvements for canals and, later, railroads to aid the flow of goods. These pressures led to the proliferation of state-chartered corporations during the 1820s that provided opportunity and protection for investors in projects that, while privately owned, served public purposes.[19] In keeping with this trend, in the session of 1824–25, the legislature of the Commonwealth of Massachusetts incorporated a company to construct a hotel building worth up to $500,000; however, three years passed before Eliot took action. In 1826, a *Boston Newsletter* editorialist deplored the lack of proper accommodations in the city, stating that "public houses of entertainment, although seemingly numerous, are scarcely adequate to the purpose: hence boarding houses of respectability are glutted with visitors."[20] He added that another hotel of equal stature to the old Boston Exchange Coffee House would be welcome.

The Investors

In May 1828, Eliot proposed a plan for financing the structure, agreed to by the other major investors, who included Thomas Handasyd Perkins, James Perkins, Andrew Eliot Belknap, and Samuel Atkins Eliot.[21] Even these few names invoke the investment pattern of about forty interrelated families that dominated Boston corporate activity from the late eighteenth century onward in both the commercial and philanthropic realms. This consolidation of capital among kin networks translated into financial, cultural, and political leadership roles for the merchant and industrial elite in antebellum Boston.[22] With the Eliot and Perkins families prominently at the project's helm, not only was a subscription for a $100,000 loan easily filled by early June, but the project also received the community's enthusiastic support. The register of investors consisted of eighty-seven individuals and thirty businesses, representing a total of 144 Boston entrepreneurs. Of these, many hailed from readily recognizable Boston families such as the Cabots, the Appletons, the Perkinses, the Eliots, the Lawrences, the Quincys, the Searses, the Lowells, and the Brookses. Fifty-two served as directors of fourteen of the fifteen Boston banks. The investors included the mayor, Josiah Quincy; future mayors, governors, and congressmen; and ten of forty Boston members of the Massachusetts House of Representatives. Of the entrepreneurs, nearly two-thirds were merchants engaged in either overseas or domestic trade or dry goods.[23]

While the list was heavy with those of established fortunes, such as merchant Peter Chardon Brooks, Boston's richest citizen, and Thomas Handasyd Perkins, the "merchant prince," it also included the names of Boston's new textile industrial elites, such as the Appletons, Lowells, and Lawrences, and represented a marriage between older and emerging economic forces in the city. Whereas the decade of the 1820s featured power conflicts in seaboard cities between existing commercial interests and the new manufacturers, in Boston the rifts were nearly inconsequential. Family networks eased the merchants into industrial investments, and the two groups shared a modern entrepreneurial outlook that emphasized the primacy of capitalism and a willingness to innovate. The Boston elite invested in large-scale initiatives such as bridges, wharves, building construction, canals, and early railroads that, in aiding transportation and trade, served both public and private interests.

Despite the seeming ease of financing the project, investors still regarded the hotel as a risky investment. Nowhere in the United Sates, and some said the world, had a hotel of this magnitude or design been proposed. As a June 15, 1828, *Boston Daily Advertise*r article noted, "Similar establishments through the country ha[d] proved uniformly unsuccessful." However, the hotel's principals possessed those characteristics necessary for successful capitalist ventures: resourcefulness, innovation, willingness to assume risk, and the ability to supply or obtain capital.[24] The wealthy merchants of Boston, through their banking activities and experience with development projects, wielded tremendous power and influence that proved useful in real estate operations involving building permits, incorporation charters, and property assessments. Increasing the number of investors not only moderated the degree of risk but also enlarged the circle of influence. The hotel became important, not just as an investment and a city necessity but also as a public representation of the personal, business, and civic ethos of a class of men.

The proprietors, however civic-minded and enthusiastic about their undertaking, nonetheless harbored concerns for the financial health of the project.[25] It would have been out of character for them to do otherwise. In 1895, when developers finally demolished the Tremont, William Havard Eliot's son, Dr. Samuel Eliot, wrote a short letter to the Bostonian Society to be read at a commemorative lecture. In it he wrote, "It ought to be generally known in Boston, first, that the hotel was of a style entirely new, not only in this country, but in Europe; second, that it was an enterprise in the public interest rather than for the advantage of the proprietors." Eliot pointed out that the investors did not intend to profit from the enterprise, that indeed, they had leased the building for a nominal amount of rent for its first ten years. Finally, he noted, "The subscribers to a loan in aid of the building largely failed to make good their promises, and left the burden of the undertaking to the proprietors, especially to their leader, William H. Eliot, who assumed the responsibility of completing the building."[26] Despite the passage of fifty-seven years, Samuel Eliot's brief and prickly letter evoked the bitterness that emerged early on. It referred to a power struggle between his father and the Boston City Council. Even though the community endorsed and supported the project, a debate over a proposed bounty to be paid the corporation highlighted changing

ideas about public and private development and the government's role in supporting projects that, while purporting to be in the public interest, also fell within the
realm of private investment.

The Debate over Tax Abatement

In early June 1828, William Havard Eliot petitioned the city government for the
sum of $500 to be paid annually for ten years to himself and his associates as
proprietors of the hotel. The resolution passed the initial reading in committee
unanimously and with enthusiasm, and it concluded by praising the project and
urging the city council to support it. The amount of $500 equaled that which
would have been levied against the hotel for annual taxes. In effect, Eliot asked for a
tax abatement that not only acknowledged the risk he had assumed in the public's
interest by financing the hotel but also substantiated the city council's support of
the project.[27] As was the custom, the several semi-weekly newspapers reported the
proceedings from the Monday night council meeting, sparking a lively and entirely
unanticipated public debate concerning the role of municipal support of private
enterprise.[28]

A member of Boston's Common Council challenged the propriety of contributing to what had been put forth (in his view) as a speculative project catering
only to the wealthiest class of people. He drew on the idea of precedent to support
his position, asserting that the nascent iron, cotton, and wool manufacturers had
not benefited from commensurate aid from the city coffers. He directed his argument against the increasing power of early industrialists and, in doing so, voiced
his opposition to government support for large-scale private enterprise. Thus, he
cast the hotel project into the category of private investment rather than civic enterprise. The critic dismissed the logic of the "trickle down theory" *avant la lettre,*
a scenario suggested by supporters in which wealthy southerners, seduced by the
hotel's luxuriousness and comfort, would linger longer in the city and spend more
money. This invoked, then, the debate about luxury and its role in driving the
emerging capitalist economy. Did a project such as this, directed at the wealthy few,
constitute an appropriate objective for public support?

Eventually, this council member revealed his primary concern, that of the "danger of the precedent" that underwriting such a project might introduce. While
he conceded the worthiness of the hotel project, he asked, "Who shall defend the
treasury from the assaults of the middling, the poorer, and the numerous classes of
society, all of whom abound in projects that promise to increase both the capital

and population of the city?" While private men could reject subsequent pleas for funding, the "body politic" needed to adhere to more rational procedures. Patrician Whigs such as Eliot regarded themselves as helmsmen of the nation's economy and fully expected government support in the form of special privileges. However, the realities of strained government budgets combined with changing ideas about democratic power and the public good asserted themselves in this dissenter's voice. The point of his opposition was not to denigrate the hotel project but to protect the city treasury from claims of other, equally entitled classes of people who, in the climate of Jacksonian ideology, "could not have been turned away without injustice or the most gross inconsistency."[29]

An exchange of views followed over the course of the next several weeks, with the opposition persevering in its point of view, buoyed by the certain knowledge that whatever the proposal's outcome, plans for the hotel would move forward. The Common Council defeated motions to postpone the issue indefinitely, and eventually the decision rested on the opinion of the city solicitor, Charles P. Curtis. Taking a narrow interpretation of the city's general powers for granting money, Curtis stated (in what was called a "sensible and manly speech"), "The object referred to in the proposed resolve, is not one of those which have been provided for by any special act of the Legislature."[30] In other words, the city of Boston had explicit authority from the legislature to assess taxes for the purpose of such expenditures as street lamps and the nightly watch, but not for hotels. The editor of the *Commercial Gazette* supported Curtis, praising his "powerful arguments against yielding to the solicitations of private interest, as opposed to the public." The ongoing high-profile controversy in Boston, the state legislature, and the courts concerning the Charles River Bridge had sharpened the debate over the nature of private and public interest.[31] The solicitor argued that improvements such as streetlights unambiguously represented the public good, but that the hotel's potential for generating profits for its investors clouded the issue of public benefit. Thus, a complex aggregation of attitudes involving class, politics, capitalist development, and power were at work in the discussions, resulting in a stunning and wholly surprising blow to Eliot and his partners.

Clearly disturbed by the public setback, Eliot, in his widely reported toast at the celebratory dinner that followed the cornerstone ceremonies, grumbled: "Tremont House—That liberal spirit which has enabled the projectors of the Hotel to commence this work.—May no *legal impediment* ever occur to prevent its exercise on similar occasions. Tremont House—May the *legal impediment* erected to oppose the stream of public patronage prove but a temporary dam to increase the force of the stream of private enterprise."[32] The force of Eliot's waterpower metaphor, significant on several levels, given the importance of the water-dependent textile industry and early manufacturing enterprises in Massachusetts and New England, expressed not only his regret but also his position that government's proper role was to provide an environment hospitable to economic investment.

Laying the Cornerstone

Excavation for the hotel began on June 17, 1828. The cornerstone was laid in a festive ceremony on the "ever memorable" fourth of July, a day, as one of Boston's early chroniclers noted, afforded "millions of Freemen an opportunity to express with gratitude and with one voice, the numberless blessings Independence has brought in her train to this republic."[33] Indeed, on the very same day to the south, similar festivities were marking the first construction shovel on a steam railroad in Baltimore and inaugurating the Baltimore and Ohio Canal in Washington, D.C. Such commemorations brought investors, politicians, workingmen, and private citizens together to acknowledge just how significant internal improvements were to the economic and political health of the young nation.[34] Bostonians similarly regarded the Tremont as an important link in the nation's burgeoning transportation network.

Despite the preceding controversy over the tax abatement issue, the laying of the cornerstone was an occasion of considerable pomp, symbolizing the relationship in the new nation between free citizenry and free enterprise. In what one historian calls the characteristic genre of the nineteenth-century civic ceremony, an early morning parade of infantry companies, accompanied by thundering cannon and merry bells, preceded a procession of hotel investors and members of the Massachusetts Charitable Mechanics Association (MCMA), an organization of Boston's tradesmen.[35] Participating at the invitation of Eliot, an honorary member of the MCMA, the procession began at the rebuilt (but significantly smaller) Exchange Coffee House and made its way to the site of the future hotel, at the corner of Tremont and Beacon Streets.[36] An inscribed silver plate and a sealed glass case holding a parchment bearing the names of all the investors were placed in a hole cut into the stone.[37]

During the ceremony, "in the presence of a great number of spectators," Major Benjamin Russell read aloud the names of all the subscribers and the amounts of their loans. Samuel T. Armstrong, president of the Mechanics Association, gave a short address, followed by another delivered by Eliot. Drawing on heartwarming and patriotic references to the Pilgrims, Benjamin Franklin, and President Washington, the speeches also introduced a class-consciousness into the ceremony. Eliot addressed the tradesmen: "The proprietors . . . felt that there was a particular propriety in requesting you to assist in commencing this enterprise because no class of their fellow-citizens has a more direct interest in every measure which promotes the advancement of the community in refinement and the growth and improvement of the city. If the execution of this work shall prove in any degree beneficial to our city, we are happy to believe that you will be among the first to receive a share of the advantages which will accrue from it."[38] What sounds a bit like condescension toward the artisans is perhaps a frank acknowledgement of social hierarchy, illustrated graphically by the parade order, in which the mechanics followed the investors. Many Brahmins remembered early-nineteenth-century Boston as a "small homogenous Yankee town where hierarchy, neighborliness and order prevailed."[39] Among the list of investors, only seven, perhaps eight at most, held membership in the Mechanics Association.[40] Armstrong himself eventually became mayor of Boston and filled other political offices, including lieutenant governor. Though he accumulated considerable wealth over the course of his lifetime, Daniel Webster yet referred to him as "one of the common people."[41] This kind of class rigidity buttressed the implication in Eliot's speech that while the mechanics (or tradesmen) might be the "bones and sinews of society," they would not, in all likelihood, constitute the class of men who would patronize the hotel.

The Architecture

Situated on the southwest corner of Beacon Street and Common Street (now Tremont Street), the hotel took its name from Boston's first English name, Trimountain.[42] In 1828, the area surrounding the site was a mix of upper-class residential homes and important community buildings. One guidebook, recognizing the elite's influence on the city's financial health, called the area "the Heart of the City."[43] The hotel displaced three pre-Revolutionary homes and a large garden. To the south, the site adjoined the Granary Burying Ground, where Samuel Adams, John Hancock, Paul Revere, and the victims of the Boston Massacre were buried, a cemetery that one Bostonian described in 1895 as "beautiful before modern ideas had destroyed its effect."[44] He was, no doubt, referring to the granite retaining wall and iron fence built by architect Solomon Willard in the 1830s. Completing the block to the south stood (and stands) the Park Street Church (Congregational) with its magnificent wooden spire, directly across Park Street from the Common.

Stately mansions—many of them belonging to the hotel's investors—lined Beacon Street from the hotel to the gold-domed State House. Nearly all the investors lived less than a half-mile from the corner of Tremont and Beacon.[45] The Tremont Theatre, designed by the hotel's architect, Isaiah Rogers, faced the hotel directly across Tremont Street. King's Chapel, the wealthy Unitarian church, stood (and remains) diagonally across from the hotel site, on the northeast corner of Tremont Street and School Street.[46]

The site itself represented the social status and power of its sponsors. For example, in the immediate neighborhood, King's Chapel's Unitarian congregants were among the city's wealthiest and socially preeminent citizens. Both the new State House and the improvements made to the Common raised property values, ensuring the stability of the wealthy residential area and encouraging new development of commensurate stature. Located in the midst of the elites' homes, the hotel reinforced spatial and territorial boundaries, housed their guests and business associates, and functioned as the venue for public dinners, celebrations, and casual associations. At the same time, it served to represent the achievement of refinement and cultural distinction for this elite class of men.

American architect Isaiah Rogers, born in 1800, designed the Tremont House. He was the third architect, after Alexander Parris and Solomon Willard, to pioneer Greek Revival architecture in Boston. Largely self-taught and extremely talented, Rogers entered Willard's Boston office in 1822. By 1826, he was practicing independently. Rogers designed the Tremont Theatre in 1827, but Tremont House was his first nationally important work. The Boston hotel's success led to many more commissions, including Maine's Bangor House (1834), Astor House (New York City, 1834–36), Burnet House (Cincinnati, 1849–50), Battle House (Mobile, Alabama, 1852–53), Maxwell House (Nashville, 1862–65, of coffee fame), and Galt House (Louisville, 1865). Although Rogers's career was varied and celebrated and not in the least limited to hotels, architects and historians nonetheless crowned him "the father of the American hotel."[47] Rogers himself attributed his success to Eliot's patronage. In December 1840, nine years after Eliot's sudden death at the age of thirty-five from influenza, Rogers acknowledged his debt to Eliot in his diary as he reflected on a recent visit to Eliot's widow and children. He wrote, "He by his aid in my circumstances was the main spring to all my after success," concluding, "William H. Eliot was my best friend."[48]

In 1830, Eliot published *A Description of Tremont House,* an elegant book describing the hotel, complete with architectural plans and detailed descriptions, that helped spread Rogers's fame throughout the national architectural community.

Its publication served as another avenue through which Boston's cultural network expanded its influence beyond the political sphere while suffering through the unfavorable national political climate precipitated by the rise of Jacksonian democrats. Typically, this process moved through religious, educational, and cultural institutions. However, by looking at the hotel's design as a representation of Boston's elite values, ones that viewed commerce, development, innovation, and elite guidance as the nation's life force, Eliot's publication and dissemination of the hotel's plans fit well within this model. In fact, as one historian observed, "It was typical of Boston that William Havard Eliot published a description of this first of its kind which cited classical authority for every detail and remarked that Doric simplicity and New England granite were made for each other."[49] The book's publication made these design ideas widely available to like-minded developers who, like Eliot, recognized that an institution of this caliber was essential to a community's vigorous commercial health. Eliot's focus on the building's architectural detail firmly linked the hotel to a Whig vision of the nation's future.

With the early building going on in Washington, D.C., American architecture began to change from the staid Federalist style popular in the late eighteenth century toward a renewed emphasis on classicism and monumentality. Ancient Rome and Greece provided republican ideological inspiration for American Classic Revival architecture, which assumed a classical, yet distinctive, form in a country seeking to express abstract values through its architecture. Qualities such as simplicity, rich colors, and generous interior spaces echoed the American continent's physical geography and lent national character to the classical model. After the War of 1812, Boston's three leading architects—Parris, Willard, and Rogers—advanced Greek Revival architecture, the later phase of Classic Revival. Parris's 1824 Quincy Market, a monumental structure built entirely of Quincy granite, was Boston's first large public Greek Revival building. Classic porticos supported by four monolithic columns ornamented each end of the 535-foot-long building.[50]

The Tremont House faithfully reproduced classic Greek Design from sources such as *Antiquities of Attica* and James Stuart and Nicholas Revett's *Antiquities of Athens*. The typical feature of the Greek Revival was the temple front, characterized by its stately columns and dignified pediment. Certainly, the centerpiece of the Tremont's 205-foot-long white granite front was its thirty-seven-foot-tall Doric portico, patterned after the Doric Portico at Athens, with four twenty-foot columns, each weighing twenty tons. Greek design held specific significance for America's first modern hotel. On one level, Greek Revival was *the* contemporary style of architecture, the latest evolutionary development of the classic revival, executed by "a new crop of modern-minded professional architects."[51] A self-consciously modern building such as the Tremont demanded the most modern architectural style. During this period of nation building, architects cultivated a simple national public architecture, one that looked for, as one observer stated, "the simplest style of architecture . . . which above all others, has assumed as the basis of its institutions, the utmost simplicity in all the forms of its governments."[52] Whether or not the rhetoric accurately represented the complex reality of layers of local, state, and national governments, Greek Revival architecture sought to symbolize the citizen's relationship with his or her state—that of a republican.

Most interesting was the application of the Greek temple form to buildings serving practical purposes, notably banks (Second Bank of the United States, Philadelphia, 1817–24), market houses (Quincy Market) and, by 1829, hotels. In the American political economy of the early republic, the politics of government were in symbiosis with a capitalist economy. An architecture that symbolized democracy's purest ideals also honored capitalism's great achievements. Therefore, it was no great leap to apply the architectural form of classic sacred shrines to the cathedrals of American commerce: banks, markets, and hotels. In this context, hotels "gave identity to the city not unlike the identity European cities derived from their cathedrals."[53] Behind the Tremont's classic facade existed a busy world of commerce, capitalist production and consumption, technology, and display that thrived on a Whig political philosophy of reward based on merit and industry in partnership with the beneficence of government.

The Hotel

A guest, after passing under the towering portico, entered the hotel through massive sixteen-foot-tall folding doors that glided open on hinges designed to counteract the pull of the doors' great weight. From there, a flight of ten steps led to a second set of doors that opened onto a circular rotunda, its domed stained-glass ceiling supported by ten Ionic half-columns. The use of stained glass, by invoking the spirituality associated with the Middle Ages, created an aura of reverence within a commercial space symbolic of the industrial era. This elegant room was the very first hotel lobby. Prior to this, hotel guests entered directly into a hotel's barroom, as at Barnum's, where the hotelkeeper transacted the hotel's business. At the Tremont, parlors that received guests upon arrival, an office, and a porter's room for luggage encircled and opened up from the rotunda. This design not only segregated arrivals from the busy barroom but also privatized the act of waiting and further refined and dispersed to specialized spaces functions such as procuring rooms and receiving and storing luggage. This mitigated the hectic bustle of arrivals and departures and lent a more genteel atmosphere to the establishment.

The large main public rooms stretched across the hotel's front and looked out onto the street. These divided, to the left and right of the central rotunda, into spaces defined by a gendered geography. Separate receiving rooms for men and women adjoined the rotunda/lobby to the right and left, respectively, when facing the hotel. A long corridor running parallel to the hotel's front gave access first to the ladies' receiving room, continuing left to the ladies' drawing and dining rooms. The gentlemen's drawing and reading rooms similarly led from their receiving room to the right. The gentlemen's drawing room served as the bar. The adjacent library subscribed to national and European newspapers, of which guests could avail themselves at will. For a small membership fee, members of the general public could also have access to the library. Six large rooms on the second story corresponding to those on the main building's first floor provided space for private clubs and parties.

Two brick-clad wings projected back from the main front at angles that fit the unusually shaped lot. The north wing housed the immense dining room, the most elaborate of all the rooms with a capacity of two hundred guests. The south wing held two floors of family apartments. This continued the hotel's gendered compartmentalization. The main dining room extended from the men's side of the building and, as a rule, was used by men, unless temporarily given over to host a ball. Women lodged in the family apartments, located above the women's parlors and dining room. The remaining floors, three in the north wing, two in the south wing, and one across the main building, divided into individual guest bedrooms. Rooms on the fourth floor above the main building housed servants. The bathing rooms (for which there was a separate public entrance), housekeeper's apartments, laundry, and larder occupied the south wing's basement story, while the kitchen and servants' hall occupied the north basement wing. Shops and offices, accessible from the street by separate private entrances, were located in the lower story of the main building on Tremont Street. The hotel had 170 rooms that combined to occupy 12,849 square feet.[54]

Innovations

Everything about the hotel suggested its innovative nature. The granite used for the facade and columns came from the quarries at Quincy, nine miles distant.

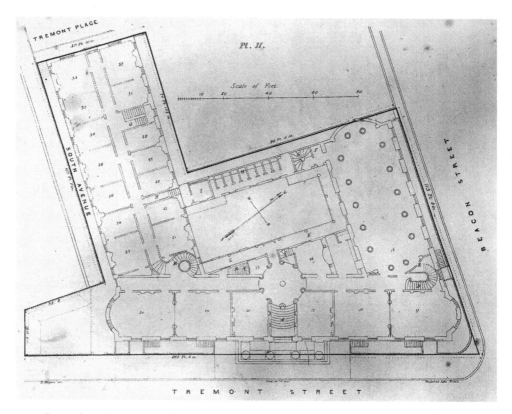

Plate 2 from *Description of Tremont House*. The plan of the first floor shows the receiving rooms (12, 13) on either side of the vestibule (A), the ladies' dining room (19), the ladies' parlor (20), the gentlemen's drawing room (16), the reading room (17), the main dining room (18), smaller private parlors (22, 28, 33, 34), and bedchambers (24–27, 29–32). The privies are at "H," and a washroom at "L." *Courtesy, American Antiquarian Society.*

New quarrying techniques and advances in transportation facilitated the use of the rich granite formations and encouraged a new trend in building that incorporated the use of massive slabs of stone.[55] The early exploitation of Quincy's granite and the difficulty in transporting large monolithic stones precipitated an interest in railroads by Thomas Handasyd Perkins and other "visionaries." For example, each of the columns for the Boston Court House required a team of sixty-five yoke of oxen and twelve horses to move them from Quincy to Boston.[56] The Quincy Market columns were transported on specially designed carriages suspended beneath the axles of nine-foot-diameter wheels.[57] The Granite Railway that Perkins and the quarrier Gridley Bryant built in 1826 to transport granite for the Bunker

Hill Monument also transported the Tremont's columns. The railway enjoyed the reputation of being the nation's first railroad. With this system of cars and rails, one horse could draw eight to twelve tons of granite in a single day at the rate of three miles per hour.[58] Thus, the granite facade created interest through the quarrying processes and transportation innovations that it utilized and was celebrated in newspapers, city histories, and stranger's guides.

The building's design incorporated several pathbreaking innovations. In addition to the newly conceived rotunda, office, and baggage rooms, single and double bedrooms distinguished the Tremont from other hotels where a multitude of guests shared sleeping quarters. Patent locks on guest room doors enhanced this new accommodation to privacy, adding security for guests and their belongings. These locks were keyed individually so that no key opened more than one lock.[59] Each key was attached to an iron bar nearly six inches long and an inch wide, a size and weight that discouraged guests from walking away with them. Guests left their keys at the office before leaving the hotel. The Tremont House arranged for the stabling of horses to take place at a location separate from the hotel, a departure from convention that usually placed stables adjacent to the hotel or in the central courtyard. The Tremont contracted for private stabling under independent management about four city blocks away. This eliminated the clatter of hooves and prevented unavoidable and unpleasant animal odors from marring the building's sense of gentility. Because of changes in traveling patterns, many visitors arrived in Boston either by stagecoach or steamboat, reducing the immediate need for stabling.[60]

The incorporation of indoor plumbing into the building's design constituted the most notable innovation at the hotel. Rainwater reservoirs in the courtyard and on Tremont Place (a small street that ran behind the hotel and connected to Beacon Street) supplied the basement laundry, kitchen, and bathing rooms. The hotel made the eight bathing rooms available for a fee to both guests and Bostonians. In 1840, for example, a bath cost thirty-eight cents. Harvard historian Samuel Eliot Morison recalled in his memoir that his grandmother and her siblings bathed weekly at the Tremont, having no "plumbing of any description" in their great Charles Bulfinch–designed mansion on Beacon Street.[61] A third large reservoir in the south wing's cellar supplied water to the greatest luxury of all, the battery of eight enclosed water closets located in a first floor passage on the courtyard's west side, between the north and south wings. Two cisterns in the attic story caught rainwater from the roof and supplied water for use in the bedchambers. Lead pipes enclosed in "boxes" surrounded by pulverized charcoal carried waste water to cesspools in the cellar where it passed into the common sewer. [62] While these plumbing improvements sound elementary, even through the 1860s only the very wealthy enjoyed running water, toilets, and tubs in their homes. Indoor plumbing did not become commonplace for all classes of people until well into the twentieth century. [63]

Gaslight also lent an aura of modernity to the hotel, following the example of Baltimore's City Hotel. Gas lamps lit all the public rooms, while whale-oil lamps lit the bedchambers. As the hotel courted a genteel clientele, its gas lamps created a setting for sophisticated society. Their soft and playful light enhanced the theatrical setting for elegant display. [64] Early lithographs of the hotel's exterior show eight gas lamps on the street, four each in front of the hotel and the Tremont Theatre across the street, this at a time when Boston street lighting was in its infancy. [65] The early use of gaslight bolstered both the hotel's aesthetics and its image as a site of modernity and refinement.

The call bell system in use at the Tremont was the first use of a system patented by Seth Fuller, a Boston bell hanger. [66] Fuller's remote signaling device replaced the tradition of guests ringing bells from their rooms, a custom that created an incessant disturbance for all guests. Bell pulls located in the individual bed chambers connected to a battery of bells mounted above the office counter within an entablature that housed a wood box fifty-seven feet long, one foot high, and six inches deep. When a guest needed a "rotunda man" (the bellman's forerunner), he or she pulled the cord in the bedchamber. The pulled cord rotated a cylinder, in turn causing a small hammer to ring a clock bell twice. The bell's sound alerted the office manager. The vibrations from the revolution of the cylinder shook a spindle that held a card printed with a number corresponding to the room from which the summons originated. Upon noticing which card was "shaking," the office manager would know where to send the messenger. The system was the first of many of increasing sophistication, leading to the room telephone that enabled hotel guests to communicate with the front office from remote locations. [67]

Finally, the proprietors understandably regarded the building's fireproofing to be a major concern. Hotel fires always constituted a serious threat, especially to those hotels built of wood. The Tremont's vulnerability intensified because of the building's size and the correspondingly large number of people expected to reside in it at any given time. Open fireplaces, cooking ranges and ovens, whale-oil lamps and gas burners, not to mention guests' smoking habits, all created significant fire hazards. Four wide staircases located at intervals throughout the house, three of which extended from the attic to the ground floor, enabled the rapid evacuation

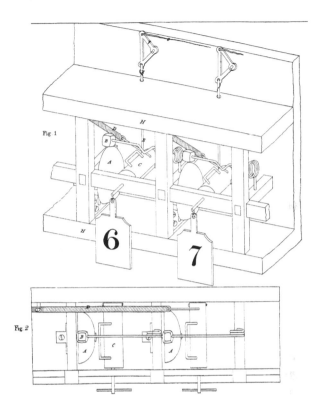

Plate 31 from *Description of Tremont House*. This diagram shows the action for the call bell system as designed by Boston bell-maker Seth Fuller. When a guest pulled on the cord in the guest room, the attached wire (D) rotated the cylinder (C) that simultaneously shook the room number and caused the hammer (B) to strike a bell (A). The bell alerted the desk manager to turn to see which of the numbered hang tags was shaking in order to send a rotunda man to the correct room. *Courtesy, American Antiquarian Society.*

of guests and servants and served as the first line of defense. Rogers prescribed the extensive use of granite, brick, and slate to limit the amount of combustible material and placed interior firewalls between sections of the hotel. In addition, a layer of coarse mortar encased each room above the ground floor. Upper-story cisterns held a water supply purportedly sufficient to extinguish any fire and able to be directed through hoses. The hotel's site, surrounded by streets and open ground, mitigated the danger of fire spreading both to and from other buildings. In 1852, when the Tremont Temple (formerly the Tremont Theatre)—across Tremont Street from the hotel—burned, the hotel emerged unscathed. The hotel's slate and copper roof protected it from the "immense clusters of cinders" spreading everywhere.[68] An engraving in *Gleason's Pictorial Drawing-Room Companion* that accompanied a dramatic description of the conflagration showed crowds of men watching the fire while perched on the hotel roof and balcony as well as under the portico.

These innovations, while perhaps sounding elementary to twenty-first-century readers, placed the hotel in the forefront of early nineteenth-century building technology. By displaying their own familiarity with the latest mechanical systems, the proprietors crafted an image for themselves that placed them among those who had great hopes for the "Mechanical Age." At the same time, the hotel's technological aspects integrated function with the building's aesthetics, adapting historic architectural form and decorative detail to the practicalities of a modern, commercial building. This dialectic between the new and the old presented itself in the relationship between the hotel's progressive technological systems and its classical monumental architecture. These aspects combined within a larger conceptual scheme that included rich, luxurious interiors and amenities; a high and attentive level of service and management; and a beautiful neighborhood that encompassed the world of Boston's elite. This concept was successful and earned the hotel a national reputation from its first days. The Tremont became the standard against which other houses judged themselves and that newer, more modern hotels sought to eclipse.

Opening Ceremonies

The Tremont House opened on October 16, 1829, nearly sixteen months to the day after excavation of the hotel site began. A festive dinner attended by 120 "gentlemen" marked the occasion, and all the Boston newspapers reported on the activities expansively.[69] In the hotel's dining room, Mayor Josiah Quincy presided over the evening's entertainment, which included a lavish feast of forty-six dishes followed by a program of celebratory toasts, songs, and poems. Daniel Webster, Joseph Story, Edward Everett, members of Congress, and most of the prominent Boston merchants attended, with the unfortunate and ironic exception of the Eliot family, who missed the event because of the grave illness of a family member. At the meal's conclusion, Quincy, as presiding officer, offered toasts to the new hotel and the Eliot family, reaffirming the extent of the family's beneficence. "The family of Eliot," Quincy declared, "The history of our *Church,* our *University,* our *Hospital,* and our *Hotel,* bear ample testimony of their *piety,* their *love of learning,* their *benevolence,* and their *public spirit.*"[70] The toast validated the Eliot family's role as one of the Boston families who served as community stewards. Edward Everett continued the theme in his characteristically lengthy oration that followed the many and varied after-dinner toasts.

Several speakers congratulated the "mighty power of the Mechanics," whose separate dinner took place the following night. Everett's remarks, however, were a paean to the merchant class and the temple they had just completed for their mutual worship. After a toast to the absent Col. Thomas Handasyd Perkins, the "bright ornament" of the city's merchantry, Everett expressed his pleasure with what he regarded as the balanced interdependence between the various classes of America's prosperous society. In his eyes, the farmer, the manufacturer, the mechanic, and the merchant were "happily adapted to each other." However, Everett professed a special fondness for the merchant class, claiming "without the least disparagement to other callings," that there was "something favorable to liberal and expanded views, in the pursuits of the merchant."

Everett characterized the merchant as a risk-taker whose personal financial rewards flowed out to the community to produce an "ample harvest of prosperity." His description of the merchant evoked both romanticism and virility, as he rendered a portrait of adventurous and enterprising men dealing with large fortunes, often put at stake in a single venture. "There is probably scarce a gentleman who hears me that has not more than once seen a Canton ship, itself worth thirty or forty thousand dollars, with a hundred thousand dollars in specie on board, dropping down Four Point Channel without a cent of insurance." The success of these bold undertakings depended to a certain degree on "the smiles of a good Providence" but more so on an "intelligent, sagacious, and far-reaching calculation." The fortunes derived from such enterprise were being put to good use to found and support manufacturing interests as well as to forge advances in farming and animal husbandry.

In recounting the Boston merchants' contributions to the city's public institutions, Everett added the Tremont House to the list of Harvard, Andover, the Botanical Garden, Massachusetts General Hospital, and the Athenaeum, all beneficiaries of merchant benevolence. Most of these institutions, like the Tremont, functioned exclusively for the elite class and served as vehicles for Boston's worthies to cement their authority, especially as the populism of Jacksonian politics surged. Everett's following toasts, to the merchants of Boston, the unfinished Bunker Hill Monument, internal improvements ("the nation's wealth and glory, may Massachusetts enjoy both!"), the Tremont's proprietors ("may those who assume responsibilities for the public good, find the public willing to make exertions in *their* favor"), and the manufacturers ("may their water rights soon become water privileges") enumerated, indeed, a political agenda for the Whig party.

The following evening, October 17, 1829, 130 mechanics (that is, tradesmen), most of them members of the Massachusetts Charitable Mechanic Association, gathered for a second celebration at the Tremont House.[71] Samuel T. Armstrong presided over the dinner, with Mayor Quincy, Edward Everett, and Daniel Webster again in attendance. The *Boston Courier*, the city's working-class paper, was the only newspaper that reported on the mechanics' dinner. The *Courier*'s editor, Joseph T. Buckingham, was a longtime member of the MCMA. This occasion of the class-bound, separate dinners is an ironic commentary on the "age of the common man." The proprietors had paid tribute to the tradesmen through their toasts, yet the mechanics could not attend the dinner to hear them. The tradesmen celebrated their project's completion in the hotel's dining room, a venue that most would probably not enter again. Yet each group testified to the comfortable and beneficial dependence of one upon the other. The mechanics raised their glasses to "the mutual good feelings and confidence between the merchants and mechanics, the projectors and the laborers of this day."[72] Other toasts, through their language, made class and political differences clear. Take, for example, "The Union—A cask of twenty-four staves—bound by hoops of white oak and hickory—there is no danger of its bursting, while the dregs can work out at the bung," which incorporated metaphors of trade while also referring favorably to Andrew Jackson—"Old Hickory"—decidedly *persona non grata* at a Whig dinner party.

The Hospes Letter

The same newspaper issues that reported the proprietors' opening dinner also printed an exchange of letters initiated by a correspondent writing under the name of Hospes, using as a pseudonym the Latin word for both "host" and "guest" from which such words as *hospitable* and *hospice* derive. Hospes addressed the public from the pages of the *Daily Advertiser* with what he considered useful suggestions for the new hotel's management. First, Hospes hoped that the hotel would not serve as a lounge or drinking place for "idle and dissipated" young men.[73] To discourage the congregation of unruly drunken youths, he urged that no food be served. Hospes then recommended keeping crowds of idlers away from the front entrance so that women and children alighting from stagecoaches would not be forced to make their way "through a crowd of gazers." Indeed, incurring the gaze of strange men, and worse, being forced to speak to them in order to secure passage, violated strict rules of etiquette meant to protect women from rudeness and incivility in urban settings.[74]

Additionally, Hospes wished that the Tremont House would "never be disgraced by a pile of dirty slippers shoveled up in the corner of the office" and that the abomination of smoking be "wholly banished." In the nineteenth century, the custom prevailed that travelers with muddy boots removed them upon entering a hotel. Urban streets, whether paved or not, were typically marred by animal droppings, open sewers, dirt, and mud, all of which threatened footwear and the hems of long skirts. Management provided pairs of "public" slippers for those not traveling with slippers of their own. Servants removed and cleaned the dirty boots and returned them to their owners. Hospes—probably with good cause—found the pile of common slippers distasteful. As for smoking, Hospes presaged (or perhaps initiated) the secondhand smoke issue: "Now my lungs will not bear the smoke of tobacco, particularly after it has passed through the lungs and throat and nostrils of another man." He added, "I contend that no man has a right, under color of seeking his own pleasure to fill the atmosphere I aim to breathe, with the smoke of a plant that poisons me." Men's smoking and their extensive use of chewing tobacco often created offense, particularly through the habit of spitting in all public places, indeed, even, as later recounted by Dickens, on the carpets in the president's mansion.[75]

Many of Hospes's suggestions both amazed and amused the newspapers' editors, exposing the correspondent to ridicule, but they also elicited some revealing commentary. In particular, the *Commercial Gazette*'s editor argued that one of the main goals for building the Tremont was to attract wealthy southern travelers. Because "every citizen who has traveled beyond the limits of Massachusetts" understood that these men smoked, not only after dinner but after breakfast and lunch as well, prohibiting tobacco use constituted a "dangerous innovation." The important concern was the patronage of "these gentlemen with their families, who are liberal in their expenditures, genteel in their persons and accomplished in their manners; who possess as much refinement and intelligence and moral worth as the most polished classes in New-England." It was important not to drive them away. This argument constituted a tacit defense of slavery, whose labor enabled tobacco production and the planter lifestyle that measured up to that of "the most polished classes in New-England." As for the other suggestions, such as those directed at the mess of slippers and public food service, supporters averred that "none but genteel company will visit Tremont Hotel," thus negating any need for formal rules. Guests possessing genteel manners would simply know how to behave.

The editor of the *American Traveller* was not quite as thoughtful, although the thrust of his argument against Hospes was similar. He chose to focus on the question raised by the specter of idle and dissipated young men. Taking considerable civic offense and satiric liberty, he avowed that "no city in the whole world . . . is more exempt from bar-room dissipation than our own. What reputed respectable Hotel did this sage *Hospes* ever enter, and behold a set of 'idle and dissipated young men' lounging away their time . . . in drinking and smoking?" He declared, "Our bar-rooms exhibit an almost studied sobriety and order."[76] Letter writers conducted most of the brief flurry of debate in similar mock indignation. However, the discussion is useful because it articulates both the public's expectations for the hotel—what kinds of things it deemed necessary in order to fulfill community goals—as well as the frank and unselfconscious acknowledgment that the building was an elite institution and a studied contrast to the somewhat unruly and uncontrollable world outside its walls.

Inside the Hotel

Almost no descriptions exist that detail the hotel's interior decoration beyond some general comments published in travelers' accounts and the plates in Eliot's *Description of Tremont House* that depict the architectural embellishments. In the early nineteenth century, commercial parlors or "publicly accessible rooms of state" required certain components that served both as extensions of personal display and as backdrops for refined behavioral conventions. These included carpets, window draperies with lace curtains, suites of reception sofas and chairs, a center table, a piano, and other decorative objects, such as pier mirrors, urns, and framed pictures.[77] Eliot's book listed the cabinetmakers, upholsterers, and importers of carpets, mirrors, and lighting devices who supplied the Tremont's furnishings. These commercial enterprises most likely also furnished elite Boston homes. Henry Tudor, an English visitor, commented on the Tremont's "better style of furniture."[78] An engraving of the hotel on period guest bills shows tied-back and fringed draperies with curtains billowing through the open parlor windows.[79]

Another source that helps depict life at the Tremont is a little 1833 two-volume book of stories, *Sayings and Doings at the Tremont House,* published pseudonymously by an enterprising British satirist writing under the name Costard Sly. Asked by his editor, Dr. Zachary Philemon Vangrifter, to supply a description of the hotel, Sly responded, "But what need of a Preface? . . . Every American know[s] the TREMONT HOUSE; and every foreigner, who has visited it, must in candor, admit, and that for comfort, good cheer, and the extent of its accommodations, it is not surpassed, if equaled, by any similar establishment." A physical description was unnecessary, not only because "everyone" knew the Tremont but also because the kinds of social conventions and intrigues that the stories related could only take place in an environment befitting upper-class manners and activities. Indeed, Sly scolded Vangrifter by saying, "If you do not find yourself breathing an atmosphere of GOOD-FELLOWSHIP,—you must either be a very so-so sort of person yourself, or—or you have no nostrils!!!" [80]

The settings of the various stories conveyed a picture of the hotel's milieu in addition to its physical description. The stories themselves narrated tales of business association, mistaken identity, courtship, and love and often touched on upperclass pursuits such as gardening, fashion, trout fishing, hunting, and sightseeing in the countryside. Sly described his characters as "well-bred" and "gentle mannered." One young lady on the eastern social circuit described the object of her affections as one who "drove those fine black horses," evoking wealth and discernment on the part of both the subject and the observer. A description of a scene at the front entrance gave a different appreciation of the one anticipated by Hospes, who feared the crowds of idle and dissipated young men. Sly set the stage: "Several gentlemen were standing on the steps;—some leaning against the pillars of that noble portico;—some smoking,—All listless, happy—small-change rattling persons, whose self-complacent looks might have challenged a comparison with those of any community of monks in the world." In the dining room, waiters maneuvered "to and fro, amidst a running fire of champaigne [*sic*] corks," and "everybody who has been in the Tremont House" remembered parlor No. 87 as a "very pleasant sitting room." A late evening of drinking and smoking in the men's drawing room by a cosmopolitan gathering of acquaintances that resulted from the coincidence of their stay at the hotel summoned a vision of "huge jugs of whiskey-punch and brandy-toddy;—innumerable glasses, and an immense dish of cigars." Additional stories sketched the hotel lives of mixed generations of upper-class women dressed in finery, playing the piano in the parlors, receiving visitors, and gossiping about dances, beaux, and potential mates. [81]

The commentary of foreign travelers to the United States often dwelled on the dining room and its activities. The Tremont's dining room measured seventy-three by thirty-one feet, with a fifteen-foot-high ceiling. Ionic columns framed the area occupied by the two long dining tables, and six large multi-paned windows looked out on Beacon Street. French doors led to the interior court piazza. Two open fire-places warmed the room, with added heat supplied by an air furnace that captured heat from the cooking fire below and delivered it to the dining room. Eliot chose the dining room for the hotel's most elaborate decoration, seeing no "impropriety in surrounding its occupants with cheerful and tasteful objects." In the 1830s, the Irish actor Tyrone Power remarked on the tables' perfect appointments. [82]

The daily rate of two dollars included four meals: 7:30 a.m. breakfast, 3:00 p.m. dinner; 6:00 p.m. tea, and a supper served promptly at 9:00 p.m.[83] English visitors in particular lampooned the *table d'hôte* system in which guests ate meals communally at appointed hours, selecting food from an overwhelming number of prepared dishes presented to them by a cadre of livery-clad waiters marching in military-style cadence. The "crash of a gong" announced each meal, calling the eager guests to the dining room. Ladies with male company ate together in the smaller ladies' dining room. Those with private apartments could order meals at will.[84] Irish immigrant "lads" served as waiters, a situation that elicited some commentary. Observers judged the services of these young boys well but noted that immigrants filled these jobs because free-born Americans perceived positions of service as being undignified in a democratic country.[85]

While most visitors agreed on the food's excellent quality, they were universally astounded by the speed at which it was consumed. James Boardman, another British traveler, commented that "the dispatch at meals, particularly among the mercantile classes, is almost incredible to those who have not witnessed it," adding that five to ten minutes sufficed for most meals. As an Englishman, Boardman always found himself "quite unable to keep pace at meals with the Americans."[86] An editorial from the *National Intelligencer* noted that "the process of bolting would seem to be done by steam."[87] Henry Tudor, while enjoying his soup in a leisurely way, noticed that not only had all the diners disappeared but the food had left the table as well. After laying "embargo" to two platters of meat, Tudor, as "the lonely unit out of 150 well-dined and departed persons," left the table shamefacedly, resolving to eat double-time the following day.[88] Tudor attributed this speedy manner of eating to the "economizing diligence of the American people." Their haste at meals, particularly at midday, when many local businessmen boarded at the hotel, enabled them to return to work that much more quickly. This observation illustrates another way in which the hotel integrated itself into the mercantile community. As a product of commerce as well as a setting for it, the hotel provided a basic

View of the Tremont House dining room from the cover of the Tremont Quadrilles. Nineteenth-century sheet music covers commonly featured local buildings and attractions. Here, fancily dressed patrons dance down the center of the Tremont's dining room to the music of an orchestra perched on top of the colonnade at the far end of the room. This is one of a very few interior images of the hotel. *Courtesy, American Antiquarian Society.*

service, drawing its practitioners into its midst and enabling easy and convenient communication among them.

Travelers consistently gave the Tremont House their highest commendations. Boardman called it "the largest, and certainly the most elegant building of the kind in America" where one could "far[e] sumptuously everyday, and repos[e] in comfort at night."[89] Godfrey Vigne agreed, describing the Tremont as "by far the best hotel in the States."[90] And despite Tudor's discomfiting acclimation to the dinner table, he nonetheless enthusiastically endorsed the hotel as "a splendid establishment" that could scarcely be "exceeded in point of beauty, convenience, and excellent arrangement."[91] As late as 1852, in a description of a commemorative banquet held at the Tremont for a retiring commander of the Boston City Guard, *Gleason's Pictorial* referred to the magnificent appearance of the banquet hall and the wonderful entertainment at "this favorite house."[92] From the very beginning, the Tremont earned unconditional acclamation and thus served as a model for emulation throughout the nation.

A Place in History

By any measure, the Tremont House was a successful enterprise. Despite the proprietors' original fears, the hotel quickly developed into a financially sound investment. According to contemporary accounts, the house filled to capacity by the end of the first night, and many bachelor merchants or those with small households made their homes there over the years.[93] Up to one hundred new visitors registered at the hotel each day. The long list of famous guests, including Charles Dickens, Ralph Waldo Emerson, Daniel Webster, Andrew Jackson, Abraham Lincoln, Alexis de Tocqueville, Fanny Kemble, and even John Wilkes Booth, in addition to the countless lesser-known foreign travelers, testified to its world reputation. Dwight Boyden, the Tremont's first manager, retired after seven years with a small fortune of $150,000.[94]

City guides, directories, and travelers' accounts provide evidence that the hotel fulfilled the hopes of both the promoters and the city in terms of providing exemplary accommodations that enjoyed widespread recognition and generated civic pride. In a burst of enthusiasm, the city guidebook, Bowen's *Picture of Boston*, printed a comprehensive description of the hotel in its 1829 edition, including a full account of the laying of the cornerstone with the complete text of Armstrong's address. Subsequent editions prominently featured an engraving of the hotel in a Tremont Street scene opposite the table of contents, with an additional engraving accompanying a lengthy description inside. While the book listed about fifty other Boston hotels, boarding houses, and taverns, none of the several other descriptions compared in length or detail, at least through the decade of the 1830s, to those given the Tremont. In a quite different arena, *Parley's Magazine for Children and Youth*, which began publication in Boston in 1833, featured a picture and description of the Tremont House in its first issue, indicating a continuing pride in the building and highlighting the spirit of genteel respectability its publishers hoped to foster among its young readers.[95]

The Tremont House's place in history derived from the layering of modern ideas over traditional forms. Its proprietors and the city leaders who gave the project their support regarded the hotel as both a representation of modern commercial success and an important component of Boston's present and future ranking in the global marketplace. Isaiah Rogers incorporated the necessary elements of a first-class hotel and superimposed modern ideas of luxury upon them. These included innovations in spatial design and technology such as indoor plumbing, remote communication, and patent locks. These expressions of technological luxury embodied ideas about material prosperity as a symbol of progress, the role of luxury as an engine driving capitalist enterprise, and the energy of the emerging industrial age. Interpreting the hotel as a product of a new technologically framed culture prompted Tyrone Power to describe the Tremont as a quietly regulated machine.[96]

Housed within a building of monumental proportions and style, the hotel filled its institutional role and also became a city landmark. Its architecture represented the Whig entrepreneurial spirit that emphasized private investment supported by government policy in the name of public progress. The same developments in technology, manufacturing, and capitalist structure that defined the hotel also produced a class of people with rising levels of disposable income that compelled them to demand accommodations commensurate with their station in life. The Tremont provided services for the community, most obviously the furnishing of food and housing in a class-specific environment for both travelers and city residents. It served as a social center for club meetings, political dinners, and balls. For example, in election years the Massachusetts Electoral College held celebratory dinners at the Tremont.[97] The "Saturday Club" of Ralph Waldo Emerson, Henry Wadsworth Longfellow, Oliver Wendell Holmes, Asa Gray, Charles Eliot Norton, and Benjamin Peirce met there.[98] Exchanges of information took place in the parlors, library, barroom, and dining room. Commercial enterprises such as the bathing rooms, stores, post office, barber shop, and stage lines accommodated travelers and Bostonians alike. The concentration of these functions at one site gave rise to the metaphor of the hotel as a "city within a city."

As a central gathering place for Boston society, the Tremont House became a key institutional setting for public discourse on the wide range of issues facing a nation still working out key ideas about politics, capitalism, and society. While ostensibly dominated by the Boston elite, the Tremont welcomed a steady stream of visitors from across the nation and Europe and served as a marketplace of ideas and commerce. The elaborate settings of its public rooms dominated and shaped an emerging public sphere that included a stylized parlor society. These provided an environment for what horrified Britons and Europeans regarded as a promiscuous mixing of social classes that characterized the growth of the American bourgeoisie. Proximity did not erase social distinctions. Rather, the opportunities afforded by the public rooms gave rise to both a broadening of civil society and an ironic breakdown of private family life, as such functions as dining, courting, and socialization took place in public. Thus, the hotel hosted a constant flux of private individuals engaged in several realms of the nation's work: the political, the commercial, and the social. These activities became even more public as national mass-market newspapers and periodicals chronicled the rise of the buildings as well as the activities within.

The Tremont also served as a symbol for the city of Boston. In much the same way that mansion houses represented the personal achievement and refinement of the wealthy elite, the Tremont spoke to the nation about Boston. The expansive granite front, the four towering columns, and the broad staircases that swept its guests into its midst distinguished the hotel from all others in the city, representing to the world the combined accomplishments of the Boston commercial community. The building's well-publicized $200,000 price tag provided a quantitative standard of investment that implied a corresponding qualitative achievement of decorative sophistication, capitalist endeavor, technological proficiency, craftsmanship, and refined social behavior. The publication and distribution of Eliot's book added to the public construction of the Tremont House as a production worthy of emulation and of Boston as its beaming, prideful progenitor.[99]

The Tremont and other hotels of the same period appeared, especially to foreign travelers accustomed to a rigid, institutionalized social hierarchy, as democratic and egalitarian institutions. While these perceptions hardly accounted for the social and economic restraints that in fact limited patronage of the hotels to those having both financial means and cultural acumen, in some ways this was so. The hotel's Greek Revival architecture sought to invoke and reify a democratic political structure that allowed capitalist enterprise to flourish. The interior design mimicked the relationship of citizens to the geography of their nation. The massive but beautiful building belonged in its totality to all guests, just as the nation belonged to its citizens, but was divided into many small individual rooms that maintained the primacy and integrity of the individual. These, of course, were idealized visions. The same economic structure that enabled the production of great wealth also resulted in widening gaps between the rich and poor as well as a class-stratified society whose informal yet palpable demarcations restricted access to places like Tremont to all but the wealthy.

While the nature of the hotel, as well as its location, cost, and luxury, clearly marked it as an elite institution, as a symbol of the city's prosperity and national stature, the Tremont instilled pride in many Bostonians. All who strolled along Tremont Street could appreciate the building's exterior beauty and grandeur. The lithograph of its facade in Eliot's book depicts fine carriages and elegantly dressed people passing by. However, on the opposite side of the street, two tradesmen, clearly identified by their clothing, caps, and carts, chat comfortably and gaze across the street at the hotel. Sheet music for the "Mechanicks Quick Step," a composition dedicated to the Mechanicks Rifle Company, used the hotel as a backdrop for the line of uniformed rifle corpsmen pictured on the cover. The mechanics' symbol, a bent muscular arm, sleeve rolled high, holding a sledgehammer, floats above the hotel's image. This connection between the hotel and the men who both built and defended their state indicates a pride of communal ownership that transcended both class and politics.[100] Whereas Eliot's lithograph represents a romanticized construction of the Whig universe, the Mechanicks' picture evokes an honest feeling of participation in nation building.

The moniker of a "palace of the public," first posited by the *National Intelligencer,* suggests another way in which hotels assumed a democratic character. European royal palaces continued to be constructed through the mid-nineteenth century.[101] Hotels built with rooms numbering in the hundreds were the only comparable building type in the United States. In contrast to the codified exclusivity of aristocratic royal palaces, large city hotels allowed in "the people," at least those who could pay the rent. At the same time, this was precisely what limited their egalitarian nature. The cost of two dollars per day effectively sifted out the middle and lower classes, preserving the hotel as a province of the wealthy.[102]

This, however, was not at all inconsistent with the ideology of the Boston elite, who believed wholeheartedly in the egalitarian myth and the idea of the "self-made man." While obviously not all people shared equally in America's abundance, the myth held that equality of opportunity allowed those possessing the proper work ethic to succeed and move upward through the social classes.[103] For the Whigs who built the Tremont, the transformation of labor into capital by thrifty industrious democrats would result in the accumulation of wealth, at which time barriers to the hotel would cease to exist. While this was unlikely for the vast majority of immigrants or working-class laborers, belief in the possibility allowed the proprietors to endorse the hotel project as a necessary enterprise beneficial to all.

The many printed descriptions of the Tremont House, save Eliot's book, did not include the technological elements of the building. The importance of these systems lay in their contribution to a new definition of luxury that included technologies and reinforced the idea of luxury as progress. Through its novel mechanical systems, the hotel came to be seen as a technological artifact, one that could be endlessly improved upon to create even more pampered and luxurious surroundings. With the Tremont House, technology represented the innovative character of its investors. By 1834, when the Astor House opened in New York City, technology had become an attraction in its own right, one that defined the building. The tradition of modern design and technological competitiveness that had its roots in the Tremont House did not only create a new technological aesthetic, it also limited the hotel's days as the nation's foremost hostelry, as it eventually fell victim to the obsolescence that modern development ensured.

....

CHAPTER THREE

THE PROLIFERATION OF ANTEBELLUM
HOTELS, 1830–1860

....

"Every Thing Is on a Gigantic Scale"

T HE LUXURY HOTEL concept seemed to tap into a well of American enthusiasm, as urban developers proposed hotel projects throughout the United
States and journalists and writers chronicled their rise. As hotels proliferated in
American cities, major American and British periodicals regularly published descriptions and commentaries about hotels that recognized and reinforced their
centrality to urban social and cultural life for local communities and the traveling
public. Foreign visitors took special notice of the buildings, not only because they
differed so from English and Continental accommodations but also because they
appeared to represent key aspects of an apparently distinctive American character,
particularly through their size, inventiveness, accessibility, and what commentators often referred to as "gregariousness," code for promiscuous public socializing.
As one British journalist observed about New York City hotels in 1857, "Every thing
is on a gigantic scale, in character with the land of the Mississippi and Niagara."[1]
Three key themes emerge from the antebellum commentaries.

First, the hotel became a prominent venue for political and social events for
men and women as the buildings' large communal rooms evolved into increasingly
elaborate staging areas for a growing public parlor society whose activities ranged
from public mating rituals to large state dinners. The expansion of social activities
away from private parlors to commercial parlors in hotels, steamboats, and later
department stores found a particular resonance in the hotel where both locals
and travelers convened in fashionable array. Certain hotels became known for accommodating particular groups of people. For example, for decades New York's
Astor House retained a reputation as a meeting place for politicians, particularly
Whigs. Westerners were known to frequent the St. Nicholas, and Californians the

Metropolitan. The New-York emerged as a democratic stronghold to offset the Astor House's predilection for Whigs, and the Fifth Avenue Hotel had a reputation for hosting stockbrokers and gold speculators. Newspaper accounts, magazine articles, special dinner menus, diaries, and hotel account books and records all detail the bewildering variety of affairs that took place in the luxurious environment of hotel dining rooms and parlors.[2]

Second, even though the basic format for successive hotels remained literally cast in stone, developers and architects embellished on the design with complex technological systems that redefined notions of comfort, service, and luxury. This escalation in decoration, technology, size, and cost began to suggest that certain consequences might ensue from unbridled development. The steady succession of new construction continually set increasingly—and often alarmingly—greater standards for size, opulent decoration, and the application of mechanical invention. As a result, the newer, grander buildings rendered "older" buildings obsolete within two or three decades, contributing to the continuous cycle of tearing down and building up characteristic of nineteenth- and twentieth-century cities.[3] More than one commentator in the 1830s imagined that the 1836 Astor House would serve as a monument "for centuries to come," when, in fact, in less than twenty years it would already yield its preeminence to the newer buildings uptown.[4] This dizzying tempo of development was characteristic of the nineteenth-century ideology of progress, which viewed such advancement as evidence for the continuing perfectibility of civilization.

Finally, as the leadership in finance and commerce began to relocate to New York City, so did the model for hotel design. Because New York served as a major port for domestic and international travel and shipping, the demand for fine accommodations there increased dramatically during these years. This need received an additional boost from New York's 1853 Crystal Palace Exhibition, an industrial fair that followed on the example of London's highly successful Crystal Palace Exhibition two years before. New York's effort drew 5,272 exhibitors to the city as well as over one million visitors. Beginning with John Jacob Astor's Astor House (1834–36, demolished 1913) and culminating with the splashy Fifth Avenue Hotel (1856–59, demolished 1908), hotel entrepreneurs participated in an escalation of standards that a 1930s historian characterized as a game, "like the annual blooming of new models of motor cars and radio sets."[5] Smaller cities could support only one rendition of a "monster hotel," as the very large hotels came to be called. New York had enough traffic and people to support many.

Emerging technologies and technological systems fueled and defined this development. Remarkable changes in transportation resulted in increasingly larger numbers of travelers, creating a corresponding need for accommodations. The mechanization of many industries that supplied hotels with opulent furnishings, such as textile, carpet, and furniture manufacturing, helped to satisfy the clamor for luxury products on a heretofore unprecedented scale. Within the hotel itself, the application of steam power to the laundry, to the kitchen, to the movement of goods, and to the heating system, as well as improvements in plumbing and communication, created new definitions of comfort and luxury that in turn fed the hotel patrons' expectations for continued inventiveness to respond to their needs and wants.

Hotel Sociability

Philip Hone's diaries, meticulously written between the years 1828 and 1851, are a goldmine for historians interested in New York politics and social conventions.[6] As one of the wealthiest and most prominent men in New York City, Hone lived at the center of that city's most notable society. Quite literally a "big" Whig, Hone may have been, as one editor of his diaries stated, "the last man to be able to know personally every one of importance in the United States."[7] His engaging diaries are replete with descriptions of meetings with the people who dominate U.S. history textbooks. Many of these meetings and events took place at hotels throughout the country and in New York City, most notably the City Hotel and Astor House, bastions of Whig gatherings. These references are important because they illustrate how crucial hotels were not only to the natural course of Hone's daily activities but to the life of the nation as well. As the settings for dinners, political rallies, balls, club meetings, celebrations, auctions, parade reviews—all in addition to overnight accommodations—hotels satisfied the requirements of a wide range of urban functions and centralized them so as to create a social, political, and commercial hub.

As a person at the center of New York's social and political world, Hone attended most of the city's prominent events. In 1837, he described the Booksellers' Dinner for 277 men as "the greatest dinner I was ever at, with the exception perhaps of that given to Washington Irving on his return from Europe." Held at the City Hotel, the Booksellers' festivities were "gotten up, arranged, and conducted in admirable style." Later that same year, Hone described a banquet at the Astor House in honor of John Bell of Tennessee with about 220 men in attendance. The guests arrived at 7:30 in the evening and ended up staying all night. There were "speeches upon speeches" and endless toasts. Finally, at two in the morning, Daniel Webster rose to speak. Urged on by the boisterous company, Webster kept at it until four in the morning. Many smaller personal dinners were consistent features of Hone's life as well, such as the one with Webster for "a select knot of four and twenty Whigs" at the Astor House in 1842, and a festive evening sampling the gustatory virtues of Merrimack salmon, again with Webster at the Astor House.[8]

Hone commented on the growth of various towns that he passed through on his travels, frequently mentioning the existence of a good hotel as a measure of a locale's progress. In 1833, he remarked on the development of several small towns, especially that of Amsterdam, New York. "When I knew it formerly," Hone wrote, "it was a mean, inconsiderable Dutch village: now it is a thriving town with large stove manufactories, etc., and an excellent hotel." July 1842 found Hone traveling through Binghamton, New York, reacting with the same excitement: "Instead of a dirty depot of lumber which I expected to find, here is one of the handsomest towns I ever saw. . . . There is a long street of business with brick stores, spacious hotels, a magnificent court house, academy, female seminary, and such other buildings as mark the progress of wealth and enterprise." For Hone, a comfortable hotel in the center of town affirmed his own entrepreneurial spirit and symbolized a level of success befitting the animated character of what he termed the "go-ahead" age.[9]

In addition to the hotel serving as a key element of a city's robust economy and a significant social venue, the antebellum luxury hotel was also an important setting for political activity. Hone attended national political conventions and presidential inaugurations, for which the hotel served as the setting for significant auxiliary events. As an example, President Martin Van Buren visited President-Elect William Henry Harrison in Harrison's rooms at Gadsby's Hotel in Washington, D.C. prior to his 1841 presidential inauguration. In addition to turning its dining room into a giant dormitory, Gadsby's erected a temporary building in its courtyard in which to feed the "immense mass of animated Whig matter" that had congregated for the inauguration. Hone's description of the 1844 Whig nominating convention in Baltimore reveled in the continuous hubbub at Barnum's: "The *affair* begins to subside. The masses are breaking apart and leaving Baltimore by every practicable conveyance. The crowd is still great about Barnum's." Other entries recorded smaller Whig meetings at the Astor House, political speeches delivered from hotel balconies, receptions for men such as Henry Clay, Webster, and visiting presidents, and even candidate meetings for Hone himself, during his unsuccessful bid for the New York State Senate.[10]

Hone's diary is valuable because it unselfconsciously conveys both the quotidian life of the antebellum social and political elite and the manner in which the urban luxury hotel served that constituency. Most Americans, however, learned about these activities through the major middle-class mass periodicals that recounted events and activities at the major hotels. Many of these articles displayed dramatic engravings that showed the urban hotel functioning as a hub of activity. *Gleason's Pictorial Drawing-Room Companion,* a Boston weekly magazine catering to a national middle-class audience that commenced publication in 1851, is a particularly strong source for the 1850s. Indeed, its offices were very close to the Tremont House. In 1851, Louis Kossuth, the Hungarian revolutionary hero, elicited great enthusiasm as he toured the United States. An engraving of his reception in front of New York's Irving House (named for New York's favorite son, Washington Irving) shows the street packed with hundreds of men, including the participants of a torchlight procession. Kossuth appears on the balcony of the hotel's front portico, under a regalia of banners extending three stories above him. According to the accompanying text, Kossuth was "compelled many times to appear on the balcony and say a few words to the whole souled multitude, who seemed crazy to have him ever before them."[11]

That same year *Gleason's* carried similar engravings of equally exuberant crowds in American cities where, in one example, a vast multitude filled the square in front of Boston's Revere House to receive President Millard Fillmore.[12] Another engraving recreated the scene of Daniel Webster addressing "an assembly of citizens" in front of the Revere House as a consequence of his being denied a venue at Faneuil Hall. As the crowd of men doffed their tall black hats to their beloved orator, adoring women hung out the hotel's windows waving and tossing handkerchiefs.[13] Still another issue depicted Webster atop the front portico of Philadelphia's United States Hotel addressing a well-dressed crowd of men and women in the street below, while onlookers peered out the hotel's open windows.[14] These examples demonstrate the way in which the hotel served as a central gathering place for public activities that offered a degree of protection to notables as they participated in crowd-drawing events. After standing on the balcony or at a window of the hotel, famous persons could easily retreat into the privacy and protection of their private rooms.

An 1862 *Atlantic Monthly* article by Nathaniel Hawthorne describing a tour of Virginia battlefields and military camps further illustrates the way that hotels of the era served as a key venue for the exchange between private interests and informal governmental structures. In the article, Hawthorne claimed that Washington D.C.'s Willard Hotel "may be much more justly called the center of Washington

Daniel Webster, speaking to crowds in front of the Revere House, Boston. While one or two women appear in the street crowd, the hotel itself serves as a protective barrier separating the hordes of women leaning out of the windows from the crowd of men below. *Gleason's Pictorial Drawing Room Companion,* vol. 1, no. 3, May 17, 1851, 37. *Courtesy, American Antiquarian Society.*

and the Union than either the Capitol, the White House or the State Department." Willard's had already earned a reputation as a mecca for Union officers and as a staging ground for politicians. As Hawthorne noted, "Everybody may be seen there." Hawthorne described the multitudes that gathered at the hotel to conduct the nation's business: "You exchange nods with governors of sovereign States. You elbow illustrious men, and tread on the toes of generals; you hear statesmen and orators speaking in their familiar tones. You are mixed up with office-seekers, wire-pullers, inventors, artists, poets, prosers (including editors, army-correspondents, *attachés* of foreign journals and long-winded talkers), clerks, diplomatists, mail-contractors, railway-directors, until your own identity is lost among them."[15] In addition to recreating for readers the air of bustle and intrigue at the hotel, Hawthorne's vignette also captured the mixing of classes that so disturbed and fascinated foreign visitors used to the more formal separation of classes characteristic of monarchical societies. Even the president-elect stayed at the Willard. The hotel's 1861 ledger contains the entry for "Hon. A. Lincoln & Family," who resided there from February 23, 1861, until Lincoln's first presidential inauguration on April 4, 1861. He occupied parlors 6 and 8 and settled his bill for $773.75 on April 19, 1861, using funds from his first paycheck as president of the United States.[16] Future incarnations of the Willard Hotel served in the same way. At one point it became a temporary White House for Calvin Coolidge, who lived there as vice president and remained a resident there while he transitioned to the presidency after President Warren G. Harding's death.[17]

In addition to being a place for these kinds of public political events, the urban luxury hotel became a social center as well. One traveling Frenchman noted that the great hotels of America produced more marriages than private society. Linking hotels to the economic health of the country in a somewhat unusual way, he noted, "These establishments, therefore, have a great importance in a country where the increase of the population is considered as one of the principal elements of its prosperity."[18] As an illustration of how matches might occur, a short story published in an 1857 issue of *Putnam's Monthly* described a four-year courtship conducted during the annual visit to a New York hotel by a Georgia plantation owner and his daughter on their way to Saratoga Springs. The first year, the lovesick narrator merely gazed at his intended and plotted ways to secure her acquaintance. The second year, he managed an abbreviated introduction to the young woman's father. The third year, in conversation once more with the father, he learned that the planter knew the young man's father, but, to the suitor's horror, reminisced that the boy's father had treated the planter badly in business. Hopes dashed, the fourth year's visit brought further unhappiness with the disclosure that the young woman had married. As the story progressed, the narrator revealed the many ways that strangers mingled at the hotel, in the dining room, barrooms, billiard rooms, lobby, all of which provided opportunities for alliances and dalliances. The story also highlighted the regularity with which travelers returned to the same hotel year after year. The young man depended on that constancy to learn about the planter and his daughter through a network of friends and hotel clerks. He was able to use these connections to gain an introduction. While this particular story failed to result in the hoped-for union, it nonetheless depicts a social setting in which strangers commingled and made new acquaintances and affiliations, a setting very different from one based on private entertainments and closed social circles.[19]

William Chambers, the Scottish editor, teasingly called the long corridors ad-

The caption for this drawing by caricaturist Thomas Nast reads, "Abraham Lincoln in Willard's Hotel in Washington, on the eve of his inauguration, sitting by the fireplace, friendless and solitary, holding newspaper; March 1861." *Courtesy, Library of Congress.*

joining the parlors the "flirtation-galleries" because of their "qualities as places of general resort and conversation."[20] Here, elaborately dressed ladies displayed themselves, talking with each other or their gentlemen callers. Hotels such as the Astor House sponsored fancy dress balls for their guests' entertainment. These events were the highlight of a hotel stay. In an editorial deploring the idea of private hotel dining being experimented with at a new uptown hotel, the *New York Weekly Mirror* commented: "The going to the Astor, and dining with two hundred people, and sitting in full dress in a splendid drawing-room with plenty of company—is the charm of going to the city! ... A regal drawing-room at her service, with superb couches, piano and drapery, and costing no more than if she stayed in her bed-room—plenty of eyes to dress for if not to become acquainted with, and very likely a 'hop' and a band of music—bless my soul, says the country-lady, I hope they'll never think to improving away all that?"[21] Chambers noted that the Astor House swarmed "like a hive." The setting of indescribable luxury, where rich velvet, lace, satin, gilding, carpets, and mirrors combined to create what he called "an Elysium of princely drawing-rooms and boudoirs," constructed a theater for romance and social relations within an expanding network of capitalist courtship and enterprise.

New Hotels, 1830–1860

Magazine and newspaper editors in both the United States and abroad reported on the construction of new luxury hotels in America's major cities as important civic and national events. Local newspapers described new construction in depth, and in effect, made the atmosphere of great social change more concrete. Foreign periodicals such as London's premier architectural journal, *The Builder*, and London and Edinburgh's *Chambers's Journal* kept their readers abreast of hotel development in the United States. Newspapers typically reprinted local accounts of hotel openings in cities other than their own. Frank Gleason of *Gleason's Pictorial* in particular viewed the hotel system as one of the features of the country and, in "rejoicing at their proliferation throughout the country," appreciated these first class hotels as "one of the advancements of civilization and refinement."[22]

An 1863 article in *The Builder* describing the Lindell, a new St. Louis hotel, chronicled the lineage of luxury hotels in the United States beginning with the 1836 Astor House, which the article said had held its ground as "*the*" hotel for twenty years. The author recorded in chronological order the great hotels that had built upon one another's excesses in size, décor, and modern improvements: New Orleans's St. Charles; St. Louis's Planters, Cincinnati's Burnet House; Chicago's Tremont and Sherman houses; Memphis's Gayosa; and finally the large New York City hotels of the 1850s, the Metropolitan, the St. Nicholas, and the Fifth Avenue. Each of these hotels "excelled," "eclipsed," "overtopped," "surpassed," and "cast in the shade" the best hotel of the immediate past, demonstrating the progressive competitiveness and the national diffusion of the luxury hotel idea as originally executed by Boston's Tremont House.[23] As each of these hotels became costlier, investments grew riskier, requiring more sophisticated financing and creating an even greater dependence on luxury innovation to attract a large enough share of business.

The New York City hotels of this period demonstrate the escalating tempo of development and the enthusiasm that attended their construction. John Jacob Astor and his son, William B. Astor, built the Astor House from 1834 to 1836 as a joint project, choosing to build on the elder Astor's home site, a section of Broadway across from City Hall Park, between Vesey and Barclay Streets. While this was still a residential neighborhood, it was rapidly turning into a commercial area where the city's elites came to conduct business and to be seen in an ever-constant parade of activity. One editor noted that "crowds of beauty and fashion, domestic and imported, fill this part of the promenade of Broadway, for Astor's Hotel is on the *fashionable* side of Broadway, and here you are sure to find the *elite* of the commercial metropolis."[24] Astor chose Isaiah Rogers to be the architect. With the Astor House, Rogers designed an aggrandized version of his earlier Tremont House. The New York hotel was a spectacular six stories compared to the Tremont's four. It contained nearly twice as many rooms, larger than most in the Tremont, and cost twice as much to build.[25] Yet the floor plan and the four Doric columns and Quincy granite facade imitated that of its predecessor.

Many national newspapers—Baltimore's *Niles' Register,* Boston's *American Traveller,* and Philadelphia's *Atkinson's Casket,* to name a few—and local city papers gave detailed reports on the Astor House's construction and opening. The *New Yorker* characterized the interior furnishings as being of a "style of unostentatious richness and severe simplicity," the furniture custom-made from "strikingly handsome," highly polished black walnut. The paper pronounced the dining and drawing-rooms as "unequalled."[26] Marble columns and blue and white marble mosaic floors added to the elegant atmosphere, as did the enormous framed wall mirrors and carpets, which the *American Traveller* effusively labeled "imperial," owing to the mirrors' ornate gilding and carpets' lush pile.[27] The *New Yorker* described the hotel as "thoroughly and characteristically *New York,*" indicating a style associated with excess and a certain cultural superiority derived from New York's commercial preeminence.

In addition to the larger size and rich décor, the Astor House incorporated dramatic improvements and new mechanical systems into its design. Unlike other gaslit buildings that depended on often unreliable commercial gas manufactories, the hotel had its own gas manufacturing plant to supply the gas-fueled lighting, which continued to garner its share of attention. In its report on the hotel's opening, the New York *Constellation* observed, "The house was lighted by this gas everybody is discussing." After noting that the gas had ironically given out during a cotillion due to the quantity being consumed, however, the paper warned, "Gas is a handsome light, but liable at all times to give the company the slip; and it is illy calculated for the ordinary use of the family."[28] The hotel served as a site to showcase technologies that had not yet made the transition to the domestic household. The rotary steam engine located in the basement drove force pumps to raise water to

The Astor House, located on Broadway at City Hall Park, persevered as one of New York City's landmark buildings until its demolition in 1913. This early engraving captures the hotel's fashionable location and the bustle of an early-nineteenth-century street. *Warshaw Collection of Business Americana—Hotels, Archives Center, National Museum of American History, Smithsonian Institution.*

all floors and provide power to the various pieces of machinery in the kitchen and laundry. A mechanical clothes dryer used heat conveyed by steam pipes to a special room to dry guests' clothes and hotel linens. Despite its commercial nature, many of the hotel's functions replicated household ones, thus making their mechanization seem even more dramatic and fantastic.

Attic cisterns supplied water to the seventeen bathing rooms and two showers located on the ground level of the Barclay Street wing.[29] Plumbing extended throughout the house, with water closets and hot and cold running water on all floors, further privatizing grooming and body management. In this way, technological improvements enabled the emerging parlor society characteristic of antebellum America to execute its preparations privately, thus obscuring the personal mechanics necessary to create a genteel appearance. As historians and sociologists have demonstrated, rigid systems of etiquette enjoined against any adjustments to personal appearance in public settings. Similarly, the hotel maintained a front and back of the house that both concealed the mechanics of hotel work and created a genteel public atmosphere for socialization. Bathrooms and water closets in the upper guest room corridors adjacent to sleeping rooms provided hotel patrons with a "back of the house," where they could prepare themselves for public society in much the same way actors prepare themselves backstage for performances.[30]

The considerable power of the Astor's steam engine elicited a sense of marvel. As the *New Yorker* facetiously noted, "We believe it does not yet stipulate any assistance in bed-making, sweeping rooms, dusting furniture, attending on guest [*sic*], &c., &c., but in the onward march of improvement, we may expect all this to follow in good time."[31] Other equipment that lent an atmosphere of ever-increasing technical refinement and self-sufficiency included a printing press to print the daily menus, a bell system, and patent locks, all of which added to the hotel's reputation as "the greatest establishment of the kind in the world." The hotel elaborated on the number of specialized rooms and included a grand dining room for gentlemen, a number of smaller dining rooms for women and accompanying men, smoking rooms, a reading-room, and a bar-room. In addition, the hotel housed a cafe with an attached cloak-room as well as retail areas for tailors, boot-makers, barbers, and hair-dressers; an apothecary; and an umbrella-maker. An oyster cellar remained open until 4:00 a.m.

The visiting Frenchman described the Astor House, not surprisingly, as "a city in the midst of a great city."[32] The hotel's reputation was so widespread, both in the nation and overseas, that no hotel could ask for finer praise than being known as the "Astor" of its town or city. Over the years, of the dozens of parlors, No. 11 emerged as the Whig and Republican headquarters. In 1875, the *New York Times* reported that the plans for all Whig political campaigns, conventions, and candidate lists had taken place in Parlor No. 11, which was at all times available to Thurlow Weed, one of New York's great political bosses.[33] Thus, the hotel remained a key site at the intersection of commercial, social, and political life for decades, even as other hotels sought to capture its place among city institutions.

Several hotels were built in New York City throughout the 1840s, but none except perhaps the Irving House (1848) seriously challenged the Astor House until the Metropolitan opened in September 1852. The Metropolitan was one of a number of enormous hotels built to accommodate visitors to the 1853 New York Crystal Palace Exhibition. Reputed to cost nearly one million dollars, the Metropolitan could accommodate six hundred guests (one thousand if necessary) in a setting of resplendent magnificence said to exceed that of all other New York hotels.[34] As befitted comments on all new hotel construction, the *New York Herald* claimed, "No expense ha[d] been spared in rearing the edifice, or in fitting it out with all the modern improvements." Located on Broadway at Prince Street, the Metropolitan offered the usual assortment of ground floor shops, richly decorated public rooms and parlors, and single guest rooms and apartment suites with private baths. As was common, detailed descriptions of the opening ceremonies, the interior, and the hotel's equipment filled the papers, reinforcing the importance that the buildings held for the city. Writers took special notice of the newly invented spring mattresses overlaid by hair mattresses that furnished all the beds and lent a new and welcomed degree of comfort to accommodations. The boilers and steam engine occupied the "Engineer's Room." Steam heated all the public rooms and passages, radiating from pipes in the walls. The lighting, plumbing, and heating systems used twelve miles of gas and water pipes. *Chambers's Journal* asserted that the Metropolitan's laundry routinely processed four thousand articles per day, with the capability to wash, dry, iron, and deliver a piece of laundry in fifteen minutes. These "astonishing details" were rendered credible, according to Chambers, by the "well-known tendency of the Americans to conduct operations on the factory or large system more extensively even than is practiced in England."[35]

The Metropolitan retained its lead for only a few short months. In early January 1853, the St. Nicholas Hotel opened, also on Broadway, between Broome and Spring Streets, the first hotel that assuredly cost over one million dollars.[36] The newspaper and magazine reports of its furnishing read almost as an accountant's ledger: curtains in the ladies' parlors, eight hundred dollars each; reception room draperies, one thousand dollars each; the grand piano, fifteen hundred dollars; mirrors in the barber shop, eight thousand dollars; rosewood barber chairs, three thousand dollars; a chandelier, one thousand dollars; and on and on. In the modern world of 1853, money served as the metric by which material values could be assessed and compared. Enumerating the cost of the furnishings reassured patrons that their hotel ranked highly within the world of goods. The *New York Times* referred to the St. Nicholas as the "Hotel *par excellence*" of the day, for the "extent of accommodation, completeness of arrangement, costliness and chaste elegance of decoration, and combination of all modern improvements."[37]

Descriptions pointed to equally lavish expenditures for both the women's and men's spaces. Phalon's Hair-cutting Saloon at the St. Nicholas, the city's foremost barber shop, cost nearly $20,000 to outfit and included brilliant chandeliers, thirty massive mirrors, frescoed ceilings, and the aforementioned carved rosewood chairs. Each "artiste," as the barbers were called, wore a black velvet coat and worked at a station adorned by magnificent washstands and statuary. Edward Phalon's specialty was his talent in adapting the "length, and parting, and dressing of each man's hair to the shape of the head, the contour and expression of the face, and the spirit of the character."[38] The St. Nicholas created luxurious public spaces for

Ballroom of the Metropolitan Hotel. In 1860, the official delegation from Japan toured the United States, signing the first treaty of commerce between the two countries. Every city they visited sponsored elaborate balls to welcome the ninety-three Japanese visitors. This double-page illustration from *Harper's Weekly* draws attention not only to the elaborately dressed guests but also to the decorations that augmented the already highly ornate ballroom. *Harper's Weekly,* June 30, 1860, 408–9. *Provided courtesy HarpWeek.*

both men and women, spaces meant to convey both "home-like comfort" as well as "unequalled magnitude."[39]

The six-story white marble hotel was composed of three buildings connected by passageways on each of the stories. For fire protection, separate buildings housed the laundry and kitchens. Steam heated the hotel and flowed to all parts of the building through "metallic pipes." Gas and Croton water coursed through the fifteen miles of piping that threaded its way through the hotel walls, creating an infrastructure that brought new levels of technological luxury to the hotel's customers. Every bedroom had hot and cold running water, but suites had the unprecedented luxury of private baths and water closets. Extensive provisions existed for fire protection, including fireplugs, hoses, and alarms, as well as water reservoirs and pumps. A six-horsepower steam engine drove the machinery for the laundry and kitchen. The laundry featured both washing and drying machines capable of cleansing five thousand pieces of laundry per day. The kitchen's

The St. Nicholas Hotel, the first hotel to cost over one million dollars, occupied a large block on Broadway, shown here with its storefronts, which attracted the street's busy traffic. In the lower right corner, a circus-like team of eight festooned horses pulls a coach advertising Phalon's Perfumer, the fancy barber shop located at the hotel. *Collection of the New-York Historical Society.*

"new inventions" included jacketed steam kettles whose steam heat boiled liquids, eliminating the need for at least one large cooking fire. An "electro-magnetic" annunciator (a guest room panel that connected to the front desk) improved hotel communication. This new electrical system replaced the mechanical bell-wire system first heralded at Boston's Tremont House that, while path-breaking at the time, eventually proved inadequate because of the wires' tendency to stretch out and malfunction.[40]

The St. Nicholas overwhelmed New York society with its cost, décor, and level of technological sophistication. The *Daily Tribune* critiqued the drive within the hotel industry to supersede all competitors by commenting, "Whatever of splendid and gorgeous in this way money could procure seems to have been obtained for this hotel with a view to outdo every other in elegance and splendor." It further predicted a "new era" for metropolitan hotels whereby "henceforth they must be furnished without regard to cost." The St. Nicholas's management felt compelled to address these criticisms in an 1856 promotional booklet. Acknowledging that the project "was considered an experiment of doubtful expediency" and that critics had expressed fears that the hotel exceeded market demands, management could justify the size and cost. The booklet claimed that even in a short time, the hotel had demonstrated that "a thousand guests can be as well and commodiously lodged under one single roof as any smaller number, and that order, security, and all the comforts and luxuries of social life can be enjoyed in greater perfection in a colossal hotel than in one of inferior capacity."[41] The hotel was making claims for systemization and economies of scale that would later underpin the mass production strategies promoted by E. M. Statler and his contemporaries. Thirty-one-year-old John Vessey, a British landholder traveling in the United States, described the St. Nicholas as the "greatest hotel in Christendom," noting that "considering the number of guests at these hotels there is less bustle and confusion than a stranger would suppose," an indication that the hotel's systemized management had the desired effect.[42]

The *Daily Tribune*'s critical attitude contrasted pointedly with the superlatives that most other commentators lavished on the hotel. On the one hand, the reporter was bothered by what he considered to be compromises in the definition of luxury. He claimed that the upper-story single rooms had insufficient ventilation and were located too many floors away from the baths. This situation necessitated inexpedient treks and jarred with expectations for convenience and comfort. However, his critique also moralized against the decorating, which he claimed evoked gaudy North River steamboats rather than "real" elegance and good taste, "such as a gentleman of high culture, refinement and love of art would exhibit in fitting up a palace for his own use." In this way, he distinguished between a perceived pretense to style and that which was the product of genuine—that is, European—refined taste. This critique was part of a broader and ongoing commentary that questioned the intensifying consumption patterns through which middle-class parvenus sought to achieve social identity.[43]

Putnam's Monthly echoed these sentiments. It faulted the hotel's good looks, asserting that its newness paled when compared to Europe's weathered palaces and observed that "an over new look does not properly become a palace." As evidence, the writer noted that he and his companion were startled by their own inelegant reflection in the sumptuous mirrors, wryly observing that "it would have seemed but natural that in such splendor we too should be splendid." Truly, everyone looked like intruders, exposed by their "artificial manners and overlabored dress." A subsequent visit by the writer to a missionary's home in the poor, densely populated, crime-ridden Five Points neighborhood further exposed the moral perplexities and contradictions that the hotel represented for him.[44]

Writers christened these enormous buildings "monster hotels." With six grandly proportioned stories, the five daunting flights of stairs to rooms that travelers commonly referred to as "sky-parlors" seemed long past due to be conquered by an inventive mind. The hotel assigned these rooms to men traveling or living alone, reserving the more comfortable and better-located rooms and suites for women and families. One British traveler wearily claimed, "If anything would cure the most confirmed bachelor, it would be a tour in the United States."[45] The *Daily Tribune* advised, "There is steam power at hand and there are ingenious brains enough to invent an elegant and convenient apparatus to convey skyward the upward bound, and earthward the descending, without such excessive labor of mortal muscle." This call to arms for a passenger elevator demanded not just a practical mechanical solution, but an elegant one as well. It forcefully merged technology and aesthetics and further redefined luxury as a release from physical exertion. Indeed, climbing stairs was a laborious undertaking, especially considering the soaring fifteen- to twenty-foot ceiling heights, the discomfiting hot weather, and, for women, the weight of heavy costumes, which could include as many as seven to nine shape-defining petticoats. Before the invention of the tall building, the desire to conquer vertical space was a luxurious concept that intensified the guest's passive role. Soon enough, the 1859 Fifth Avenue Hotel answered the challenge by providing a passenger elevator.

Yet another immense white marble structure, the Fifth Avenue Hotel occupied over an acre of land on Fifth Avenue where it intersected Broadway between 23rd and 24th Streets across from Madison Square. Elisha Graves Otis had exhibited a passenger elevator at the New York Crystal Palace Exhibition in 1853, but it was not until 1857 that Otis installed one of his elevators in the E. V. Haughwout and Co. store at the corner of Broadway and Broome Street in New York City, an establishment famous the world over for its chinaware, chandeliers, mirrors, and other accoutrements of the rich.[46] When the Fifth Avenue hotel opened, it incorporated a different machine, a vertical screw railway patented by Otis Tufts, a mechanical engineer from Boston.[47] At the vertical railway "station" off the main hallway, guests entered a "little parlor," an ornate cab with plush seats, which carried them to the upper floors. The hotel elevator ushered in a new era that mitigated the compromises guests made by choosing pricier lower rooms that, while eliminating the need to climb stairs several times a day, were disturbingly noisy and dusty. Upper floor rooms, once normally forced on men traveling alone to save women the physical effort of climbing stairs, now became attractive for their cleaner air and distance from the unpleasantness of the streets below and increased the opportunity for de facto social stratification based on cost.[48]

During its construction, the hotel and its owner, Amos R. Eno—one of the city's wealthiest men—endured criticism because of its distance from the Broadway hotel district further south. Critics called the building "Eno's Folly." Once opened, however, the Fifth Avenue garnered accolades as the "best specimen we can offer of the possibilities of hotel luxury." The grand entrance hall measured an incredible 165 feet by 27 feet, with fifteen-foot ceilings. The business office, post office, telegraph and railway ticket offices, newspaper and book stand, and theater ticket office, stock and exchange telegraph, carriage and package offices, coat-room, billiard room, barber shop, and various smaller rooms as well as the passenger elevator were all accessible from the main hallway. These services, in addition to the dining facilities, eight public and 120 private parlors, and over five hundred guest rooms combined to produce the most elaborate hotel in the world.[49]

Eno contracted with Paran Stevens to be the proprietor, responsible for furnishing and managing the hotel. Stevens had already cemented a reputation for himself as the proprietor of Boston's Revere House (1847), the Tremont, and the Battle House (1852) of Mobile, Alabama. Stevens hired management companies to run the daily affairs of his hotels. For the Fifth Avenue, he engaged Hitchcock, Darling, & Co., who, along with an assistant hired away from the St. Nicholas, had learned hotel management from Stevens. Highly regarded in Boston and elsewhere, Stevens had perfected a system for managing "monster" hotels, and one hotel directory claimed he had "done more for the business interests of Boston . . . than any other man in Massachusetts."[50] By the time Stevens acquired the Fifth Avenue's lease, he had become known as the "Napoleon of hotel-keepers."[51] Under Stevens, the Fifth Avenue developed a worldwide reputation as the safest, healthiest, and "most comfortable hotel in the world."[52]

In September 1859, as elite New Yorkers returned to the city from their summer country pursuits, the city's hotels filled with visitors eager to enjoy the "fall season." The *New-York Traveller and United States Hotel Directory* claimed that hundreds were unsuccessful at obtaining rooms at the larger hotels. As attractions in their own right, the hotels themselves accounted for the hoards of visitors, as the comforts they offered were not "surpassed, if equalled, in any city in the world." The dining tables and the magnificence of the rooms at the class of hotel that included the Astor, New-York, Metropolitan, and St. Nicholas exceeded those of royalty. "Warm and cold baths are always at hand," the paper boasted, "laundries are provided of improved construction, furnishing a clean garment, while the owner is being shaved." The new Fifth Avenue Hotel had "put aside the lamp of Aladdin in wonders of dispatch, and richness of supply."[53] Hotel development in New York City had reached a level of inimitable standards for both technological and ornamental luxuries that promised and delivered a fairy-tale lifestyle to a class of wealthy cosmopolitan Americans.

Symbols of Civilization

During the thirty years between 1829 and 1859, standards for size, opulence, and technological comfort in America's luxury hotels rose dramatically. The basic design imagined by Isaiah Rogers remained in place. Shops, baths, and service areas filled the ground floor, surmounted by a floor with offices, entryways, and public areas, followed by one or more floors of parlors and private suites, and then several floors of guest rooms in various arrangements to suit individual requirements. However, the number and kinds of services and rooms greatly increased, as did the style of decoration, which became enormously more elaborate and costly. As America proceeded through its industrial revolution, technological improvements multiplied, too, as capitalists mined the capabilities of industrialization for their personal comfort and needs.

While New York City emerged as a center and standard-bearer for hotel development, the same forces of escalation acted on hotel development throughout the nation. Every major city boasted of its first-class hotel, its "princely establishment" or "noble and imposing structure," each of which was invariably presented as "one of the most extensive and elegant affairs of the kind in this country."[54] Anthony Trollope, in his 1862 book *North America,* devoted a separate chapter to American hotels because he found them to be "a great national feature in themselves," in the same way that the legislature or judiciary or literature might represent a national culture. "Any falling off in them," he stated drolly, "would strike the community as forcibly as a change in the constitution, or an alteration in the franchise." Amused by the way hotels figured so prominently in city after city, Trollope found that American hotels did not compare easily to those of other countries, that they appeared to be an institution unto themselves. Distinguished by their great size and proliferation, American hotels existed everywhere, built by speculators with the certainty that they would draw population, in the same way the railroads did in England and Europe. Moreover, Trollope despaired of the way young married couples, wedded to each other but not to occupations or geographic locale, set up "housekeeping" in hotels and thus were able to uproot their lives at any given moment for new and different opportunities.[55] Even if not particularly complimentary, Trollope's essay nevertheless confirmed how significant hotels were to American culture.

Domingo Faustino Sarmiento, an Argentinean statesman who traveled through the United States in 1847, mused that American hotels stood as substitutes for churches in a society that valued individualism. "More money," he claimed, "has been put into the Astor House than into any church in that city." New Orleans's St. Charles, built in 1837, served as the inspiration for Sarmiento's speculations. As he approached the city from the north, he could see, from the deck of his steamboat, the gilded dome of the St. Charles on the horizon, which called up his memory of St. Peter's in Rome. Fascinated by the association, Sarmiento postulated that whereas European cities derived their identities from cathedrals, American cities turned to their hotels: "Where the importance of the individual reaches the heights it has in American democracy, the temple's power diminishes in proportion to the multiplication of sects, and the hotel inherits the dome of the ancient tabernacle and takes on the aspect of the baths of emperors." In America, the hotel had appropriated the veneration formally reserved for religious and imperial shrines. At the St. Charles, a "Jupiter-sized" statue of George Washington—the father of the country—guarded the entrance, and a thousand gas jets lit the interior.

The dining room accommodated seven hundred guests, and the luxuriousness of the colossal public rooms succeeded in persuading Sarmiento's indifferent friend of the advantages of a democratic government. "I am converted," the friend exclaimed. "Through the intercession of Saint Charles I now believe in the Republic, in democracy, in everything. I pardon the Puritans, even that one who was eating tomato sauce straight, with the tip of his knife, and before the soup. Everything should be pardoned a people who raise up monuments to the dining room and crown their kitchens with domes like this one!" This enthusiastic but double-edged accolade was typical of many foreign observers who found themselves awed by the hotel but bemused and bewildered by the American customs and manners within. Surrounded by gas lighting, enormous public areas, and the company of seven hundred fellow visitors at the St. Charles, Sarmiento concluded that "if you want to know if a machine, invention, or a social doctrine is useful and can be applied or developed in the future, you must test it on the touchstone of Yankee knowhow." Hotels, Sarmiento asserted, played an extremely important role in a dynamic society such as the United States, "with an active life and a future."[56]

The enthusiasm that these mammoth buildings generated was dampened by

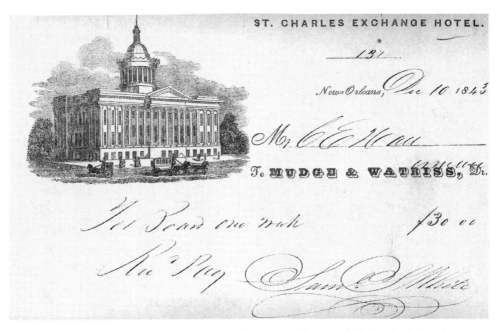

The gilded dome of the St. Charles Hotel in New Orleans, which inspired Domingo Faustino Sarmiento to muse on the role of American hotels as the "touchstone of Yankee know-how." *Warshaw Collection of Business Americana—Hotels, Archives Center, National Museum of American History, Smithsonian Institution.*

an unsettling temporality. An 1838 editorial in the *New-Yorker*, describing renovations to the American House, commented that the hotel had "risen suddenly from the obscurity into which it had been cast by its colossal neighbor," referring to the nearby Astor House. The American House's new management planned to stucco the hotel's exterior in imitation granite and build additional stories so as to harmonize with the Astor House's exterior. All this was meant to entice the return of the hotel's clientele. Yet, by 1861, George Sala described the Astor House as a "mere quiet second-class house" and, more astoundingly, the eight-year-old St. Nicholas as having "fallen into the yellow leaf."[57] The possibilities presented by increasingly available luxury fixtures and furnishings and the rapid emergence of new and refined technologies promising higher levels of comfort severely limited the expected lifespan of luxury hotels. The idea of creative destruction, through which old development gives way to new in the name of progress, found its quintessential instrument in the modern luxury hotel.[58] Despite the remarkable level of resources devoted to each new establishment, new technologies and the availability of ever more elaborate decorations insured that the "older" hotels would be forever locked in a game of "catch up" that they had little chance of winning.

As the emblem of the "go-ahead" society, developers, newspapers, pundits, and city officials embraced the luxury hotel as representing all that was progressive and successful about American society. They compared its outré decoration to the most luxurious of Europe's palaces, as if to prove the young nation's achievement of old-world sophistication. At the same time, the hotel's advanced technology trumpeted all that an inventive young country could promise. A short, effusive 1859 article in *Ballou's Pictorial Drawing Room Companion* affirmed that "hotel life is one of the most striking characteristics of American society, and our countrymen have certainly reached the acme of luxury and comfort in the vast public houses every large city and town can boast." Not surprisingly, its description of Boston's American House included the usual claims that "no expense has been spared" and that the hotel was "one of the largest in the world."

Taking an interesting twist, though, the article called attention to an anthropological description and a series of drawings at the top of the same page that featured a tribe of east African Abyssinians. As the nearly naked Abyssinians feasted on raw meat (described as "warm . . . eaten while smoking and palpitating"), their African eyes "sparkled with delight" as they tore the beef to pieces "with ferocious eagerness." The hotel article observed, "It is a little curious to compare the style of living at these fine hotels with the barbarism delineated in the preceding sketch. We have brought the extremes of life in contrast on the same page."[59] Certainly, references to the Abyssinians invoked comparisons to America's own "savages" currently being subdued on its own continent. By contrasting these examples of premodern life with the continually expanding extravagant extremes of America's luxury hotels, *Ballou's* reaffirmed for many Americans the idea of the luxury hotel as the nation's most significant standard of civilization.

CHAPTER FOUR

THE CONTINENTAL HOTEL, PHILADELPHIA, 1860

"In Reference to the Building of a Monster Hotel"

I N 1857, Philadelphia businessmen worried. New York City's booming growth as a port and commercial center contrasted sharply with Philadelphia's decline. Even before that year's devastating economic panic further eroded Philadelphia's position, a number of editorials in the city's newspapers urged local businessmen to sell their city's attributes to an otherwise oblivious national marketplace. "If we do our duty," the *Public Ledger* advised, "New York must sink into a second rank as a distributing market."[1] In this climate of competition and anxiety, a group of business leaders proposed constructing a first-class hotel. While certainly not a novel idea by this time, the combination of the nation's adoption of luxury hotels as symbols of successful cosmopolitan urban culture with New York's particularly enthusiastic predilection for building them presented a group of entrepreneurial Philadelphians with a sense of pressing need.

The story of Philadelphia's Continental Hotel is instructive because of the public debate that raged in the pages of the city's newspapers. For three months, the hotel project suffered unrelenting criticism from a threatened hotelkeeper, who, under the pseudonym "Franklin," wrote a series of letters to the editor of the *Philadelphia Evening Journal.*[2] At the time, the Franklin House was a popular hotel at 55 South Fourth Street in Philadelphia, near the location of the proposed hotel. Most likely, Franklin was the proprietor of this hotel. His attack provoked responses from interested parties, and the resulting tenor of irascibility persevered among certain of the shareholders throughout the building's construction. These letters, which addressed immediate concern over the proposed hotel, expressed a broader anxiety over a rapidly changing way of life. The arguments articulated attitudes about progress, competition, capitalism, luxury, technology, and the growing power of large corporations and the ways that these historical processes shaped the conditions of individuals acutely aware of living in times of momentous change.

By the late 1850s, America's large urban centers had changed dramatically. Ongoing industrialization, centralization of the commercial downtown, separation of work from home, growing stratification of social classes, rapid population growth, increased immigration, congested streets, and a host of other changes were indications of rapidly vanishing communal patterns of urban life. One needed only to look at Chestnut Street, the center of Philadelphia's fashionable society and the site of the proposed hotel, to witness this process. Once a mixture of public buildings, stately homes, and the magnificent chestnut trees that gave the street its name, the rapid development of a central retailing district there created what one observer derisively called a "wilderness of brick, mortar and brownstone." Concerns that several citizens voiced about the already unmanageable crowds of carriages on the street seemed to represent deeper worries about rapid growth and looming urban woes.[3] The controversy surrounding the Continental Hotel project is significant because it reveals how individuals and communities confronted the historical process of modern development.

Mid-Nineteenth Century Philadelphia

Sidney George Fisher, lifelong Philadelphian and noted diarist, wrote in 1839, "Returned to Philad: as I always do, with the conviction that dull, monotonous & humdrum as it is, it is the most *comfortable* & desirable place for a residence in this country."[4] Of New York City, he observed, "There is all the vulgarity, meanness, & ostentatious parade of parvenuism. Wealth is the only thing which admits, & it will admit a shoe-black, poverty the only thing which excludes & it would exclude grace, wit & worth with the blood of the Howards."[5] As Fisher observed, cities, like people, have personalities. And while a certain degree of haughty local pride might be expected from a third-generation Philadelphian, Fisher's assessments did not vary in substance from those of other observers and critics.

Antebellum commentary about both cities reiterated the contrast between solid, staid Philadelphia, remarkable for its endless, regular street grid, and its northern competitor, with all its noisy vitality. Frances Trollope, writing in *Domestic Manners of the Americans,* found Philadelphia to be a beautiful city, but one of "extreme and almost wearisome regularity." The predictability and monotony of its street pattern served as a metaphor for the tone of Philadelphia's society. For example, the lack of nighttime activity in particular disturbed Trollope, as "scarcely a sound is heard; hardly a voice or a wheel breaks the stillness." An after-theater excursion found Trollope and her companions alone in the streets despite the mild weather and early hour. She wrote, "We alone seemed alive in this great city." By comparison, Trollope described New York as "a lovely and noble city . . . much superior to every other in the Union." Trollope noted that in New York the shops remained open late in the evening and "all the population seem as much alive as in London or Paris."[6]

Dickens, too, drew similar contrasts between the two cities, noting with amusement the colorful dress of New York ladies. "What various parasols!" he exclaimed. "What rainbow silks and satins! what pinking of thin stockings, and pinching of thin shoes, and fluttering of ribbons and silk tassels, and display of rich cloaks with gaudy hoods and linings!"[7] By contrast, George Foster, a contributor to the *New York Tribune*, described the characteristics of Philadelphia female beauty as "a wholesome, exquisitely colored placidity of countenance, a roundness and regularity of features, a well-developed yet not overcharged contour of form, and a confident, unconcerned carriage."[8] Once again, the "regularity of features" of the Philadelphia women echoed the city's characteristic street grid, about which, after walking for two hours, Dickens observed that he "would have given the world for a crooked street."[9]

"In New York," Foster declared in 1848, "every inhabitant goes as if an invisible somebody was after him with a sharp stick: in Philadelphia we take our time."[10] Statistics supported this difference in approach, one that resulted, by the 1850s, in Philadelphia dropping to second place as measured by population, commerce, and manufacturing. Several historians have explored the various routes by which Philadelphia lost is primacy in the world of commerce. Chief among them were the influence of Philadelphia's Quaker origins, the fragmentation and divisiveness among the city's elites, the conservatism of the city's businessmen, as seen in their unwillingness to expand through the adoption of modern financing and their focus on individual wealth and achievement at the expense of communal needs. Foster's impressions support these academic analyses. "Capitalists and merchants in Philadelphia act and think only for themselves," he grumbled. Indeed, Foster claimed that if local businessmen had worked in concert toward economic goals, then Philadelphia's more southern location would serve to boost its advantages in the spring, offering an earlier opening of shipping channels. The early start would help to secure Philadelphia's preeminence with western and southern traders and encourage them to linger in the city, rather than spending their holidays in New York. It must have been all the more galling to the elites, knowing that travelers had to bypass Philadelphia in order to reach New York. Foster remained hopeful, noting, "All this will be remedied . . . and shortly, too."[11]

Foster's prediction proved too optimistic. In 1857, civic leaders continued to lament their place as the nation's second city. Edwin T. Freedley, author of business treatises and all-around commercial cheerleader for Philadelphia, published *Philadelphia and Its Manufactures,* a five-hundred-page book that described the region's suitability for investment and manufacturing and favorably portrayed the city's various business interests at length. It also calculated Philadelphia's secondary position in the "commercial firmament," going so far as to attribute Philadelphia's fall to a conspiracy perpetrated by her rivals. "Her enemies have industriously circulated, far and wide, reports which, if unexplained, must prove detrimental to her interests; and the declension of her foreign commerce has been to them a harp of a thousand strings." By contrast, Philadelphia's friends, in the absence of hard data to counteract the inter-urban chicanery or a coordinated plan to combat it, unwittingly allowed the deception to prevail.[12]

Philadelphia had certainly suffered its share of setbacks. Those geographic features that had made Philadelphia so attractive as a trading center in the eighteenth century conspired against the city when trade extended westward into the growing nation. The opening of the Erie Canal in 1825 and the Baltimore and Ohio Railroad in the 1830s diverted internal trade from Philadelphia, for which the Allegheny Mountains created a formidable engineering challenge. The sharp curtailment of foreign commerce as a result of the late-1820s tariff controversy and the failure of the Bank of the United States of Pennsylvania—the national bank's successor— threatened to leave the city, as Freedley dramatically postulated, "a mere speck on the horizon."[13] Editorials urged businessmen and merchants to use their summer vacation travels as a chance to talk up Philadelphia and win the trade of the West. In banking, railroads, and commerce, one correspondent warned, "If we wish to emulate the vast increase of New York, we must have the same facilities."[14]

Hence, various business leaders met in March 1857 at the Board of Trade rooms to discuss advancing the new hotel project.[15] By the time of this initial public meeting, supporters had already subscribed $200,000 toward the hotel. The meeting's setting, at the Board of Trade, demonstrated the board's ongoing determination to aid Philadelphia's economic development through a policy of broad-based participation and cooperation by the entire business community. Even so, despite the enthusiastic core of business leaders supporting the proposal, Franklin's protest threatened the project's backing and exposed the increasingly complex infrastructure of modern power in antebellum Philadelphia.

The Hotel Folly

Accounts of the plans for the new hotel project began to appear in Philadelphia's several newspapers at the beginning of March 1857. For the first two weeks, letters and reports about the hotel topped the front page. Fueled by controversy that entangled some of the city's wealthiest businessmen and high-ranking politicians, the discussion quickly became at once both personal and theoretical. For example, critics attacked in equal measure the self-interest of specific suppliers, such as those for carpets or china, as well as the logic and power of monster corporations. By early April, Franklin published for sale a bound volume of his first set of seven letters entitled collectively *The Hotel Folly, A Series of Letters in Reference to the Building of a Monster Hotel in the City of Philadelphia by a Corporation*. The book itself was designed to draw readers to it. A single word, a gold-embossed "Franklin," adorned a highly unusual, vividly jewel-toned turquoise cover. Dedicating the volume to "an impartial and justice loving community," Franklin had published the letters in order "to preserve them for future reference, in the belief that TIME will eventually demonstrate the TRUTH."[16]

These critical letters were long, wordy, and repetitive, filled with conjecture, suppositions, dramatic accusations and portents and peppered with quotations from Benjamin Franklin's *Poor Richard's Almanac*. Some of his overheated rhetoric expressed typical—and perhaps justified—concerns that any new project would generate, such as the danger to Chestnut Street from a potential fire, worries about the impact of increased crowds on an already busy thoroughfare, and financial concerns about profitability. For example, we know that the Girard House, Philadelphia's largest and most luxurious hotel at the time, stood empty during the Civil War years due in part to an oversupply of hotel rooms brought on by the new hotel, as well as by the reduced numbers of visitors from the South.[17] As a hotel proprietor, Franklin eventually suffered from the Continental's opening even as the community benefited. Despite the city's great hopes, Philadelphia failed to regain its commercial lead over New York City. In this era of pervasive change, personal and communal needs conflicted with one another, and the celebrated march of progress masked individual injury. In this way, Franklin's struggle foreshadowed battles that contemporary cities face when new office buildings court the tenants of older ones in the name of progress and urban development.

Franklin made several points in his extended arguments. In direct contradiction to the opinions of the hotel committee and city boosters, Franklin claimed that the hotel would not be good for the city of Philadelphia. Franklin observed that the already adequate supply of comfortable, well-managed establishments rarely exceeded a two-thirds occupancy rate, requiring "some extraordinary excitement [to] render them full to their utmost capacity." He rejected the argument that a mammoth hotel by itself would draw visitors to the city as "an idle fallacy, without any foundation in reason or probability."[18] In addition, Franklin accused the investors of pandering to "ideas of grandeur," without considering the disgrace that the inevitable failure of the project would bring to the city.[19] Finally, echoing a similar argument voiced in 1828 against the Tremont House's proprietors, Franklin derided the idea of a project conceived to create jobs: "Now that we have a new Opera House just finished, what would the stockholders think if merely to give jobs to builders and contractors, another Opera House should be put up at a cost of a million of dollars."[20] Franklin was taking issue with still-current arguments that justify large-scale development projects intended for elite patronage based on their potential to generate work for the lower classes.

Most of these emotion-laden arguments were not supported with facts by either side. However, it is possible to assess the status of Philadelphia's hotels in 1857. Chestnut Street reigned as the center for Philadelphia's carriage trade, and, as would be expected, the city's best hotels aligned themselves on this street or very close to it. In addition to the several banks and the State House, many of Philadelphia's most prominent commercial establishments, those selling drugs, fancy dry goods, jewelry, and carpeting, lined the street.[21] Thus, in response to a letter from a writer calling himself "Fair Play," who suggested that competition might keep existing hoteliers on their toes, Franklin retorted with sarcasm, "But who are these Philadelphia Hotel keepers who do not attend to their business? I assume, of course, that they manage houses on Chestnut street. The dignified Penn Manorites would scarcely stoop so low as to seek to regulate Hotels elsewhere."[22] If Franklin were indeed the proprietor of the Franklin House, his hotel was located about five blocks east of the proposed Continental site, just a third of a block from the corner of Chestnut on Fourth Street, and in the opposite direction of new development. Franklin recognized that only Chestnut Street hotels mattered in this conversation and took umbrage at the implication that he might not be worthy of the new competition.

During the 1850s, no fewer than twelve hotels conducted business on the blocks of Chestnut between Second and Tenth Streets. Most were modest in size, typically four or five stories and perhaps three to eight bays wide. The United States Hotel was demolished in 1856 after being sold to the Bank of Pennsylvania, but the addition of the St. Lawrence and the St. Albans hotels more than compensated for the lost rooms.[23] The Girard House (1852) and the La Pierre (1853), built during the corresponding boom in New York City, constituted the city's largest and most modern hotels. John McArthur Jr., who served as the architect for both buildings, later designed the Continental Hotel. Both enjoyed fine reputations as first-class hotels "fully equal to any in the country."[24] However, even with velvet carpets, handsome tapestries, and all the other accoutrements of luxury, neither of these establishments approached in magnitude that of the "monster" hotels. While Franklin's assessment of Philadelphia's hotel needs in terms of number of rooms might have been accurate, the proponents of the scheme sought the cachet that derived from extravagant size. The only way to compete with New York was through sheer scale.

Franklin aimed his sharpest remarks at the concept of the corporation. Most of the letters contained wild conjectures regarding the cost of the land, bricks, marble, and other expenses that a monster hotel would generate by virtue of its size and extravagance. His objections to the huge amount of capital believed necessary to build the hotel directly related to his fear of the structure for capitalization, the state-chartered corporation. Although the corporation was not a new method for financing hotels, most moderately sized buildings could be financed through private partnerships. But the power and limited liability represented by the public corporation frightened Franklin. In the mid-nineteenth century, Philadelphia was a conservative manufacturing center, with most enterprises owned privately, in proprietorships and partnerships. In this context, the threat of competing with a large corporation with financial power and legal protection intimidated those whose livelihood depended on playing by the old rules.

Periodic financial panics, too, had chastened entrepreneurs, who witnessed the fall of several prominent Philadelphia fortunes. The phrase, "Hotel Folly," alluded to a past Philadelphia debacle, Morris's Folly, an outrageously large Chestnut Street mansion built by Robert Morris, a signer of the Declaration of Independence. Morris was a wealthy merchant accused of profiteering during the Revolutionary War who went bankrupt in 1797 and spent several years of his life in debtors' prison. Pierre L'Enfant designed the residence, and the public held L'Enfant partly to blame for Morris's problems because of his deceptive and severe underestimation of the project's expense. Because the house cost so much to build, no one could afford to buy it, so the creditors had it dismantled. Morris's Folly became the subject of ridicule and contempt.[25] Franklin played on the still-vivid historical analogy to create doubt about the hotel project and question the wisdom of advocating luxury development as an appropriate vehicle for securing widespread benefits.

The La Pierre Hotel, on Broad at the corner of Chestnut Street, Philadelphia. The La Pierre Hotel was one of the hotels that stood to suffer from the Continental's competition. *Warshaw Collection of Business Americana—Hotels, Archives Center, National Museum of American History, Smithsonian Institution.*

Perhaps overestimating his own importance, Franklin accused those launching the corporate project of intending to put the city's other hotels out of business. Unjustifiably brought into being, the corporation would "compete for an insufficient business with those [hotels] already erect, which may be compelled, in the struggle, to fight hard against a monster corporation, that seems to have been originated with no other design than to crush them."[26] The hotel business, Franklin claimed, did not differ from other businesses by which men supported themselves and their families and gradually accumulated wealth. Indeed, more than in other industries, a hotel's success depended on the personal characters of those who ran it. Franklin deplored the growing number of corporations, those "bodies without souls," particularly because the fundamental corporate characteristic—limited liability—obviated all sense of personal responsibility. If the project failed, the protection of the corporation would prevent creditors from collecting what rightfully would be due to them.

With a nod to Adam Smith, Franklin asserted, "This business . . . should be left as free as the vital air to individual competition."[27] His reaction to the broader development of the modern business enterprise revealed confusion and an uneasiness with the changing world around him. Franklin's views were more in line with those that the incorporators of the Tremont House might have expressed. Traditional doctrine held that corporations should be chartered for the public good. While this could be and was interpreted broadly, beneficiaries of earlier corporate charters believed that the duplication of services or the gain of one group at the expense of others put investments at risk, as Franklin suggested. The new corporation toward which Franklin directed his venom had emerged out of the older structure, but it represented private profit-oriented commercial interests and accepted the disruption of established businesses as the inevitable cost of progress.[28]

Paran Stevens, the manager of five nationally recognized hotels, including New York's Fifth Avenue, and the designated manager for the new Continental, also suffered under Franklin's pen. Dripping with sarcasm, Franklin chastised the "ubiquitous Col. Stevens, who now, with a skill unparalleled even in the history of remarkable feats, manages *five Hotels* perfectly and really wants a *sixth,* to give him enough business to attend to." Even though the practice of separating hotel management and ownership had been employed for decades, the concept of a managerial system that transcended a single unit offended and threatened the hands-on local proprietary operation that Franklin understood.

The thirteen letters published by the *Philadelphia Evening Journal* also focused on the issue of self-interest and, by extension, the inherent power of great wealth. In contrast to the Jacksonian myth of egalitarianism and the Whig ideology of the self-made man, an increase in the stratification of social classes marked by little social mobility characterized the socioeconomic relations of the mid-nineteenth century. The luxury hotel, with its rigid class and gender segregation, served as a monument to this reality. In 1860, in Philadelphia, the top 1 percent of the population owned 50 percent of the city's wealth. The upper 10 percent controlled 89 percent, and, conversely, the bottom 80 percent possessed just 3 percent of the wealth.[29] The very real possibility that a few powerful investors could marshal the city's financial and political forces to crush competition justifiably triggered Franklin's fears. Franklin defined the ringleaders as those who expected to profit, not just as successful shareholders but also through monopolizing the contracts for building and furnishing the hotel. Its first proposed name, the Penn Manor, allowed Franklin to tweak the financial aristocracy by referring to the shareholders as Penn Manorites. He accused the Manorites of conspiring to dupe hapless trusting mechanics and tradesmen into exchanging their labor for shares. Yet another class of foolish subscribers, the "good easy capitalists" who allowed themselves to be persuaded to invest, stood to lose as the Manorites used their money to finance the Manorites' own lucrative contracts, increasing their wealth regardless of the project's outcome.[30]

The list of shareholders included both men of substantial wealth and tradesmen whose livelihood stood to be enhanced through profitable contracts.[31] From the handwritten list of 213 investors, about eighty are identifiable by occupation and residence through the city directory. As would be expected, nearly all those identified lived in downtown areas designated as being fashionable or respectable. The 1860 U.S. Census of Manufactures Schedule of Philadelphia County, which lists—among other things—manufacturing enterprises and their capital investment, included the names of twenty-seven of these investors. Of them, four companies were capitalized between $5,000 and $15,000, eleven were worth between $20,000 and $70,000, and twelve between $100,000 and $700,000. Even the smallest of firms represented an accumulation of wealth that masked greater personal fortune. For example, William Rice, the printer, with a reported manufacturing value of $7,000, purchased five shares in the hotel in cash at $500 each, suggesting a net worth greatly in excess of that shown on the census. At least fourteen firms that invested a total of $41,500 in the hotel eventually secured contracts supplying marble, clocks, chandeliers, mirrors, pianos, china, stone, steam boilers, ornamental iron, carpets, bedding, tinware, and plumbing.[32]

Another writer engaged the issues of "self-dealing" in the pages of the *Sunday Dispatch*. He found self-interest at work on both sides: "One would believe that there was nothing but pure philanthropy on the other side, and the persons who are trying to inveigle our unthinking citizens into making investments . . . are public spirited, and free from all sordid motives." He went on to cite the ways in which several of the "philanthropists" stood to gain: John Rice, the president and general contractor, would receive 5 percent of the project's cost; Eli Price owned land directly across the street from the new hotel, land that would presumably increase in value; William Kerr, purveyor of fine china, "has not the least idea that any china or crockery is to be used"; while James Orne expected the contract for carpeting, as did William Struthers for marble work. In the latter example, a $5,000 investment stood to be rewarded by a contract ranging from $91,000 to $120,000. Each of these, and many other lesser tradesmen, waited with anticipation for the "glorious profit" offered by the new hotel project.[33]

The critics assumed this position of righteous indignation in part due to the sullied history of John Rice, the hotel corporation's president. Prior to 1854, the City of Philadelphia occupied a small two-square-mile rectangle of land between the Delaware and Schuylkill rivers, surrounded by suburbs and small independent communities. In that year, the city and county consolidated, forming the coterminous City and County of Philadelphia. The new entity absorbed the surrounding communities, increasing its size to 125 square miles. Also at that time, people began to press for improvements to the city's produce and meat markets, which then consisted of congregations of sheds in the middle of city streets. Certain city officials and others, including John Rice, proposed a plan to construct "large and well-ventilated" buildings where shoppers and vendors could transact business without obstructing traffic.[34] The night before the consolidation bill was to take effect, the City Council surreptitiously called a special meeting to authorize the purchase of large lots of land at inflated prices, with the specification that, without securing other bids, Rice would build new market houses.[35]

The "Rice Job," as the scheme came to be known, suffered considerable public condemnation, and the city council members involved lost their positions at the hands of Philadelphia voters. Rice eventually built the new market houses, but they failed to attract either the shoppers or the vendors, the "fashion of going for marketing to Market Street ... [being] so deeply rooted in popular practice."[36] To further compound the whole affair, the city repurchased the deserted market houses from Rice, bought other land, and then gave Rice a new contract to build more market houses. Needless to say, Franklin recounted the entire sequence of events in great detail. "The Market House variations may be played over again, with some new fantasies, to 'a very pretty tune.' The President will perform, and the stockholders will 'pay the piper,'" warned Franklin.[37]

The published letters on the hotel controversy functioned almost as a hearing on the modern age. Even as he marshaled statistics, history, and Poor Richard's adages to support his arguments against the corporation and its impersonality, Franklin still embraced Philadelphia's larger need to secure an advantage in world and national markets. In fact, Franklin authored the quotation cited earlier: "If we wish to emulate the vast increase of New York, we must have the same facilities." He exhorted Philadelphians to undertake any measure necessary to remain competitive, asserting, "If New York employs lightning for the purpose of conveyances, we cannot rely upon steam, but must go in for electricity."[38] In other words, Franklin accepted the ideology of progress and the reality of a modern capitalistic economy, yet he turned to the traditional structure of his community in the naive hope of protecting his own personal resources from the power of the impersonal "soulless" corporation. He agreed with the need for development but favored regulation rather than unrestrained competition. Franklin embodied the ambivalence associated with modernity that pitted the process of development against tradition and favored the present risk-takers over the past ones.

Dissension in the Ranks

Franklin's opposition to the hotel plan—persistent though ineffective—was pitched against an inevitable outcome. The subscription and building of the hotel proceeded on schedule, despite its share of problems. One of the incorporators, Solomon K. Hoxsie, a builder, but not the builder of the hotel, continued Franklin's role as devil's advocate, but from within the firm. In September 1857, he wrote a lengthy letter to the company officers detailing alleged mismanagement of funds by the directors. Echoing many of Franklin's warnings, Hoxsie expressed resentment toward the directors for not being responsive to the shareholders. He specifically accused the directors of purchasing a smaller piece of land for the hotel at a price more than $35,000 above that previously agreed on for the larger lot. After private inquiries resulted in his being reprimanded for the crime of "impertinent curiosity," he addressed the directors by saying, "I do not assent—Gentlemen, to the position that your characters place you above the necessity of explaining such of your actions as seem to be in opposition to the interest of those you represent."[39] While the actions of the directors appeared to justify Franklin's characterizations of the detached monster corporation, Hoxsie's involvement and his questioning of the directors' actions demonstrated his and others' active participation in the direction and functioning of the organization.

On another front, Hoxsie's letter disparaged Paran Stevens, the manager, whose lavish tastes, Hoxsie charged, would result in runaway costs for which the shareholders would then be liable. Hoxsie's comments revealed uneasiness with New York standards and a certain disdain for the kind of luxurious extremes threatened by Stevens's stewardship. "What sort of marble mantles will be fine enough for Mr. Stevens," Hoxsie asked. "Are they to cost $3000 or must he have them to cost $15,000? ... The gas fixtures—are they to be in every room? Are the chandeliers to be magnificent? Who is to pay for them, the Company or Mr. Stevens?"[40] Hoxsie expressed concern that Stevens's luster was blinding the directors from their proper fiduciary responsibilities. So he assumed the role of corporate watchdog. "You cannot expect the stockholders patiently to bear the contempt with which you seem to treat them," he warned. "And I for one, now give you notice that until the whole plan is settled and the cost of finishing the building reliably estimated and approved at a meeting of stockholders, I shall refuse to pay a dollar more

upon my subscription." True to his word, Hoxsie exercised his power by withhold-
ing payment for his shares. Records show that as of March 1859, he had not yet
completed paying his subscription installments.[41] Yet Hoxsie's objections assailed
given assumptions—at least in New York—that each new luxury hotel, in order to
compete, needed to be constructed without regard to cost.

Hoxsie's criticism of Stevens foreshadowed an irksome controversy over the
new hotel's proposed elevator. The elevator's expense forced the corporation to
question the changing nature of luxury, its relationship to technological innovation and the insuperable imperative to forge ahead, literally at any cost. Hotels had
had steam-powered baggage hoists for many years, but the Fifth Avenue Hotel was
the first to have a passenger elevator.[42] This machine, the vertical railway elevator,
designed and patented by Otis Tufts of Boston, worked on a different principle
than the hoists. Tufts's elevator eliminated reliance on "running rigging," with
its deadly unpredictability of chains and ropes. Instead, it employed the ancient
principle of the screw that, through its solidity, assured passenger safety.[43] A ten-
foot-square vertical space extended throughout all the floors of the building, with
openings on each floor. A beautifully appointed passenger car or "covered room"
formed the "nut of the screw." This extended from top to bottom, so that as the
screw revolved, the car either ascended or descended. A guideway at one corner of
the well prevented the car from turning around with the screw.

The description of the Fifth Avenue's elevator conveys the magnitude of the
machine and the complex mechanical knowhow necessary to create it. The screw
itself was ninety feet long and eighteen inches in diameter across the threads. It
weighed more than eight tons. Manufactured in gun iron in several lengths, these
pieces were fitted together using six-inch-diameter wrought-iron screws. At its
bottom, the vertical screw rested on sixteen antifriction rollers, which allowed it
to revolve easily, despite the great weight of the machine. In the building's basement, steam-driven gears rotated the screw, causing the car to ascend. At the top
of the hotel, automatic mechanisms switched the car to a piston-controlled gravity descent.[44] As an article in the *New York Times* aptly observed, the going up was
not the problem. Rather, "to prevent the weight from falling, if machinery breaks
down, if malicious persons cut the belts, if the operator neglects his duty, or if
bunglers in fright or ignorance, pull the wrong rope or do the wrong thing—this
is the momentous consideration."[45]

Stevens had arranged to have the same elevator installed in the Continental
Hotel, but the $11,000 cost sent the directors into a state of shock. In early 1860,
the managers complained to Stevens, "When the plan of our ascending chamber

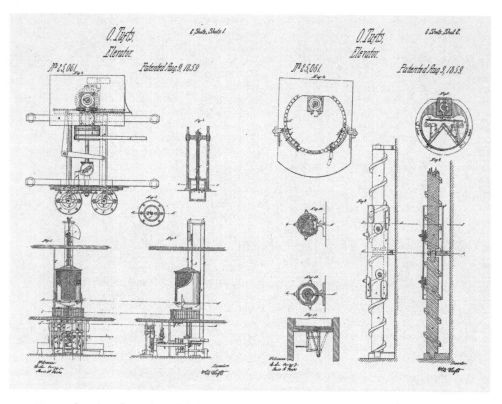

Patent drawings for Otis Tufts's "Elevator or Hoisting Apparatus for Hotels," or the Vertical Railway Elevator. Figure 1 shows the front of the machine, the passenger car affixed to the vertical screw (a) by means of a nut (e). Steam power lifted the car to the hotel's upper stories, while gravity assisted its controlled descent. Tufts's goal was to design a passenger elevator "free from the extreme and ordinary dangers, of suspension upon chains, ropes, or cords." United States Patent No. 25,061.

first suggested itself to our mind for this purpose, we never supposed it would have given us much trouble to have decided on this plan, as it has since done." Even after they realized that the elevator was not "as practicable" as first suggested, the men expressed their willingness to defer to Stevens's judgment. However, the price tag forced them to reconsider. The corporation called in a committee of three engineers to survey Tufts's machine and to recommend an alternative, which they found in a design by William Sellers, a prominent Philadelphia machine tool manufacturer. The engineers not only surveyed the various plans but also traveled to New York to see the Tufts elevator in operation. They determined that Sellers's machine "will make less noise, . . . it will not get out of order, [it will] be less expensive in repairs, require less power, can be run at a much greater rate of speed and be equally safe." In addition, Sellers agreed to guarantee his machine so that after a year or two, if the hotel were not entirely satisfied with it, he would take it out at no cost to the company. Sellers offered to install the elevator for $6,000 cash plus $2,500 in stock, for a cash savings of $5,000. Not only did Sellers's offer appear to make better sense both technologically and financially, his willingness to receive partial payment in stock demonstrated his faith in the hotel project more broadly.[46]

At the Fifth Avenue Hotel, the elevator proved expensive to operate and, because it frequently required repair, expensive to maintain. Movement in both directions was extremely slow and jolting stops at the various floors marred any sense of enjoyment passengers might have anticipated. A trip in the hotel elevator could prove unsettling at best.[47] Even though Sellers's design addressed these shortcomings, cost significantly less money, and came with a no-risk guarantee, the directors were still unable to override Paran Stevens's commitment to the Tufts elevator. With thirty shares, Stevens owned the largest block of stock, and as manager, he wielded great power. In this same letter, the managers revealed their subservient relationship to Stevens. Calling attention to the importance with which they regarded his stewardship, they noted, "It has been our constant care and desire to have you entirely pleased in the plan and construction of our building so that you might enter on your career in Philadelphia feeling that you had one of the best arranged houses ever constructed for a Hotel purpose."[48] The directors' goal to build the country's most palatial hotel and their dependence on Stevens to fulfill that goal forced a decision that ran counter to technical and financial logic. The determining factor appeared to be Stevens's firm managerial power. When the hotel opened in February 1860, the elevator issue remained unresolved. Even a year later, after the company adopted the Tufts design, the Continental's elevator was still in the process of being built. The elevator contributed to a redefinition of luxury. It released guests from the hard work of climbing stairs. The car's ornate interior integrated this new kind of technological luxury into a traditional aesthetic universe by superimposing elegant furnishings onto a mechanical contrivance. Despite the machine's imperfect performance, it nevertheless represented luxury and progress. Once it had been introduced, no hotel with global pretensions could afford not to have one.

The Continental Hotel

The hotel opened to the public on February 16, 1860. It occupied the lot on the southeast corner of Chestnut and Ninth Streets, the present site of the Benjamin

Franklin Hotel, an ironic choice of names given the historical controversy faced by its predecessor. John McArthur Jr., born in 1823, designed the Continental. An immigrant from Scotland to Philadelphia at the age of ten, McArthur trained first as a builder and then as an architect. He designed many of Philadelphia's important buildings. His best-known work is the Philadelphia City Hall (1874–1901). McArthur's reputation distinguished itself for the strength and structural character of his buildings rather than for their artistic merit.[49]

Chestnut Street was a bustling retail center, known as the "Broadway of Philadelphia," ever defining itself in relationship to New York.[50] In 1848, Foster described it as "the very witching hour of fashionable going-forth, and all that is pretentious, aristocratic and lovely in Philadelphia."[51] All the elegant shops were located along the street, including Townsend Sharpless and Sons, L. J. Levy Dry Goods, and Thomas W. Evans. Stores selling watches, hats, trimmings, musical instruments, toys, and many other goods lined the block between Eighth and Ninth Streets.[52] Many of these stores expanded into newly constructed buildings at the same time that the hotel plans developed in the spring of 1857. Concerns about traffic congestion surfaced during the public debate about the hotel. A carriage required twenty-five feet of parking space, so that a fifty-foot storefront only accommodated two customers. Already by this time, the mayor had stationed police at Eighth and Chestnut to escort lady shoppers across the busy street.[53] McArthur designed the hotel so that carriages and baggage wagons arrived at a Ninth Street entrance, thus relieving some of the anticipated bottlenecks on Chestnut.

The Chestnut Street entrance, however, was as grand as its address required. Three spacious arches defined the entryway beneath a massive stone portico supported by eight "noble" columns, each crowned by richly carved capitals.[54] The hotel's front extended 170 feet on Chestnut and 235 feet along Ninth Street. The six-story building (eight stories at the rear) expressed the commercial *palazzo* style of architecture that had rendered New York City the pacesetter for midcentury architectural achievement. One observer credited the grand new buildings with converting New York from a city of "tough granite dowagers" into a "city of palaces."[55] Palace architecture, with its richly ornamented exteriors, broad horizontal lines, and forthright invocation of wealth and aristocratic values, gave outward form to the new corporate era and stood in contrast to the neoclassical Greek Revival style of the early republic. Such architecture frankly celebrated the supremacy of economic commercial power and spoke of the way in which the hotel fit into the matrix of the city's commercial activity. Its facade promised a corresponding interior of incomparable luxury. The corporation intended the cast-iron orna-

"'The Continental' Schottisch," sheet music, 1860. The "Continental Schottisch" was composed to celebrate the hotel's opening. A schottisch was a nineteenth-century ballroom dance, much like a polka. The hotel's image shows how massive the building was in comparison to the Chestnut Street establishments just to its left. The *Library Company of Philadelphia*.

mentation, the sandstone fronts, and the brick exterior to announce Philadelphia's achievement of both fashionable and commercial success.

Retail stores selling hats and furs, men's furnishings, children's clothing, ladies' shoes, drugs, perfumes, and cigars lined both street fronts of the hotel, drawing it seamlessly into the commercial temper of the street. Men entered through the main doors directly into a ground floor business exchange that occupied a portion of an immense central hallway extending 185 feet down the median. Service

CONTINENTAL HOTEL · PHILADELPHIA · 1857

JOHN M. McARTHUR JR. ARCHITECT

· PLAN OF GROUND FLOOR ·

NINTH STREET

SANSOM STREET

CHESTNUT STREET

areas arrayed themselves along the long corridor. These included the general business offices, a men's washroom, a coat room, a package room, a telegraph office, a newsstand, a baggage room, a writing room, a smoking saloon, a billiard room, a gentleman's cafe, a hairdresser, bathing rooms, a post office, and travel offices. Separate ladies' entrances on both Chestnut and Ninth Streets diverted women from the male-dominated activity found there and up to the first floor. In its advertising brochure, the Continental claimed that the hotel offered everything a person could want or need: "Here is house rent, and servants, fuel, post-office, tradesmen's marts and carriages; and railroads at the door, connecting with every branch of road in the country.—Everything of life—free from care, and at small cost."[56] In short, the hotel provided nearly everything necessary for a comfortable and convenient, if transient, life. A complete city existed under a single roof, providing centralized services and a protective insularity from the gritty realities of an increasingly chaotic urban life. Without a trace of irony, the hotel boasted, "A family of two, or ten people, may be located at the Continental, and never find it necessary to leave the precincts of their apartments for months."

Visitors reached the parlors on the first floor by ascending the grand stairway, a work of stone masonry purported to be the only self-supporting staircase in the country. Nine feet wide, the center flight rose clear of the walls from the ground floor to a landing and then returned to the second floor in two side flights supported by the walls. It was a majestic piece of artisanship, wainscoted with Italian marble, that swept patrons into the grand promenade hall. An open balcony on the second story of the stairway gave ladies "the opportunity of viewing the busy and enlivened scene below." The design encoded roles of women as observers of men actively conducting the business of life in the hotel exchange below. It reified those aspects of separate sphere ideology in which women observed the public activity of their men from the sidelines and also served as moral watchdogs. The staircase served as a parade route along which guests ascended to the more dignified and magnificent rooms above. Adjoining the corridor could be found the elaborately decorated parlors, private dining rooms, the tea room, the main dining room, and private suites of guest rooms. Even amid this newly constructed splendor, men

Plan of Ground Floor, Continental Hotel. Notice the smaller ladies' entrances to the left of the larger entrances on both Chestnut and Ninth Streets. The elevator is to the left of the Ninth Street main entrance, near the baggage room, which had its own hoist. The men's washroom was situated conveniently between the business exchange and the saloon. *Courtesy, The Historical Society of Pennsylvania.*

Grand Staircase, Continental Hotel, from *A Traveler's Sketch*. From the staircase, the women could watch the men congregating and conducting their affairs in the business exchange below. *Courtesy, Hagley Museum and Library.*

continued their habit of spitting and smoking—much like animals marking their territory. It reminded one visiting British sportsman of sitting beneath a large rookery at night, "the dirt from the lips of men, and the same from the rooks, in noise, quality and quantity, are very similar."[57]

The second, third, and fourth floors divided into private bedrooms and suites, several of which encompassed private baths and water closets. Most of the guests, however, shared hall baths and water closets. Each floor had eighty-four rooms, the majority of them singles, but a fair number could be combined with adjacent rooms to form suites. Eleven suites on each floor had private bath accommodations. The remaining sixty or so rooms shared three hall facilities, each with two water closets only, no baths; of these, two were designated for women, one for men. The second, third, and fourth floors were identical in layout. About half of the fifth floor was allocated to long, narrow servants' bedrooms. The other rooms were small single bedrooms—undoubtedly what the traveling Englishman referred to in 1857 when he described the "sky-parlors" at the St. Nicholas. This floor housed no facilities for the help and only one tiny corner water closet serving about fifty guest rooms.[58]

When the hotel opened, its promoters published a thirty-four-page booklet that described the Continental and promoted its value to the city. *A Traveler's Sketch* reveals what the proprietors considered important for the traveling and local public to know about the hotel. The "magnitude and perfect arrangements" of the hotel were designed to "contribute to social elevation and usefulness." Craftsmanship, technology, and artistry—a mix of old and new talents—combined to produce an institution that not only answered a practical need but also served as a social and cultural model for the community. *Godey's Lady's Book* echoed these sentiments, declaring the hotel to be a standard bearer for the "alliance between the useful and the ornamental" and praising the Continental as a steward for "lessons of simplicity and truth."[59] *Godey's* blessed this house for exuding genuine refinement and avoiding the sin of superficial slavishness to fashion, "where too much time and cost are lavished on fantastic extravagance." In this "Palace Home for the Traveller," everything was "appropriate and . . . in the right place." With a circulation of 150,000 and an estimated readership of one million, *Godey's* opinion mattered, although in this particular case, it was not entirely free of vested interest. Published in Philadelphia, the magazine's publisher, Louis A. Godey, owned two shares of hotel stock.

The small four-by-six-inch booklet continued with boosterism that thumped the tub for Philadelphia's geographic advantages, situated as it was between the eastern states and the nation's political center, Washington, D.C. This sensibility of "middleness" between the over-the-top garishness of New York City and the machinations of national politics to the south further extended itself to Philadelphia's temperate climate, access to major railroad lines, and reserved yet discriminating social character. Thus, the hotel recommended itself to travelers seeking an "air of stability and comfort."[60] Again, *Godey's*—America's arbiter of middle-class respectability—affirmed the hotel's, and, by extension, the city's, quiet good taste by

Business Exchange, Continental Hotel, from *A Traveler's Sketch*. The diamond-shaped and checkered marble floors were a common design element in hotel lobbies throughout the nineteenth century and recalled similarly patterned floors at Versailles. *Courtesy, Hagley Museum and Library.*

noting, "The furniture has nothing gaudy, not a daub of vulgar finery or tinsel show is seen about it; yet its elegance and beauty exceed any description we could give."[61] Several pages in *The Traveller's Sketch* described the size and location of the various rooms as well as the elevator and grand stairway. A generous number of engravings of the main public rooms provided visual access to their décor and scale.

It is telling that the proprietors devoted nine of the sixteen pages that described the building to the hotel's "Working Department" and the "Mechanical Elements." Except for a few lines on the five acres of carpeting used throughout the building and a brief mention of the Italian marble used for flooring and wainscoting, the booklet does not detail the hotel's furnishings or decoration at all. Instead, it focuses overwhelmingly on the technological arrangements. As one architectural historian noted about mid-nineteenth-century hotels, "Much of their interest lies in the fact that the efficiency of their plans and the development of their mechanical plants . . . placed them in the vanguard of the architectural development of the period."[62] The building's machinery constituted more of an aesthetic than its decoration. Given that Paran Stevens furnished both the Fifth Avenue and the Continental within a year's time of one another, it is safe to assume that the Continental's decorations followed what *Harper's Weekly* characterized as "the usual hotel style," albeit in Philadelphia's characteristically restrained manner.[63] But the escalating complexity of its mechanical plant elicited unabashed excitement and concrete evidence of modern progressive achievement, supporting the *Public Ledger*'s claim in the name of "persons who are qualified to make the comparison, that it is the most complete and magnificent hotel in the country."[64] For many of the hotel's systems, the hotel guide offered the only entrée into their workings, invisible as they were within walls or relegated to areas off-limits to most guests.

Steam power drove special-purpose machinery and appliances throughout the building, creating an almost magical perception of the hotel. The kitchen's array of steam-driven equipment, in particular, drew special interest. *Godey's* urged visitors to cross the sacrosanct divide between the hotel's public spaces and service areas to inspect the kitchen: "Then the large, airy kitchen, instead of a place to be shunned . . . should be visited as a curiosity by all persons interested in the niceties of domestic economy."[65] Equipment included a twenty-two-foot steam table to keep food hot; boilers for meats, fish, and vegetables; steam-driven spits for roasting meat; plate warmers; coffee and soup urns; a coffee grinder; candy and jelly kettles; and ice-cream freezers. The last, by virtue of the regular rotary motion achieved through mechanical means, produced ice cream of an even consistency previously unattainable. In addition, a gas bread toaster, "secured an even toasting on both sides" of the bread. The "celebrated model dairy farm of Mr. Kershow" supplied the hotel with milk and cream. Here, "the cows [were] milked by the new patent milching [*sic*] machine . . . in contrast with the usual mode." Steam-powered platform elevators transported foodstuffs and supplies between the ground floor receiving rooms, the cellar storage rooms, and the second story kitchen that served the main dining room.[66]

The brochure described other mechanical systems throughout the house. The 8,700-square-foot laundry employed steam power to do the washing, wringing, and mangling. Steam heat facilitated the boiling, drying, and airing of clothes and linens so that one thousand pieces could be processed in an hour. Yet another platform elevator transported linens between the laundry and the guest floors. The plumbing and gas systems required seven miles of gas pipe. Twenty miles of bell wire and three thousand feet of speaking tubes outfitted the Jackson annunciator system by which guests could communicate from remote locations with the general office. Thirty-five thousand feet of steam tubing directed heat throughout the building, the heat being circulated by powerful steam-driven fans. The boilers, engine, and pumps were located outside the walls of the hotel, under the rear sidewalk, so as to reduce excess heat and noise in the main building. Gas lighted all the rooms.[67]

As befitted the opening of a city's new landmark, the city's newspapers also printed descriptions of the hotel. While it is clear that management fed the papers much of the editorial copy, the degree to which they discussed the hotel's technology underscores the cultural fascination with the new mechanical world unfolding before the public's eyes. For example, the *Public Ledger* described the heating and ventilation system in exquisite detail, noting the different ways in which many sections of the hotel were both heated and supplied with fresh air. At least two different articles in this one paper described the water system, kitchen, gas fixtures, and the as yet unfinished elevator. Such details served to contrast the hotel's technology with the modest mechanical standards of the average home, placing the hotel in the vanguard of national progress and constructing an image of sophistication and modernity for the city, the hotel, its proprietors, and its customers.[68]

Stories about the Continental also invoked a new language to convey the immensity and intricacy of the building: the language of numbers. Up until this time, building descriptions typically included the dimensions of a structure and often the sizes of rooms, including ceiling heights, as well as the number of rooms dedicated for specific purposes. Depictions of New York hotels throughout the 1850s called attention to the costs of the interior decorations. Recall the enumeration of the costs of furnishings at the St. Nicholas Hotel that read like an interior designer's invoice. Even the descriptions of the Fifth Avenue Hotel, opening only a year prior, focused largely on the furnishings and the building's size.

Those writing about the Continental quantified its components: seven miles of gas pipe, eleven tons of solder, twenty miles of bell wire, 450 tons of iron, twenty-five thousand feet of marble tile, fifteen thousand hinges, twenty-five tons of sash weights, one hundred gross of hat and coat pins, one thousand window frames, six million bricks.[69] Not only did these figures highlight and emphasize the immensity of the building's design, they also served to convey an increasingly rational way of viewing the world. Observers saw the hotel as a system of component parts, each of these components carrying some sort of scientific value, a rubric whose umbrella at that time covered both science and the useful arts. For example, one paper described the stairway as a "scientific piece of masonry," while another characterized the elevator as being contrived on a "new and scientific plan."[70] Measuring the hotel's pieces drew on the rationalization of manufacturing techniques and represented a first step toward the standardization of hotel construction and management. This descriptive language became part of the vernacular for communicating the extravagance and grandiosity of hotel buildings. It also offered a way to quantify luxury and thus made it possible to compare projects. Certainly, if four acres of velvet carpeting were wonderful, five acres would be that much better.

Finally, the magnitude and complexity of the technological systems altered the profile of the hotel staff. The Continental employed a small army of artisans and mechanics to maintain the building and its new technology. This represented an enormous change from hotels of the early part of the century, when service work—cleaning, preparing and serving food, and accommodating guests' personal needs—characterized the bulk of hotel labor, in addition to the artisans and craftsmen who maintained the physical structure. By midcentury, a hotel staff not only included the usual chambermaids, laundresses, waiters, and bell staff, but also machinists, plumbers, and accountants. The Continental employed up to three hundred people, a number *Chambers's Journal* agreed was necessary to take care of six hundred guests.[71] Thus, the introduction of new "labor-saving" devices resulted in a hotel requiring twice as many employees, one for every two guests, and many with specialized skills. This trend continued well into the twentieth century; one trade journal reported in 1923 that the most highly mechanized hotels required two employees or more per guest.[72]

The Continental Hotel opened with great success, and it continued to be the center of Philadelphia's social and political life for many years. In its first year, it played host to national political figures, the Japanese Embassy, the Prince of Wales, and the newly elected president, Abraham Lincoln. In February 1861, after a brief speech from the hotel's balcony to the crowds in the street below, Lincoln stationed himself, much like royalty, at the top of the grand stairway to receive his many visitors. As Franklin feared, the Continental drew business away from the other hotels. Already vacant by the beginning of the Civil War, the neighboring Girard House became a military depot. Over two thousand women worked there during the war sewing uniforms for the Union Army.[73]

The Continental Hotel's story demonstrates the cultural role the hotel assumed for the city of Philadelphia. More than just an independent business enterprise or a link in an emerging service sector, the hotel served as a high-profile vehicle for recapturing and increasing Philadelphia's share of a developing national economic market. A majority of Philadelphia's largest commercial enterprises embraced the hotel project as a means to showcase their city as a manufacturing and trade center worthy of national attention. While the proprietors engaged in the national competition by following the standard monster hotel format, they also sought to differentiate their hotel by highlighting local talent and capabilities. Pundits might have claimed that "all the first-class hotels are so similar . . . that to know one is to know all," but cities endowed their hotels with place-specific characteristics that endeared them to the communities they served.[74]

The hotel project embedded itself in the community in several ways. First, the corporation's shareholders, with the glaring exception of Paran Stevens, were for the most part Philadelphians of distinction in the business community and the trades. Rather than the clear association with a cultural elite that defined the Tremont House, this corporation lay firmly in the hands of commercial and manufacturing men who stood to gain through contracts and the association of their companies and products with a building of national stature. Instead of turning to New York suppliers to furnish the hotel, the corporation relied with confidence and pride on local businesses. Firms such as Morris and Tasker, who supplied the steam boilers and heating system, and Cornelius and Baker, whose chandeliers adorned both the Continental and the Capitol Building in Washington, D.C., enjoyed national reputations that promoted Philadelphia as a first-rate manufacturing center.

Thus, the hotel became not simply a New York impersonator but also a regional product of pride that, through its decorative and technological choices, embodied and exhibited the city's character—in a tasteful way, of course. The Continental was in every respect a commercial enterprise. From the *palazzo* architecture to the hotel's location in the heart of the city's retail district to the overwhelmingly masculine commercial nature of its ground floor, this building was about commerce and Philadelphia's ability to maintain its place in a competitive world. This is important because hotels are often thought to be little more than hyperextenuated houses. While the Continental's daily operation entailed providing domestic services, its purpose and reason for being lay in anchoring the expanding web of commercial enterprise.

The public debate of 1857 also brought the community into the hotel project. Despite their overwhelmingly negative nature, Franklin's criticisms rendered public plans and construction details that might not have been aired otherwise. The controversy contributed to the community's continued interest and the excitement generated by the hotel's opening three years later. The debate also revealed the rather amazing familiarity that members of the community had with the size, costs, materials, and construction of other Philadelphia buildings. Even though the city was growing into a major metropolis, the men engaged in the debate possessed a rich working knowledge of both local history and technical expertise. Despite others' fears of a "soul-less" corporation, the Continental's shareholders perceived themselves as active participants—even those holding a solitary share questioned the wisdom of the corporate officers from time to time. In short, the Continental's story captured a snapshot of a city in the process of change. Hands-on control by a community network peopled by interested and concerned shareholders still governed and influenced large-scale projects and corporations more characteristic of the modern world to come.

The hotel's ground floor plan in particular reveals in a material way changes that were occurring in the American city in the mid-nineteenth century. Retail stores replaced parlors on the fronts of both Chestnut and Ninth Streets, emphasizing the growing importance of consumer goods, their place in the luxury market, and their role as fuel for capitalist development. Separate restaurants for men and women represented not only the gendered realities of midcentury but also the increasing presence of both men and women in the downtown area. These restaurants also contributed to the development of a central business and shopping district whose existence depended, in part, on places for people to eat. The ladies' restaurants offered safe havens for middle- and-upper-class women, while the gentlemen's restaurants and cafes served those whose professional activities brought them into the center of town. The business exchange, saloon, and billiard room dominated the rest of the ground floor space. These male spaces excluded women and promoted a camaraderie of commercial, civic, masculine life, making concrete the ascendance of capitalist commercial culture. The personal facilities on the ground floor clearly demonstrated to whom this floor belonged. The men's washroom contained twenty-two water closets and eight urinals—there were no facilities for women.[75]

Even as the stability of the country's political situation crumbled, the Continental prevailed as an urban representation of an idealized American society. In its final form, the hotel symbolized commercial prosperity, progress, and the hierarchical structure of social relations. However, by looking more closely at the process of development, the hotel's story reveals, too, the conflicts present in the tumultuous evolution of the urban order of the long nineteenth century. The removal of the magnificent trees that gave Chestnut Street its name and the relentless encroachment of retail stores and the hotel on land that formerly held stately homes—these serve as a metaphor for the seemingly unstoppable imposition of modern life. Moreover, as the Continental's story shows, this development did not proceed without fierce debate, deliberate questioning, and a self-conscious awareness of change. As the mid-nineteenth-century commercial elite appropriated the idea of the public good to promote private capitalist ventures, its vehicle, the state-chartered corporation, perverted those ideas by advocating private profit-oriented commercial interests and accepting—indeed, promoting—the disruption of established businesses as the inevitable cost of progress. In service of the city's larger economic goals, individual entrepreneurs were forced to submit to the power of the corporation of "good easy capitalists" and their schemes to bolster Philadelphia's fortunes. Franklin was steamrolled by the consortium of business leaders who believed that progress had its necessary, if unfortunate, costs. While promoters argued that the hotel would create jobs for Philadelphia's artisans and workers, they worried little about the threat that the monster hotel represented to other hotel businesses and the livelihoods of those who ran them. These same arguments are heard in cities today, as civic leaders argue for convention centers, sports stadiums, shopping malls, hotels, industrial parks, and medical marts as ways to draw visitors and business to the city, create jobs, and stabilize faltering economies. But, as anyone who has driven by empty shopping malls and gallerias, deserted entertainment districts, and vandalized industrial parks knows, this kind of development is not necessarily an effective and long-lasting means of addressing urban problems.

....

CHAPTER FIVE

PRODUCTION AND CONSUMPTION IN AN AMERICAN PALACE, 1850–1875

....

"To Keep a Hotel"

A S HOTELS GREW to such great size during the middle half of the nineteenth century, a system of hotel management evolved with the goal of efficiently serving the enormous numbers of both transient and permanent guests whose expectations for service had escalated with the size and extravagance of the buildings. This became known in both the United States and England as "the American System." As historians of technology know well, these same words refer to the method of industrial manufacturing that emerged during the long nineteenth century that featured standardized, interchangeable parts and the organization of work into discrete segments. This manufacturing "American System" reached its ultimate configuration in Henry Ford's moving assembly line in the 1920s.[1] Contemporary observers deemed the hotel's "American System" as no less transformative, complicated, and brilliant. In particular, British critics of their nation's hotels championed the "American System," regarding it as a cultural manifestation of American character, yet another example of Yankee manufacturing prowess and knowhow that had burst onto the world stage at London's Great Exhibition of 1851.

Despite this celebration of organization and production, social critics at home attacked these developments, focusing particularly on hotel extravagance and on the tendency of people—particularly bachelors and young married couples—to establish homes in hotels. Rather than "setting up housekeeping," these young people did just the opposite, outsourcing, in effect, their domestic household management to professionals. They left the drudgery of housekeeping to the hotel staff, freeing themselves to partake in a life of purported debauchery. Such a life offered wives and mothers rewards not tied to any kind of work and directly challenged ideas about domesticity that had developed in the first half of the century, ideas that reified the cult of womanhood and her place in the home as the guardian of virtuous American family life. The anonymity of fluid social relationships that characterized hotel life created similar problems for bachelors, who, with fewer social constraints, allegedly found themselves in predicaments that threatened their reputations and hopes for social advancement. Thus, ambivalence about the emerging consumer society and the artifice of fabricating social identities through the possession and display of material goods squared off against the celebration of industrial progress that the hotel represented. The luxury hotel served as a lightning rod for these debates that existed in the wider culture, because it housed—in aggrandized fashion—both production and consumption while at the same time forcing a confrontation between commercial culture and the domestic ideal.

Critics recognized that these hotels represented big business.[2] Not only was the hotel a regular industry, but it also functioned as an extension of the marketplace, so much so that one observer described the great hall of New York City's Metropolitan as "a little Bourse," referring to the Paris stock exchange.[3] Hotel lobbies, bars, and billiard rooms served as gathering places for powerful men conducting both formal and informal business in the tradition of male sociability. Typical male activities such as drinking and conversation, particularly about women and business, were part of the worldly realm, replete with competition, bawdiness, and play. All of this stood in sharp opposition to the domestic realm of women that prized sincerity and levied on women the responsibility to seal their families off from the harsh realities of the world beyond the home.[4] In horrifying contrast, money and appearance operated as the entrée to this world of consumption, where, for four to five dollars per day, those who wished could "sniff the upper air of the very latest civilization."[5]

Women guests, while separated from actual business activity, flaunted their economic role in the luxury market by appearing conspicuously "*en grande toilette.*"[6] More than anything else, the hotel's brazen commercial nature and oversized luxurious décor placed it in the vanguard of the burgeoning middle- and upper-class consumer society and in direct opposition to traditional attitudes about domesticity and class hierarchy. Despite the private domestic nature of hotel services—incorporating sleeping accommodations, meals, bathing and toilet facilities—urban luxury hotels remained, as we have seen, commercial spaces. Within the richly decorated hotel parlors, men and women enjoyed a society divorced from the sober responsibilities associated with the domestic ideal and did so willfully, in the context of a market exchange. This hotel world differs greatly from the picture presented in analyses that characterize early nineteenth-century women as "channeling consumption into the sweet pleasure of domesticity" while men retreat into a form of "inconspicuous consumption" in order to free themselves from the "corrupting force of luxury."[7] Contemporary criticism directed at the luxury hotel unveils critiques of consumer culture that many historians have located as emerging in later nineteenth-century department stores.[8]

The hotel's engagement with capitalist culture colored observations about everything from the dining room to the administration of the hotel to the standardization of experience. British and American hotel critics wrote highly droll analyses about their experiences. Travelers and observers joined the fray, detailing their hotel adventures and musing endlessly about food and surly hotel clerks, the apparent twin nemeses of every traveler. Hotels served as the settings for fictionalized remonstrative stories in popular periodicals that served as cautionary tales about strangers, morals, class and race mixing, and the deceptive nature of appearances. While articles and stories either celebrated American hotels as the crowning national achievement of a great civilization or deplored them as a sign of pervasive degeneration, none failed to exact wonder and respect for the precise administration and management that enabled establishments of such immense size and complexity to function as they did. The phrase, "to know how to keep a hotel," permeated the American vernacular as a way to signify great administrative ability. As one editor noted, "A man who *could* keep a hotel was fit for anything from constable to Governor." To apply the common admonition, "You are a pretty good fellow, but you can't keep a hotel," was a sorry tribute indeed.[9] Still, the "American system" created a structure for the kind of highly visible consumption for which the hotel was known and separated these functions only with a highly permeable boundary that demarcated the front from the back of the house.

First Impressions

In 1861, George Augustus Sala, British journalist and foreign correspondent for the London *Daily Telegraph*, employed a well-known simile in an article on American hotels written for the London magazine *Temple Bar*. He stated, "An American hotel is to an English hotel ... what 'an elephant is to a periwinkle.'" Sala went on: "As the American has hardly any resemblance to an Englishman," neither did an American hotel show even a faint resemblance to a British one.[10] Sala believed American hotels to be at least a century ahead of English ones, which he described as "costly, cozy, secretive place[s], with fat, velvet-footed waiters." Even Anthony

Trollope, who had much to say (if not much good to say) about American hotels, complained about English hotel life as being solitary, uncomfortable, and extraordinarily expensive.[11] An article in the Edinburgh journal, *Chambers's*, described a typical English hotel sitting room as gloomy and funereal, with bedrooms attended by solemn waiters and "tart and vinegarish" chambermaids.[12]

Other writers voiced angry complaints about the dreadful British food, extra charges for necessities such as soap and candles (fees calculated by number of inches burned) and especially the extortionary tradition of feeing or tipping each of the attending employees of the hotel.[13] In his satirical tirade against English hotels, Albert Smith described the soap (for which he had to pay extra) as a "little inconvenient latherless cube of indurated composition." He asked, "Is it really soap, or cheese, or wax, or chalk, or gutta-percha, or cement, or all those things combined?"[14] Smith's hilarious forty-page invective against English hotels urged, above all, that hotels charge a fixed rate to eliminate the constant need to tip for unnecessary services such as being led to a bathroom or handed a towel or a bar of the unusable soap.

The expression of this deep dissatisfaction and the resulting comparisons between English and American hotels focused on the American System. Two characteristics in particular seemed to represent the essential "American-ness" of hotel buildings and their management, to wit, the regimented logistics that governed and occasionally tyrannized a guest's movements from arrival through departure and the hotel's room arrangements and mechanical systems that served similarly to control movement and activities throughout the hotel. Analogous to a factory that organized the division of labor within a logically arranged physical plant, the hotel's operational methods and building design worked in concert with one another to direct and manipulate the flow of people and secure their comfort. The hotel's architectural program both articulated and facilitated the systematic management of life within its walls.

In his 1861 article "American Hotels and American Food," Sala described a typical experience at a New York City hotel. He chose Broadway's New-York Hotel to serve as an example of a good fashionable family house. The Astor House, he stated frankly, had become a quiet second-class hotel (even though it survived as a prominent meeting place for politicians until it closed in 1913), the Fifth Avenue was still too new and exceptional, and the St. Nicholas, described as having "fallen into the yellow leaf," had become "rather a rowdy-place, where revolvers occasionally [went] off in the billiard room, and not always innocuously." Staying at a Broadway hotel situated a visitor in the center of city life, opening up a vista of lively activity populated by strangers from distant parts of the world, which represented both the charms and dangers inherent in hotel living.

At the railway stations and ports, a swarm of hotel trucks and hacks, independently owned but representing the different hotels, met each train and steamer to transport baggage and passengers to the various hotels. A New York Hotel truck-master gathered Sala's luggage from the railway station's baggage porter. The truck-master fastened one of two identically numbered brass plates to Sala's baggage and gave the other to Sala as a claim check. This service was not free. In Sala's words, "I also at this time . . . take his card, because I, and not the hotel, shall have to pay him; although he carries for the hotel, he is independent of it."[15] On this particular day, Sala chose to walk to the New York rather than ride with the truck-master. Not wanting to pay the "ruinously expensive" fare but also anticipating a stop for oyster soup, Sala set off on foot. His luggage would be available to him within an hour or at whatever time proved convenient, whether that day or, as he mentioned, three weeks later.

Arriving at the New York, Sala took note of the Cuban planters smoking in the entryway rocking chairs and the other well-dressed men crowding the verandahs, halls, and sitting rooms. As an 1865 *New York Times* article noted, each of the New York hotels catered to its own peculiar clientele. "No two are alike," the *Times* stated, "each has its class of custom, its habit of treatment, its own internal economy." At the Astor House, "a restless throng of politicians, officers and professional men" as well as "a buzzing crowd of merchants and traders" packed the doorsteps, rotunda, corridors, and parlors. The New York, by comparison, attracted travelers from the South, Cuba, England, and France, but visitors from New Orleans also considered it a second home because its proprietor had formerly "kept" the St. Charles, the famous New Orleans hostelry.[16] Wealthy young married couples gravitated to the Fifth Avenue, a hotel distinguished by its costly elegance and "reckless and lavish expenditure." One family reputedly paid $20,000 per year to live there, not including extras such as laundry, alcoholic beverages, and personal services. The International Hotel attracted customers from "the rural districts," while Californians and military people, including West Point cadets, stayed at the Metropolitan. Southerners patronized the LaFarge House. Thus, each hotel proprietor, while adhering to general principles of service and organization, attracted customers through individualized style, service, and theater-like atmosphere.

As Sala entered the New York, he immediately observed a crucial difference in American and English customs. He noted with some amusement that "no servant meets me; no one with cold, mechanical smile, and fawning treacherous eagerness, slides forward to take my coat, or show me a room."[17] While the absence of servility

A truck wagon hauls baggage from the railway depot to the various hotels, whose names are lettered on the various trunks. At the same time, Studley's Express is loading up. *Collection of the New-York Historical Society (79484d).*

appealed to Sala, he also found himself disconcerted that no one really cared that he had arrived. Making his way through the lengthy entrance hall, he approached the business office's massive counter to register for a room. Rather than being greeted warmly, as one might expect, Sala encountered the iconic figure of the hotel desk clerk. Midcentury American hotel clerks were renowned for their imperious indifference to their guests' arrival. As hotels grew in size, accommodating ever larger numbers of guests, the landlord became increasingly distanced from his patrons as midlevel management positions such as head clerks, assistant clerks, night clerks, bookkeepers, housekeepers, and the like became necessary to serve the hordes of people continually moving through the hotel. As owner or lessee, the landlord also belonged to a superior social class than his underlings, a class whose status derived from property ownership. The days when a proprietor such as David Barnum presided at table, personally entertaining his guests, had given way to a bureaucratic regime that left guests at the mercy of mid- and low-level wage workers, albeit ones with an uncomfortable degree of power over their customers.

Anthony Trollope described the head clerk as a "despotic arbiter" who seemed to acknowledge or ignore those seeking attention without any apparent logic. Trollope attributed this to the lack of courtesy as the natural product of democracy. "The man whom you address," Trollope stated, "has to make a battle against the state of subservience, presumed to be indicated by his position, and he does so by declaring his indifference to the person on whose wants he is paid to attend." For Britons, manners served to organize the social orders, but in America the lack

A crowd of men read the daily newspapers on the steps of the Astor House, New York. Men habitually congregated on the steps of a hotel, thus creating the need for separate ladies' entrances. *Every Saturday,* vol. 3, Whole No. 95, October 21, 1871, from the Rare Books and Manuscripts Library of The Ohio State University Libraries.

of clearly defined deferential social formalities heightened the chaotic perception the British held of American democracy.[18] An even more caustic critic blamed the clerk's demeanor on the day-to-day subjection to the constant reiteration of identical questions. Writing for *Lippincott's,* he asserted, "It is very wrong and cruel to expect much intellect in such a man." Equally disconcerting was the unpredictable vacillation "between the extremes of cold inscrutability and the warm, boundless confidential." Even when commending some clerks as "thoroughly good fellows," such praise was undermined by the claim that these clerks were without exception "stout or [had] a tendency to stoutness," appearing to "receive the questionings of the world innocuously—somewhere . . . in their superfluous fat." Perhaps a life of dining at the *table d'hôte* had unhealthy side effects.

Even the trade journals caricatured their own. The *Hotel World* described the clerks as having a power "in their words, looks and gestures" that reduced the inquiring guest "to the last verge of inferiority." This difficult posturing—a stereotype throughout the nation—presented a puzzle that conflicted with the head clerk's responsibility for the daily comfort and servicing of his guests. One rare enthusiastic defender saw him as a "guide, philosopher, and friend," as both a product of and an indispensable part of the American system. The hotel clerk's reputation served as a gatekeeper against those who yearned to experience the hotel's glittering bounties but lacked the cultural and social bearing to do so with confidence.[19] Both the clerk's behavior and the commentary about it revealed the class tension inherent in his post. He wielded considerable power over those with a higher class stature than himself yet, as Trollope suggested, did so on a selfconsciously equal footing within the egalitarian ideology that the democracy fostered.[20]

The more critical Britons deplored the mix of social classes and the hotel employees' independent nature, both of which proved confounding. In particular, the crowded hotel lobby overwhelmed travelers. One of *Putnam's* writers described the scene as an "immense host of smoking and spitting men, which surges up and down the vast hall, overflows upon the street without and up the broad staircase within, and through which he has to make his way by sheer force, in order to reach the counter behind which stands the impassive master of his life for the time during which he will stay at the house." This was in considerable contrast to British hotels, which sought to imitate a "peaceful home" where guests "never see nor hear the other guests." Having been forced to wend his way through the clamor, the as yet unidentified guest was subjected to being judged solely by dress by an obvious inferior and risked going "the way of the mechanic in his holiday suit."[21] Travelers were at the mercy not only of a hotel clerk's mood but also of his ability to read social and cultural cues accurately.

This attitude naturally carried over to the black porters and Irish servants who formed the corps of wage workers in the hotel. Their jobs were to carry baggage and attend to various needs that the guests might have, such as bringing up a newspaper or a pitcher of ice water. More than one writer noted that the black servants were more helpful than the others but often assumed a republican hubris when approached without the proper respect or at inconvenient times. Irish male servants and especially the Irish chambermaids were roundly impugned not only for their incivility, impudence, carelessness, filth, ill temper, and indolence (to name just a few characteristics) but also for their wholehearted adoption of the American "dogma that all men (and women) are free and equal."[22] Service workers, especially immigrants, regarded their jobs as one step on the "social escalator." Even if reality dictated running in place just to stay on the same tread, the American myth of equality and mobility continued to offer hope for social mobility.[23]

The American System

In 1873, writing this time in the London magazine *Belgravia,* Sala again described the American system in an essay that compared British hotels to the French, Swiss-German, and American ones. He devoted his attention primarily to American hotels, designating them as the "parent" to the French and Swiss-German versions. Sala knew that his readers were quite familiar with American hotels, based on the copious amount of material written on the subject by a steady stream of British authors and observers, including himself. Indeed, he speculated that "a good many tourists" had traveled through the United States for "the special purpose of writing about hotel life, and nothing else."

Sala's focus on American hotels sought to explain how they so perfectly represented American attributes. The hotel's distinctively American character derived from a combination of specific and singular circumstances that included the nation's great geographic expanse, the sudden development of transportation networks, the energy and cleverness of Americans, a labor scarcity that encouraged invention and a willingness to look to mechanical equipment to replace workers, and, finally, Americans' "restless vanity and ambitions."[24] Observers like Sala believed that participation in the capitalist market enabled all Americans with means to purchase freely and equally the democratic version of the aristocratic experience. Such a viewpoint created a precious irony: for a price, Americans could universally experience that which by definition is reserved for an elite few. This market-based perspective interacted with and fueled an emerging consumer-based ideology that redefined democracy in terms of equal access to market goods. Foreign commentators like Sala rarely considered the informal yet clearly understood restraints such as unfashionable dress and coarse behavior that kept Americans without "means" from patronizing the better hotels. Perhaps this was because they regarded most Americans they saw at the hotels as already being unfashionable and coarse. However, in practice, these restraints worked to ensure that a first-class hotel stay represented more than simply a market exchange. The luxury hotel provided an extravagant setting in which patrons could enjoy the cultural and social elevation that a restricted market society reinforced.[25] Thus, a stay at an elite hotel helped patrons to craft their social identity.

With the interest of reforming and improving English practices, Sala presented what he considered to be the six components of the American system. His first two, the fixed cost per day and the absence of tipping, were closely related. One of the most common complaints lodged against British hotels was the complete obfuscation of hotel charges. Travelers never knew what to expect on a hotel bill until it was presented to them. The rigid expectation that guests would tip all the various hotel employees coupled with the custom of charging for small extras such as candles incited a broad public condemnation of these practices over a period of several decades. By contrast, American hotels charged a single rate for all rooms in the house, and this rate included as many as four enormous meals per day. Guests incurred extra charges when occupying family suites, and for laundry, baths, and alcoholic beverages, but servants did not expect to be tipped. This custom—considered egregious by those expected to do the tipping—did not make headway in the United States until the late 1870s. Therefore, under most circumstances, guests in American hotels could anticipate what they would see on their account statements.[26]

This system worked well until the 1850s, when monster hotels became so huge that rooms within a hotel began to vary a great deal in size from one another. Earlier hotels such as the Tremont or the Astor House had three or four identical floors of rooms and suites, so charging the same for all rooms made a great deal of sense. However, when hotels climbed to heights of six, seven, and eight stories, architects carved out smaller rooms on the top floors, which proprietors tended to outfit with plainer furnishings. Hotel clerks often assigned men traveling alone to these rooms—the sky-parlors—because of the strenuous climb required to reach them. Elevators were, as yet, neither dependable nor widely available and served more as symbols of luxury—such as at the Fifth Avenue or the Continental—than as useful machinery. Architects were just beginning to design buildings whose height required an elevator's services. The first building for which the elevator became an integral component of the design scheme was New York's Equitable Life Assurance Building (1868–70; destroyed by fire 1912).[27] As late as 1870, a writer for *Putnam's* complained about the unreasonable custom of charging the same for all rooms. He deplored the fact that "a room furnished with splendor" on the first floor cost the same as one found "at the end of a ten minutes' ascent, in the garret, and hold[ing] merely a bed, a wash-stand, and a chair."[28] As the 1871 plans of the Continental Hotel showed, these little upper-story rooms, aligned along long corridors, were as small as 152 square feet (measuring 10'5" by 14'7"), and as many as thirty of them shared a single set of men's and women's baths and water closets.[29]

What Sala celebrated as one of the most commendable attributes of the American system was actually undergoing great change. In 1873, the *New York Times* published hotel rates for hotels in different parts of the city. The newer uptown hotels had adopted differential pricing, charging $35 per week for fifth floor rooms, $42 per week for fourth floor rooms, $56 per week for rooms on the third floor, and between $75 and $150 dollars per week for rooms on first and second floors. Thus, the lower floors continued to command higher rates despite the introduction of the elevator. Daily rates had climbed from $2 to $3 per night before the Civil War, to $4.50 to $5.00 per night afterwards.[30] With the new graduated scale of prices making headway in the best hotels, people of more modest means and others, such as commercial salesmen, could find affordable rooms, thus opening up the hotel to people farther down in the social hierarchy.[31] While these rates included a day's meals, that custom was changing as well.

The *Table D'Hôte*

One of the most widely discussed features of American hotels was the common dining table or the *table d'hôte*. Only a few things had changed since the 1820s when visitors to the United States wrote of their incredulity when faced for the first time with the astonishing *table d'hôte*. As a gong sounded throughout the hotel, masses of people stormed the dining room, took the first available seat, plunged into platters of food, ate without saying a word, and left the room. The whole process took only fifteen minutes. By the 1850s, smaller tables seating six to twelve persons began to replace the long banquet tables, establishing a more congenial and elegant atmosphere in the elaborate dining rooms. Also, newer dining customs opened the main dining room to women and children, who took their places at the table with their menfolk or sat at tables of their own. Indeed, as an 1854 Tremont House menu specifies, children occupying a seat at a regular table would be charged full price.

The bill of fare was still confoundingly long, with as many as one hundred different food items listed at both lunch and dinner. Hotels had greatly expanded mealtimes as well. A patron could breakfast anytime from six to ten in the morning, have dinner from one to five or perhaps two to four in the afternoon, partake of tea from six to nine in the evening, and enjoy supper between nine and midnight. Variations existed, of course, but the expanded periods of service were generous, as hotelkeepers tried to accommodate guests' schedules. Clearly, the occasional grouse who complained of the regimentation that prohibited him from being able to dine "at an hour that would suit his engagements" might have been

An 1854 menu from the men's dining room at Boston's Tremont House showing the broad array of food from which to choose, as well as dining "rules" concerning meal times, room service, guests, and charges for children. Hotel dining room menus typically featured a picture of the hotel and were printed daily. *Courtesy, American Antiquarian Society.*

overstating things.[32] The headwaiter escorted diners to seats that they would then occupy at meals for the length of their hotel stay. A second waiter would take the guest's order from the bill of fare and deliver small amounts of the chosen items to the table within five minutes' time.

While diners usually selected five or six different options from the menu, articles often described those who felt "bound to order everything upon the bill of fare, and try to eat their way right through."[33] This was no easy feat. Menus included a soup course, two or three choices of fish, a course of five or six boiled meats followed by as many roasted meats, and then a game course. Cold selections such as lobster, chicken, and ham salads and other cold meats came next, followed by a choice of five or six entrees such as calf's head, chicken fricassee, macaroni aux gratin, relishes and other accompaniments, plus at least a dozen vegetable dishes. A large array of pastries, fruits, nuts, and coffee finished out the meal.[34] Regional specialties often made hotel dining an adventure. Sala described Wisconsin hotels serving prairie-hen, Alabamans serving possum, Mississippians offering alligator steaks, and Arkansans presenting bear. This last made Sala nervous to think "that the bear might only last week have chawed up half a backwoodsman," creating unpleasant implications for his own diet, once removed.[35] In all cases, the amount of food available to diners was prodigious; enough, as one critic noted, at even the most modest meals to provide "abundant support for a navvy."[36]

Commentators seemed particularly fixated on the eating habits and demeanor of American women at table. Anthony Trollope characterized the way in which women studied the bill of fare as being unattractive. He deplored "the anxious study, the elaborate reading of the daily book, and then the choice proclaimed with clear articulation." Rather than issuing their outsized order in the modest "gentle whisper" he evidently expected, American women declared their choices "with the firm determination of an American heroine."[37] While Trollope objected to women's forthrightness at the table, Sala was mesmerized by the amount of food women ate in public: "That masculine yet bony authoress from New England has actually built up a sort of monument to Mrs. Beecher Stowe of slap-jacks; now she butters each layer, then pours libations of molasses on the whole, and lo! in less than ten minutes the monument is no more, and the strong-minded woman herself has stalked down-stairs and gone shopping, in defiance of all dietetic laws, human and divine."[38] Sala's jabs took aim at far more than the woman's appetite. This "masculine but bony" woman had decidedly unsexed herself by intruding into a male world, adopting manners at odds with at least British expectations and metaphorically destroying with her knife and fork the private domestic life espoused by both Harriet Beecher Stowe and her older sister, Catherine Beecher, whose popular *Treatise on Domestic Economy* served up instructions for creating the ideal family home.[39]

Travel writers gave the *table d'hôte* mixed reviews. On the one hand, many travelers appreciated not having to pay for individual meals, while others with smaller appetites or busy social schedules objected to paying for food they could not eat. The critic in *Putnam's*, for example, noted, "The general custom of charging three, four, and five dollars a day for rooms and meals has, no doubt, its advantages—to a Gargantua."[40] While some rhapsodized about the food, others complained that the dishes were served cold, stood too long during the extended meal times, floated in grease, and seemed indistinguishable from one another. One poetic connoisseur asked, "Do not the vegetables all taste alike, and does not the beef so fraternize with the mutton, for instance, that they both become demoralized?"[41] Sala, however, raved about the quality and diversity of American vegetables, and all travelers acknowledged that, whatever its faults, American hotel food—without a doubt—surpassed that of English hotels.[42]

In 1857, *Harper's Weekly* published three paragraphs on "The Decline and Fall of Hotel Life." *Harper's* was glad for the change on which it was reporting; it characterized the tendency for families to live in hotels as a worse danger to the country than the slavery question, state enmities, Mormons, and secessionists. It blamed hotel life—and particularly the *table d'hôte*—for the inability of American women to be competent mothers, wives, and reputable members of society. The article claimed that women were incapable "from physical and mental weakness" of caring for either their children or their husbands and could not "as a general rule, discharge satisfactorily any one of the functions for which they were sent into the world—can neither work, nor talk, nor cook, nor make a bed, nor form a rational judgment on passing events, nor interchange sensible ideas." For those families living in a hotel, when the men went off to work, the women were left with nothing productive to do. However, the article's main point was to report on the rise of independent eating houses—restaurants—that were encroaching on the popularity of the *table d'hôte* and, as importantly, barred women from entering. The decline of the *table d'hôte*, described as the main pillar of family hotel life, would force families to live in their own homes and ultimately compel women to "practice the high virtues of industry, economy, and usefulness."[43] The destructive influence of the *table d'hôte* and, by extension, hotel living on the family, and on women in particular, became a common theme for several decades to come.

After the Civil War, some hotels began to adopt what was called the "European

Plan," whereby guests paid for their rooms and took their meals at local restaurants or in the hotel dining room, ordering and paying for meals *à la carte*. For the most part, this resulted in higher costs per day, but many felt it was a fairer system.[44] Some scoffed at the term "European Plan," because in Britain and on the Continent, meals were often brought to one's private sitting room to be eaten in solitude or with family. Conversely, a European *table d'hôte* involved a more ceremonial seating at which all diners ate whatever the bill of fare offered. Therefore, while the method of payment derived from European practice, the experience of dining did not. Nonetheless, the *table d'hôte* persisted well into the twentieth century, with both systems often in place side by side in the same hotel. For many observers, the *table d'hôte* continued to represent republicanism and democracy in its most conspicuous and endearing form, where so much bounty was provided on equal terms for the greatest number.

Hotel Stories of Gregariousness, Fashion, and Intrigue

Visitors often summed up one of the most outstanding characteristics of American hotel life with the word *gregariousness*. Unused to the crowds and the public aspect of what they considered private life, foreign travelers often found themselves at odds with the goings on in the dining room, barrooms, and parlors. Indeed, two additional elements that Sala considered part of the American system were linked to the idea of public sociability. The distinctiveness of the ladies' drawing room and the scarcity of private sitting rooms for men traveling alone were considered uniquely American. As one writer noted rather sarcastically, in order for a hotel to be considered a "good" one, it must count its guests by the thousand. The parlors provided a setting "where [a guest] can meet large numbers of friends, and his wife and daughter can exhibit their expensive wardrobe before a critical crowd, which stands them instead of friends and acquaintances."[45] Superficial interaction with strangers, and often hundreds of them at that, in hotels, on steamboats, and on railroads as well, provided a source of wonder and consternation for foreign visitors. Accustomed to a more rigid separation between classes and a greater emphasis on privacy, British and European travelers to the United States found the constant mingling as strange and novel as any of the sites of interest they visited. The breakdown of social life as they understood it served to construct their definition of American democratic society, one that increasingly came to be defined by equal access to the capitalist market and identities forged through the display of consumer goods.

A writer for *Putnam's* described this affinity for collective behavior as a defining hallmark of the American character. The American, he claimed, was definitely a "gregarious animal." Not only did he love a crowd, he preferred to live in a crowd. Americans were even born in a crowd, for according to physicians, there were "more births of twins in the Union than in other lands." Filling schools and living in common rooms at colleges, Americans swarmed together in railways and steamboats and were "not satisfied with aught but monster meetings." Deadly steamboat and mining disasters enabled Americans to die in crowds, and even after death, the author declared, "he loves to lie amid a crowd in those enchanting cemeteries which his quaint hospitality leads him to show in every town." As an institution, the large hotel that counted its guests by the hundreds or thousands made perfect sense in a nation where men accepted the occasional sharing of rooms and beds with strangers. Sleeping in a crowd was yet another symptom of American gregariousness.[46]

This predilection for public sociability naturally lent itself to and shaped the way in which American hotels organized themselves. Private sitting rooms; small, empty entrance halls; and solitary meals—all hallmarks of British and European hostelries—found no enthusiasts among a people for whom their right to congregate was written into their Constitution. Ladies' parlors, too, were an American peculiarity. In England and Europe, hotel accommodations always offered the option of a private sitting room where one could entertain family and friends in seclusion. Therefore, there was no need for public parlors. Clearly regimented behavior based on state-defined class hierarchies precluded the random mixing of classes that one might encounter in American public parlors.

Also at work in the evolution of the ladies' parlor was the system of separate spheres that manifested itself in the hotel through the design and use of space. As the organizing principle of nineteenth-century society, one that evolved from the separation of income-producing work from the home, separate sphere ideology assigned gender-specific roles to men and women that necessarily translated into the language of the built environment. The luxury hotel both conformed to and departed from conventional expectations that built on the standard, if simplified, dichotomies between women, the home, and the private sphere; and men, their work, and the public sphere. Architects designed the hotel, in effect, using a domestic template that emphasized highly exaggerated public rooms. The wealthy culture of display that dominated the décor imparted a feminine character throughout the "house" yet worked against the prescriptions for modesty and privacy that characterized domestic ideology. However, while certain spaces, such as the ladies' parlors, suggested a certain degree of gender segregation, in fact the hotel replicated patriarchal social relations. Men had full run of the house, including the women-specific spaces—and especially the parlors and dining rooms—while women were relegated to those rooms alone.[47]

Men wielded their authority in the ground floor marketplace, but in the second floor parlors and reading rooms, women reigned. In the largest hotels, the ladies' parlors consisted of a series of elaborate drawing rooms decorated in the grandest fashion with enormous floor-to-ceiling mirrors, "heavy masses of gilt wood, rich crimson or green curtains, extremely handsome rose-wood and brocatelle suits, rich carpets, and so forth."[48] These matched sets of rosewood furniture uphol-stered with brocatelle, a heavy-figured silk and linen brocade, dominated hotel interior design for decades. As hotels increased in size, so too did the size of these parlors, which became increasingly more elaborate as well. Architects designed the corridors along the drawing rooms and parlors as wide avenues with seats and mirrors and shimmering lights, replicating, in a controlled and luxurious fashion, the city streets outside. All proscriptions that guarded against public display or unwanted contact with strangers were thrown out the proverbial window in the halls and parlors of the hotel.[49] Parading along the halls arrayed *en grande toilette*, women displayed themselves, creating an identity based on fashion and consump-tion. Young women of marriageable age adorned themselves in the latest styles, hoping to catch the attention of visiting bachelors, thereby setting themselves up as commodities in the marriage market.

A brutally acerbic commentary on the large American hotel appeared in an 1872 issue of *Lippincott's*, a nineteenth-century literary magazine published in Philadel-phia. The article moved deftly through the hotel experience, setting the tone at the outset by asking, "Is not the big hotel necessarily a failure?" The author referred to the hotel as the "Valhalla of the upholsterer," characterized the staff as slaves, denigrated the hotel clerks as fops with young wives who sang lunatic ballads, and described the inevitable lonesome guests as "yawning mid-day ghosts of the reading-room." Not leaving anyone or anything unscathed, this observer dispar-aged "that class of young men who seem to exist for no other object than that of supporting the big hotel" before finally setting his sights on women boarders and the frivolous superficiality of the ladies' parlors and dining rooms.

The writer did so through two anecdotes, both of which set up women board-ers, young and old, as pathetic creatures for whom the hotel served as an unfor-tunate influence. The hotel's elaborate interiors found their counterpart in what he contemptuously termed "vestals of the modes," a class of women of all ages who succumbed to the influence of fashion and display. In the first story, the city's

Private Parlors, Continental Hotel. These ornate parlors offered opportunities for introductions and courtship. *Courtesy, Hagley Museum and Library.*

most fashionable milliner came to the hotel daily for her meals. As he described her, "She was not young, she was not pretty, but she would not have been positively ugly if it had not been for the grand water-color style of her taste." She never spoke to anyone except when ordering food, and she lingered in the parlors after the midday dinner to—horrifyingly—pick her teeth, all the while watching the flirtatious couples and, according to this observer, suffering from hopeless longing. The writer took care to contrast her plain face with her gorgeous ensembles, the ironic dissonance being more than he could bear. She was, he noted, "at once the martyred victim and officiating priestess of fashion."

The silence, vulgar manners, unfortunate plain looks were only a partial litany of this woman's offenses. "What else," he asked, "but the service of her hollow idol had excluded her from some home-circle where for half the hotel-charges she might have had ten times the comfort and at least fifty times less loneliness?" She had violated this man's standards for womanhood by working in fashion, displaying fashion, and taking her meals publicly in a hotel draped in the latest interior fashion. The story seemed to instruct readers that the milliner had sacrificed a happy home life for the pursuit of a fashionable but superficial and emotionally empty daytime existence at the hotel. Her plainness might not have been as repellent to him had she been a married woman presiding over a more modest parlor in her own home, where, presumably, she would be surrounded by people with whom she could talk. There was no speculation on what brought her to her unfortunate life, nor was there any commendation for being a successful businesswoman who could take care of herself.

The author continued his attack against hotel boarders through a second story that involved two young ladies, both of whose sets of parents had taken up residence at the hotel because of "the suppressed fact that neither could afford to keep house in the melodramatic manner of its choice." The two girls became fast friends and, in the spirit of blood brothers (or sisters, in this case), convinced their mothers to allow them to dress alike to "make the friendship eternal." Each evening the duo waited patiently at the dining room's side doors until a suitable number of young men had assembled in the room. Then, at just the right moment, they would make their way majestically along the full length of the hall for full effect. This went on for a full five weeks, at which time "the two misses loved no more in concert, sat no more together, and dressed studiously *not* alike." The night clerk, who spent his off-duty afternoons observing the girls (suffering from unrequited love for one of the pair), reported that they had both fallen for the same mustachioed ribbon salesman. During his stay, this drummer had showered affection upon only one, disrupting the friendship but nonetheless leaving both adrift at the end of his sojourn. The girls' parents, while sympathetic to their broken hearts, refused to buy their daughters new toilettes, and so, eventually, their friendship was reconciled, each girl having missed the "check silk dress with trimmings of robin's-egg blue" that the rivalry had prevented them from wearing.[50]

These stories are instructive in terms of the hotel and the culture in which they were situated. Like the earlier article from *Harper's Weekly,* both disapproved of women's rejection of the prescribed domestic ideal. In the mid-nineteenth century, women were supposed to nurture individual family units in an inviolate domestic sphere that protected their families from the harsh, corruptive influences of the marketplace. But the hotel, in fact, *was* the marketplace, and by choosing to live there, women became part of this dangerous world by participating in and succumbing to the influence of fashion and display. The poor milliner was "plain," "ugly," and worst of all, unloved. No matter how up-to-date and beautiful, her ensembles could not save her from her pathetically lonely life. She, herself, was a metaphor for the hotel. Dressing up did not make her a "real" woman, just as the hotel's extravagant décor did not make it a home. The author regarded the hotel world as a shell, a stage set, and an impoverished substitute for a real home.

Fashion masked sincerity and threatened to become the sole indicator of a person's worth.

The two "Siamese twins of wardrobe" fared equally poorly. This story raises the issue of class and living beyond one's means. In choosing hotel life, both of the girls' families were pretending to a lifestyle they could not legitimately afford. The lure of the drawing rooms, the *table d'hôte,* and the exquisitely dressed guests proved too enticing. They discarded a simpler, more authentic, and more virtuous home life in favor of virtual wealth that their true station did not merit. The story's narrator found the hotel's material culture guilty of fostering a society where appearances ruled in an unsuccessful attempt to blur distinctions among classes. The young ladies shared the same misguided value system as their parents, and this evidenced itself through their silliness in dressing alike. These girls could only emulate the upper class, and, with poetic justice, one of them attracts the attentions of a drummer, the American symbol of pretense, transience, and commercial theatricality. In the end, the girls were left with their checked silk dresses, united in their disappointment and inability to measure up to the setting or the man.

Clothing and display carried other implications, as Americans used their clothing to construct identities for themselves. With the nation's populace becoming increasingly more mobile and fluid, people found themselves in communities and places like hotels where they were unknown. Therefore, their clothing became an important way to communicate all sorts of messages about themselves expressing taste, refinement, experience, and social class, or, often enough, a lack of all of the above. In the anonymity of the hotel, visitors with at least some means could create a fantasy life by wearing the right clothing and by paying rent. This nonproductive lifestyle based on consumption drew criticism for its excessiveness and, in equal measure, for its conscious rejection of the producer work ethic that characterized the creation and sustenance of the ideal domestic home.[51] Writing in *Harper's Weekly* in 1857, the "Man About Town" quoted a friend who was about to be married as saying that he would "rather take my wife into a ship on fire than take her to live in a hotel or boarding house." According to the "Man About Town," his friend understood the sacredness of the family home, that a married man could only enjoy life under his own roof. Nothing needed to be said about how a woman could enjoy her life, for it was well understood that domesticity and motherhood were the cornerstone of a woman's happiness.[52]

Critics targeted men who lived in hotels, as well. Bachelors seemed in particular danger from the debauchery that lurked in the many corners of the hotel, especially in the sky-parlors assigned to them, clustered high above the more respectable

HOW WE SIT IN OUR HOTEL ROOMS

Harper's Weekly, in one of its screeds against hotel living, emphasized the outlandish manners that hotels encouraged. This drawing illustrates the custom that men had of putting their feet up on the rosewood and brocatelle furniture. Other engravings on the same page showed a honeymooning wife languishing while her husband reads the paper, a man enjoying a shave, and a view from the street of the soles of men's boots propped up on the window sills, all demonstrating the bad habits engendered by hotel living. *Harper's Weekly,* December 26, 1857, 825. *Courtesy, Kelvin Smith Library of Case Western Reserve University.*

family parlors on the floors below. Disconnected from normal kinds of social control, sky-parlor inhabitants could easily be victimized by socializing with people thrown together in virtual anonymity. In 1857, *Putnam's Monthly Magazine* published a two-part first-person narrative that chronicled the life of a New York bachelor living in the sky-parlors of a large city hotel. His name was Isaac Inklespoon and his self-deprecating satire, "My Hotel," acknowledged that "his hotel" was only his to the extent to which he paid his bill on time. His shabby seventh floor room, while the size of a closet, nonetheless afforded entrée to the entire house, including the dining room, where a "tall, good-looking negro" served his meals. The waiter, the author noted, bowed "down to me as subserviently as though he were my own peculiar property." With that comment, the author placed the hotel's black workers in the same category as the other property the young man pretended to own: the furniture, curtains, silverware, and other amenities of hotel living. While the first installment focused on a story of disappointing love, it described in some detail the relationship between Inklespoon and his waiter.

"There is my waiter," he began. "And I feel at times a little in awe of him, he is such a superior kind of negro. . . . He dresses better than I . . . and more over, he is such a knowing, self-possessed fellow, with a peculiar dignity of manner, . . . and a certain quiet vein of satire lurking in the corners of his eyes, as though he could read me through and through, and thoroughly analyze my pretensions." This was the first part of a long self-mocking passage that described their daily interactions and the waiter's superior understanding of the food and the menu. More important, the passage assumed that the author and his middle-class white readership shared a particular perspective in regard to race, class, and servitude and would "get" the condescending humor and find it amusing.[53] The reader can almost see the winks.

The second installment, "Another Glimpse at My Hotel," took these racial interactions further. In this sequel, Inklespoon and others are victims to a series of petty robberies. The hotel hires a detective who masquerades as a gentleman boarder. At the same time, as Inklespoon described it, "an alarming epidemic broke out in My Hotel." This is an epidemic of lovesickness that affects not only Inklespoon's close friend, the bank clerk, but also the "Head Waiter," who has fallen in love with "some little colored girl that works about the building." Once again, Inklespoon indulged in an overwrought description of the headwaiter's respectability, noting that he was a "universal favorite with all." Unable to read or write, the waiter asks Inklespoon to pen a torrid love letter on his behalf. Eventually, the waiter and his bride marry secretly in Inklespoon's apartment because house rules prohibit marriage between employees. Despite the outward camaraderie, the author took care to let the reader in on the joke. Earlier, as the waiter is leaving the room with letter in hand, the bank clerk remains behind "to go into convulsions."

After the marriage, the detective bursts in to reveal that the waiter was the thief. He had been stealing from the residents and pawning the goods, in anticipation of deserting with his wife, who turns out to be not an employee but a slave belonging to the father of the bank clerk's fiancee. The moral lesson is clear. Inklespoon cheerfully denigrated his own social rank, yet starkly drew race and class boundaries between himself and the waiter. Inklespoon's pretensions to a lifestyle beyond his means paralleled but were in no way comparable to those displayed by the black waiter. Despite having violated class and race boundaries by becoming involved in the waiter's intrigue, Inklespoon emerges unscathed. He regains his property, and in the end, his friend, the bank clerk, becomes the butt of the joke. Almost anticlimactically, the waiter goes to jail, but friends spirit his bride to freedom. Uncharacteristically, her owner decides it is too much trouble to pursue her in the North and departs the city. That action, however, frames the bank clerk, like the waiter, as another victim of the love epidemic that has infected the hotel. They

In "Another Glimpse at My Hotel," *Putnam's* caricatures the African-American waiters as they march in with the desserts, in typical regimented formation. The better hotels in both northern and southern cities commonly hired black waiters in the dining rooms. *Putnam's Monthly*, vol. 10 (August 1875), 169. *Courtesy, Kelvin Smith Library of Case Western Reserve University.*

each lose their loves, but only the waiter ends up in jail. The story is a window into class and race relations between the guests and hotel employees. Far from clearly bounded lines of engagement characterized by strict demarcations of prescribed social distance, this story reveals close interactions across race and class, albeit with predictable power relations and outcomes. At every juncture, the reader understands the insincerity of Inklespoon's *bonhomie*.[54]

Together, these stories forcefully demonstrate the idea of a "city within a city." The social relationships inside the hotel were extensions of prevailing patterns of urban life, replicating the anxieties, prejudices, and dangers found outside its walls. Luxury hotels catered to a higher-class clientele, and many of the nineteenth-century newspapers and periodicals promoted a narrative that described a sophisticated and classbound hotel society. But, as the *Putnam's* stories and other commentaries clearly show, the guests were not necessarily paragons of virtue or the only inhabitants of the building. The front desk clerk and his minions suffered from a scathing critique suffused with class animosity. Yet another ongoing

The hotel's private parlors, with all their draperies and luxurious furniture, offered private corners for seduction, despite proscribed admonishments about the dangers. From "Another Glimpse at My Hotel," *Putnam's Monthly*, vol. 10 (August 1875), 168. *Courtesy, Kelvin Smith Library of Case Western Reserve University.*

narrative revealed the hotel's secret underworld, exposing the prostitutes, thieves, adventurers, and swindlers who practiced their craft at the city's finest hotels. In 1854, *Gleason's Pictorial Drawing-Room Companion* reprinted a little article from an unnamed New York newspaper. Entitled "Republican Institutions," it read: "Fashionable hotels are the most democratic institutions we know of. Here people meet, sleep, and eat together upon a scale of dead level, quite curious to consider. We took dinner not long ago at the ——— House, and observed a noted gambler *vis à vis* with a famous poet of an adjacent town. Alongside of a distinguished divine sat a noted pickpocket; "while two ladies, of 'no-better-than-they-should-be' notoriety were elbowing the lovely wife and daughter of a New Hampshire judge. In our hotels are mingled thief and honest man, orator and convict, virtue and vice, highwayman and hangman, legal eminence and eminent loafers, and in short, a perfect chowder of the heads and tails of society."[55]

This perfect chowder, while characterized here as emblematic of the democracy, was exactly the kind of thing that critics so abhorred. As a privately held urban institution, the hotel was a highly rationalized, carefully designed mini-technocity, meant to resemble more a platter of regimented crudités than a perfect chowder. At the same time that it valued efficiency and service, order and control, the American hotel also promoted a grand pageantry of human activity. Passions and foibles determined its character, and the comings and goings of its patrons forged its urban identity as much as the elaborate and complex design of its building.

"Tomfoolery and Gilded Misery"

As these stories demonstrate, critics deplored the luxury hotel because it seemed to encourage a wholesale desertion from traditional values. In 1873, the *New York Times* published an editorial entitled, "How We Live," with the secondary headlines trumpeting the connections among money, extravagance, and hotel life: "The Way Money Goes. Where the Extravagance Begins—Private Life and Hotel Life contrasted—The New Element in Hotel Expenditures—Some Facts and Statistics." A spate of complaints about the excessive costs of hotel living had surfaced from bachelors reeling from the recent crisis in the economy, the severity of which precipitated a six-year economic depression. The editorial, however, pointed the finger not at hotelkeepers but at the public's demand for all the accoutrements of a royal lifestyle. Fortunes made earlier in the century through industrialization and land speculation and the enormous wealth generated by the gold rush resulted in a "multiplication of wants" that, while signaling the advance of civilization, also seemed out of control.[56]

The article blamed millionaires who decorated their homes with richly carved mahogany doors, custom-dyed carpeting, central heating systems, and $1,500 lace curtains for generating a competition of display among themselves. As an example, the author constructed an elaborate model that illustrated how acquaintances exacerbated competitive consumption: "The family of A furnished their drawing-room with a delicate blue satin. A friend of B suggested lace covering to the satin, and it was adopted to the mortification of A. The C's had a friend traveling in France, who introduced *cashemire*. Old D, a man of taste, hit himself upon Gobelin tapestry, and a great decorating house has just introduced *cretonne*." With each escalation, the women, according to the article, indulged in refurnishing their homes with the wholehearted consent and underwriting of their husbands. "They wanted no urging to be kept up to the mark, and to be as good as others." For two long columns, the article enumerated the luxurious materials that went into Fifth Avenue mansions and their cost. "The millionaires of Gotham," it asserted, "were willing to befriend the man who should invent a new way of throwing away money."

The debate on luxury reared itself once more when the author cited this kind of extravagance as an inescapable and inevitable result of advancing civilization. Despite the scathing critique on such sybaritic ostentation, critics found themselves in a difficult and ambivalent place. A scaling back of spending—in this case the purchase of European goods that sent money out of the country—would result in the unemployment of those producing the luxury goods. Referring to Aesop's fable of the Belly and the Members, by which the members (the arms, legs, mouth) go on strike against the indolent self-indulgent belly only to realize that they are all interdependent, the author asserted, "Millionaires are necessary to civilization . . . if they did not exist as a class, it would be necessary to create them." The theory of development that regarded luxury as a powerful force driving the capitalist economy continued to find adherents. In 1874, Frederick William Sharon, future owner of San Francisco's Palace Hotel, addressed this idea in a junior year essay while studying at Harvard. "Thousands of workmen are employed in the manufacture of articles of luxury, and if no one encouraged their industry, would be reduced to the verge of poverty." Bringing his argument to its conclusion, Sharon claimed, "In fact, it is the absolute moral duty of the wealthy to indulge in luxuries, and thereby, spread the benefit of their wealth among all classes."[57] Personal avarice masked as public good underpinned Sharon's embrace of the luxury market, but what behooved the wealthy as a "moral duty" suggested moral vacuity for those who pretended to upper-class status.

The *New York Times* specifically criticized the adult married children of wealthy households who deliberately chose to live in hotels such as the Fifth Avenue. Hotel living enabled them to live as splendidly as they had been raised without having adequate means to do so. As one might expect, the decision was generally charged against the "fair sex," who, it was said, shrank from the exertion of housekeeping or contact with the "brutal temper of Irish handmaidens, and the phlegmatic carelessness of German frauleins." Men, needless to say, would have preferred "a house all to themselves." These couples demanded from the hotels and were willing to pay for the kinds of interior furnishings found in the wealthiest private homes at the time. They were the "new element" influencing outlandish hotel expenditures. They insisted on annual redecorating, *en suite* bathrooms, electrical bell systems, as well as the army of waiters who existed to answer their every summons. In this instance, the hotelkeepers, who normally bore the brunt of criticism for creating the scene and then charging exorbitant rates to cover their costs, were let off the proverbial hook.

The shift in mentality from frugality and restraint to gratification through spending heralded a new consumerism that not only threatened the integrity of family life, particularly in the hotel, but also encouraged a pseudo-reality of purchased experience as more and more Americans became enmeshed in the market society. The preceding stories are examples of the shrill warnings that peppered the newspapers and magazines, often in the form of morality tales. These warned readers against succumbing to the hotel's enticements and reinforced the idea that the hotel existed to house travelers, not boarders. Yet, the "hotel system" utterly depended on an enormous population of people who both rejected the critique and were willing to indulge in an abundantly rich lifestyle. These observations support the notion of a very active consumer market in place in the middle half of the nineteenth century. Consumer culture studies that link consumerism to mass-produced goods have no choice but to date it from the 1880s. Yet the hotel was a world of luxury and goods that presaged the department store of the later nineteenth century, where so many historians have located their studies of long nineteenth-century consumer culture. The characteristics of the early department stores—the architecture, opulent design, themed rooms, and ladies' lounges—had their precedent in the nation's luxury hotels. Those entrepreneurs who created lavish palaces of consumption—A. T. Stewart, Marshall Field, R. H. Macy—drew heavily on the luxurious modern hotels and their amenities for inspiration.

The litany of sins levied against the "new element" encompassed the world of luxury goods from clothing to furniture to food. The *New York Times* article and others like it suggest that this burgeoning consumerism was driven by demand, rather than being pushed by the availability of mass-produced goods. The hotel, as a purveyor of experience, enlarges, too, the categories of consumption beyond consumer goods and entertainment. Staying and living at the hotel involved a market exchange. "Money," one enthusiast wrote, "is the one irrevocable 'open sesame.'"[58] For a price, any good republican could live like a prince, be treated like royalty, and feel entitled to aristocratic luxuries.[59] American travelers and residents alike, by engaging rooms at the luxury hotel, were purchasing a life experience that either exceeded their normal station or, in the case of the wealthy, simulated an aristocracy that lay counter to their country's political ideology. In both cases, the experience was one part fantasy and one part market. And this experience was not limited to New York

City. Every large city had its monster hotel, and smaller cities enjoyed their versions as well. In this respect, the observation, "to know one is to know all," rang true.[60]

"A Marvel of Mechanical and Labor-Saving Ingenuity"

While others carried on with their jeremiads, in 1873 *Scribner's* published an unabashed tribute to the latest version of the modern American hotel.[61] This article celebrated luxury, progress, and comfort as found only in this monument to American inventiveness, creativity, and skill. Like the other articles that recognized the great expertise required to keep the "complicated machinery" running smoothly, this author noted that "we plainly perceive why '*the man who can keep a hotel*' has become the synonym for supreme and unimpeachable capacity." Even as European service was acknowledged to surpass that in the United States, the article asserted that American reliance on modern technology made the American hotel "a marvel of mechanical and labor-saving ingenuity, as well as a miracle of systematic order in the conduct of its daily life." For this author, a woman, the hotel was a "machine," so she defined luxury as mechanical genius, a representation of progress in its purest form. By focusing on the technology, the debate about artifice bred by consumerism ceased to be important. Technology and progress were unambiguous representations of civilization's advance. The male critics harped on the effect hotels had on a woman's ability to execute her domestic role, but here was a woman who celebrated technology's ability to take on the hard work of domestic life.

The entire history of the modern American luxury hotel lay within the author's living memory. She remembered her grandfather reading aloud a description of the Astor House from the *Journal of Commerce*. "To my childish imagination," she recalled, "it seemed like one palatial glitter, from beginning to end, and I thought how delightful it was to have a real palace right in New York, while my aged relative could only shake his head, and reiterate 'Visionary'; 'visionary' 'they'll sink a fortune there.'"[62] Her characterization of the hotel focused, like so many others, on the fabulous technological advances that defined the building and transformed it over its fifty-year history. Like Philip Hone before her, this author recognized the modern hotel as "the outgrowth of a new order of things."[63] It sat at the center of business, and, in addition to keeping up with the latest fashion, as the *New York Times* piece would have it, hotels also became outdated as a city's business center shifted, necessitating new development to answer the needs of a growing and changing metropolis. But, most of all, the article celebrated the building's ability to offer its clients comfort and ease through its machinery. "What is it?" she asked. "It is a massive pile of marble, and brick, and slate, and iron, with a handsome architectural front, and so vermiculated . . . with water-pipes and steam-pipes, gas-pipes and drain-pipes, speaking tubes and nerve-like bell-wires, that, viewed as a whole, it is like a living organism." The hotel lived and breathed, and as a representation of the age, she claimed it to be a "marvelous mechanical monument, to 'the whole glorious brotherhood of industrial inventors.'"[64] Joining the idea of technological invention to luxury, she acknowledged that this technological luxury came at a price, but, even so, it was well worth the cost.

The article was similar to others that extolled this urban institution. An elaborate description of the new "drum" elevator that "accomplish[ed] all the wonders of the flying carpet" not only convinced travelers of its safety and comfort but also invoked the idea of luxury with its stained-glass ceiling, gas chandelier, thick carpeting, and exotic wood paneling. Drum elevators came into use in the 1870s. They consisted of a steam-driven drum in the building's basement around which wound rope or cable that raised and lowered the elevator car. Tall buildings required very large drums and so, like the screw-elevators found in the Continental and Fifth Avenue Hotels, were impractical for the period's tallest buildings. But, like their predecessors, drum elevators brought an air of novelty and modernity to their buildings.[65] The elaborate decorations transformed the utilitarian device into a moveable parlor and once again, in joining art with technology, helped to establish ideas of technological luxury.

The article included a complete sketch of a gentleman's day and the ways that the hotel served him. The hotel offered a concentration of city services. The modern conveniences—from annunciators to the laundry, steam boilers, and kitchen—whose perfect order seemed to be run by "hidden, noiseless springs" helped the traveler make the most of every minute of his busiest day. These were "highly vitalized, energetic *live* men,—the natural chieftains,—those who . . . push forward the mighty material interests of the world, whose business brings them to the great centers."[66] These were the kind of men whose commercial enterprises helped to create an institution as glorious as the modern hotel. The hotel was both their monument and a natural habitat for them. It functioned as a gateway to the world: immediately to the life of the city, while the rest of the globe was only a step away, at the ticket counter in the lobby. Far from the dissolute hotel lifestyle castigated by men who despaired for American women's ability to keep a house, this woman's picture of the hotel offered "thinkers" and captains of industry a life blessedly free from the petty cares of daily maintenance. The two interpretations could hardly be more different.

In a personal anecdote, the author captured some of the changes that had occurred contemporaneously with the development of the hotel. When she was a
young girl, her aunt arrived via the Boston mail-coach to stay with her family while
shopping for her "wedding things." Arriving with only a small hair trunk, the aunt
needed only an additional bandbox in which to carry home her wedding bonnet.
The aunt left after three months, apologizing for the shortness of her stay, explain-
ing, "When one is going to be married, time is so precious." A generation later, this
same aunt's daughter arrived in town on the identical errand. The author asked,
"Could you have heard the perfect accent and scrupulous care with which she
used the word *trousseau*, you never would have dreamed that the English language
containe[d] its equivalent." But, more to the point, the young woman "brought
a trunk of the size of Baubier's Chicago shanty. She stayed a week at the Modern
Hotel, and within a stone's throw of our old homestead, and when she went home
she took the original 'shanty' and an additional 'Saratoga Cottage.'"[67] Baubier's
shanty referred to an early Chicago "hotel" whose enterprising owner rented out
a blanket in his shanty. As one customer fell asleep, the Frenchman would lift it
off and rent it to another. A "Saratoga Cottage" conjured up the enormous resort
hotels in Saratoga, New York, where wealthy vacationers spent the summers. The
huge trunks into which they packed all the requisite costumes were called "Sarato-
gas." Thus, the development of the modern hotel served very nicely as a metaphor
for evolving consumerism, in this instance of wedding needs.

The little anecdote represented a host of changes. The daughter—the author's
cousin—had assumed a level of refinement that her use of the word *trousseau* signified. The woman's rural home had been consumed by the expanding city. Time
had collapsed on itself so that a week was sufficient to complete a whirlwind of
tasks far in excess to that which had once taken an inadequate three months. The
young modern bride had become a professional consumer, and the transience and
impersonality of the hotel had replaced the bosom of the family while serving as
a staging ground for serious shopping. Rather than remaining embedded with the
family, the bride's hotel stay represented her immersion in the capitalist market.
The author, even as she distanced herself from the wedding story in her narrative, was equally complicit. An accomplished and published writer, she celebrated
the system that enabled her cousin to take advantage of a networked system of
transportation, hotels, and urban retail districts. The American system that emphasized order and the regimented production of services was integrated into a
larger system of production and consumption that found its most distinctive—if
contested—expression in the American hotel.

....

CHAPTER SIX

THE PALACE HOTEL, SAN FRANCISCO, 1875

....

"The Greatest Caravansary in the World"

T HE YEAR 1875 was eventful for San Francisco," observed Society of Pioneers historian John S. Hittell in his 1878 history of the city. He noted that San Franciscans were obsessed with conversation about the ongoing feverish speculation on the mining stock exchange, the political machinations surrounding the city's water supply, as well as calamities like Virginia City's destruction by fire. Topping the list of newsworthy events were the Bank of California's failure, the death of William Chapman Ralston, and the opening of the Palace Hotel.[1] These last three were tightly interwoven, and each, in its own way, were sources of astonishment for the young vibrant city's citizens. William Ralston, the Bank of California's president, was an extraordinarily energetic and beloved San Franciscan capitalist. He built the Palace Hotel to promote himself and to secure a place for San Francisco among the world's great cosmopolitan centers. Although the completed hotel contributed to the successful achievement of these goals, Ralston's extravagant conception for it exacted a dreadful price, one that not only brought down California's premier banking institution but cost Ralston his life as well. The story of the Palace Hotel is a story of one man's fascination with size, money, technological innovation, position, showmanship, and hyperbole—obsessions that fit in well with the dynamic and growing nation.

San Francisco's Palace Hotel opened in October 1875. It was intended to be, and was, the largest and most expensive hotel ever built. It represented the capstone of the monster palace hotel style, as embodied by New York's Fifth Avenue and Philadelphia's Continental. The monster hotel style also flourished in Chicago after that city's 1871 fire and developed further in England and on the Continent. As the world's largest hotel, the Palace sustained its worldwide eminence for at least a decade. Soon thereafter, the introduction of steel frame construction and advances in elevator technology relegated the palace architectural style to the heap of bulky, outmoded, nineteenth-century "piles," as critics often called them. Nonetheless, the Palace's powerful cultural iconography ensured its place as San Francisco's most beloved institution, even as more modern structures sought to threaten its ranking.

Hotel Life in San Francisco

Today, San Francisco enjoys a reputation as a city of extraordinary hotels and restaurants. While San Francisco's earliest hotels were exceptional only because they rivaled the austerity of the "Baubier's shanty" mentioned in the last chapter, the city historically relied heavily on its hotels to provide living space to both transients and permanent residents. The city experienced a dramatic population increase as the result of the Gold Rush of 1849. To accommodate the influx of miners who came to San Francisco for outfitting before heading out to the gold fields, hotels composed 7.2 percent of the city's 1850 commercial establishments, the single largest percentage of any one kind of business.[2] San Francisco's first social directory, the 1879 *Elite Directory,* unselfconsciously launched into a description of the 1850s hotels because they offered the city's only respectable housing for the small minority of women who made up San Francisco's early genteel society. For the few merchants' wives who braved living in the fledgling city in 1850 and 1851, a series of hotels "suitable for feminine entertainment" rapidly replaced one another as centers of "fashion and sociability." The St. Francis, Oriental, Tehama House, and Brann House successively "were more or less noted as abodes of wealth and elegance."[3] Interestingly enough, the discussion of these early hotels did not fall within a special section on hotel living, as was generally the case in other contemporary published directories of this kind. Rather, it was integrated into the narrative, implying that these hotels were the only places for women of means to live. A series of articles inspired by the Gold Rush in *Gleason's Drawing-Room Companion*'s 1851 and 1852 volumes depicted vividly rendered "native" Californians, miners, *vaqueros,* Indians, Chinese, and emigrants, all of which combined to draw a picture for eastern readers of wild and primitive societies. Thus, descriptions of early San Franciscan hotels as centers of gentility and elegance helped to combat images of an uncivilized frontier.[4]

As San Franciscan society "crystallized" in the early 1850s, the wealthy built homes for themselves. The directory observed that "social intercourse had the polish of older communities, mingled with the dash and freedom of the frontier."[5] Within two decades, backed by incredible fortunes accumulated through mining, real estate, and speculation, San Francisco's elite built homes that rivaled and exceeded the luxurious residences of "most European princes." Even so, San Francisco hotels maintained their hold on the city's wealthy class as appropriate places to live and gather. In a city where the boundaries between social classes were permeable and adventurers and gamblers often metamorphosed overnight into gentlemen and gentlewomen, hotels provided a readymade home that equaled or surpassed the very best elite dwellings. Many observers agreed with Hittell's assessment that attributed social mobility to "high wages, migratory habits, and bachelor life," conditions that also contributed to San Francisco's predilection for hotel living.[6] City pride reveled in the quality of the city's hotels, leading one local periodical to assert, even in pre-Palace years, that San Francisco stood "far ahead of any other city in the United States, or indeed in Europe, in point of hotel accommodations."[7] Local pride played on hyperbole and set a context for the Palace's future renown.

A first-class hotel district formed in the early 1860s on lower Montgomery Street, north of Market, between Kearney and Sansome streets. There, the Russ House, Cosmopolitan Hotel, Occidental House, and Lick House reigned as the city's leading establishments.[8] In 1870, the Grand Hotel joined them, located across Market at the corner of Market and New Montgomery. Together, these hotels accounted for eighteen hundred rooms. In 1873, echoing the same kinds of critical sentiments expressed in eastern periodicals, the *San Francisco Real Estate Circular* reported that the first-class hotels were filled with boarding families, much to the detriment of the city. Prompted by rumors of Ralston's proposed hotel project, the paper's editor buzzed that the city's extreme shortage of housing had made the hotels busier than ever. He predicted that if the new hotel materialized, it would fill immediately.[9] Some months later, again manifesting grave concern for the state of the city's domestic life, the newspaper blamed the custom of hotel living not only on the housing shortage but also on the wives who "have either been driven by impudent and incompetent servants to hotel life, or who choose it because their sole desire is to dress, gossip, show off and have an easy and lazy life."[10] As standards of hotel luxury traveled west, they inspired the same debates about luxury, fashion, consumerism, and domestic life. By 1874, San Francisco's pioneer spirit had given way to gendered interpretations of proper domesticity.

San Francisco's first-class hotels received enthusiastic endorsements from observers whose comments sounded much the same as those elicited by eastern hotel descriptions. The city's finest houses offered "elegant and comfortable accommodations" adorned by typical furnishings like French carpets, rich damask furniture, pianos, and large gilt-framed mirrors.[11] While somewhat smaller than the mammoth eastern hotels, they were not only furnished elegantly but also fitted up with technological comforts such as hot and cold running water, the usual multipurpose steam engines, speedy laundries, and mechanical kitchens. Because of the mild climate, San Francisco hotels continued to heat guest rooms with fireplace fires rather than through centralized heating systems. In 1870, San Francisco's only two elevators could be found at the Occidental and Cosmopolitan hotels. The Occidental's Otis elevator was a "pretty little room, neatly carpeted, well lighted and ventilated, with comfortable seats round the sides, [and] handsome mirrors on its walls." While some regarded the elevator a luxury, one resident re-classified it as a necessity after experiencing several days of downtime for repairs.[12]

San Francisco was no different than other large cities with several important hotels in that each of its first-class hotels attracted its own particular clientele. The Lick House, whose wealthy proprietor later endowed the Lick Observatory, had a large number of residential families who, according to one chronicler, composed "as distinct a branch of society as it is well possible to imagine; and to it the 'Occidental set' ran a lively opposition."[13] The Occidental and the Cosmopolitan shared the foreign travelers between them, including Easterners who evidently seemed foreign enough, too. At all times of the day or night, as that same writer claimed, one could see at the Occidental "twenty-five or thirty pairs of Eastern bootsoles staring at him through the plate-glass windows, while the individuals to whom they are attached sit with chairs tilted back at alarming angles." These well-heeled visitors, complete with toothpicks and cigars, "look out at the busy street, and ogle, as the Eastern tourist can, the fairer portion of the native population," recalling the Bostonians' concern for their women who were forced to make their way through the loitering hoards at the entrance to the Tremont.[14] The Cosmopolitan, on the other hand, exuded the imprint of Nevada's Washoe silver mines whose investors constituted its patrons. As newly acquired wealth from mining and real estate continued to swell elite ranks, the established hotels could no longer accommodate all who wanted to live in them. The 1870 Grand Hotel, built with capital supplied by William Ralston, alleviated this shortage and drew many of the eastern transient visitors as well. The four-story Grand covered an acre and a half. One guidebook referred to its style as the "'modern combination,' highly ornamented."[15] However, Ralston soon fell prey to the demons of the "era of hotel improvement" and, within two short years of the Grand's completion, began his plan to build the largest hotel in the world.[16]

William C. Ralston

William Chapman Ralston was born on January 12, 1826, on a farm near Plymouth in north-central Ohio. After his parents moved to Wellsville, a small Ohio River city on the state's eastern border, Ralston attended common school until he quit his studies at age fourteen to work. Eventually, he made a career clerking on Mississippi riverboats. In 1849, he traveled to Panama, where he became an agent for a steamship company that connected New York to San Francisco. Gold seekers crowded the isthmus, looking for a faster and less arduous route to California than by land. Ralston's dauntless energy and skills helped him take advantage of the opportunities and profits available to risk-takers able to make order out of the chaos that prevailed. While in Panama, he became schooled in speculation, management, ruthless competition, and finance. In 1854, his firm transferred him to San Francisco, where within a year he became a partner in a bank with Cornelius Garrison and Charles Morgan, wealthy shipbuilders who mentored Ralston in the Panamanian steamship business.[17]

Ralston and others organized the Bank of California in 1864 with Darius O. Mills as president, but Ralston was its driving force. The bank immediately became the city's leading commercial bank, and Ralston's circle of friends and partners became known as "the Bank Crowd." The Bank of California's capital drew from the astronomical fortunes earned from the Comstock Lode exploitation that began in 1859. Ralston and others profited not only from their mining investments but also through the enormous sums of money that passed through the bank and, additionally, through stock market investments that, for example, saw individual mining stock shares soar from $500 to $6,300 in little more than a year.[18] Many years later, prominent San Francisco businessman Charles Lee Tilden, speaking to the California Historical Society, sought to capture the rampant speculative spirit that permeated San Francisco at the time. He wrote, "Lots were being purchased, lots were being given away for, say, five dollars, and then sold again for, say, fifty dollars and after a while the same lot would sell for a hundred, and then five hundred, and then a thousand. . . . The same thing, too, with mining stock. The gambling spirit was everywhere."[19] Ralston stood at the center of this excitement, and his enthusiasm for his adopted city led him to become one of its wealthiest and most popular community leaders. Ralston's style and the seemingly infinite sums of money at his fingertips enabled him to approach projects without, as one historian suggested, feeling the need to keep his personal fortune separate from that of the bank.[20]

Ralston gained a reputation for supporting, both personally and monetarily, a wide variety of industrial and civic enterprises. He was particularly keen on strengthening California's manufacturing capabilities so as to gain independence from eastern interests. Toward this end, he invested in vineyards, railroads, telegraph lines, the Mission woolen mills that provided cloth for Union army uniforms, the Kimball carriage factory, a watch manufactory, a tobacco farm for producing California cigars, irrigation projects, a furniture manufactory, dry docks, and a sugar refinery. His civic projects included building the California Theater, two hotels, serving on the Board of Regents for the newly organized University of California, expanding the commercial district through the development of New Montgomery Street, and directing the future of the city's water supply.[21]

Ralston was shrewd and expected to benefit financially from these schemes. Even though many of them proved financially disastrous, recurring boom periods always seemed to rescue Ralston's often precarious standing. The twenty-five years between 1850 and 1875 were extraordinary ones in San Francisco's history, a time when opportunity seemed divinely ordained and, as the title of one of Ralston's biographies declared, "nothing seemed impossible."[22] Hubert H. Bancroft, assembling his late-nineteenth-century *Biographies of Men Important in the Building of the West*, began with Ralston, describing him as an example of "man in his most fantastic mood." Ralston's newspaper memorial helps to explain Bancroft's claim: "The advisor of capitalists, the friend of genius and enterprise, the instigator of trade, commerce, and manufactures, the builder of real estate values, the central figure of financial movements, the workingman's employer, the poorman's benefactor, the free but unostentatious giver of almost boundless charities ... he was the object of love and esteem to the entire community in which he dwelt." San Francisco's wild expansion and Ralston's flamboyant vigor and enthusiasm for extravagance framed his plans for New Montgomery Street and the Palace Hotel, plans that, even by San Francisco standards, were over the top.[23]

New Montgomery Street

The location of San Francisco's business district is nearly the same today as it was in 1850, centered on lower Montgomery and Sansome streets just north of Market. While the area north of Market is a grid where streets run due north and south or east and west, Market cuts across this checkerboard at about a fifty-five degree angle, following the original road that led inland from the bay to the early Mission Dolores. The grid south of Market runs parallel/perpendicular to Market on this slant. Not only do the streets north and south of market not line up with one another but the blocks south of Market are four times as large, making coherence nearly impossible. The business district's growing congestion in the 1860s called for some sort of plan, but the proposals extending the existing streets from the north to the south across Market Street faced difficulty because of the incompatible grids. In the mid-1860s, Ralston envisioned opening up the commercial center by extending Montgomery Street across Market and developing this area with office buildings, new hotels, and elegant shops. Ralston's first plan proposed extending Montgomery due south, bulldozing right through the middle of existing blocks, thoroughly disrupting the established grid south of Market. With great ambivalence, the city government approved the plan, but angry property owners managed to subvert it through litigation and court injunctions.[24]

By early 1868, Ralston favored a new plan that would create New Montgomery Street, a grand thoroughfare that would begin across Market, just south of Montgomery, run southeast and thus conform to the existing street pattern. Ralston's New Montgomery Street Property Company began buying property along the proposed street's site. In September 1868, a devastating—although convenient—earthquake shook up the real estate market, allowing the company to purchase much of the necessary land at reasonable prices. Once the street was in place, Ralston hoped to recoup his money by selling the lots along the new street. However, this time the depressed real estate market worked against him. Even so, the *San Francisco Real Estate Circular* expressed confidence in the plan. The proposal featured earthquake-proof construction, exceptional paving materials, exquisite street lighting, and building height uniformity, promising a street as beautiful as any in the United States. The *Circular* quoted the *Daily Evening Bulletin* that compared the new street to Paris's equally new and spectacular Boulevard de Sébastopol created by Napoleon III and Georges-Eugène Haussmann.[25] Ralston's thinking meant to establish San Francisco as one of the great cities of the world, not merely of the West or the nation.

Ralston built his Grand Hotel on the south side of Market between 2nd and New Montgomery streets. Representing an investment of over one million dollars, the Grand immediately "stepped . . . to the pinnacle of hotel fame in San Francisco."[26] The Grand was not the largest of San Francisco's hotels in terms of occupancy, but it boasted the most costly furnishings and sat on the largest and most valuable piece of hotel property. The Grand's ornate Second Empire facade with rows of arched windows and a crowning mansard roof punctuated by domed towers rose three stories but was still modest in comparison to New York and Philadelphia's monster hotels. The spacious and luxurious public rooms and the "elegantly frescoed" bar-room earned a reputation as the west coast's most richly decorated. As might be expected, the wealthiest of the "hotel-patronizing families" made their homes in Ralston's New Montgomery Street landmark.[27]

Despite the success of the new hotel and the relocation of several fine stores to its ground floor, interest in the New Montgomery Street properties continued to flag. Even though new ore discoveries in the Comstock Lode bolstered the soft economy of the early 1870s, the real estate market remained stagnant. Ralston's new Montgomery Street Property Company scheduled a land auction, but it was postponed when it failed to attract buyers. The *San Francisco Real Estate Circular,* however, continued to express confidence in the district, stating that the property "is among the most central and valuable of any in the city." In an effort to salvage his investment, Ralston began formulating plans for another hotel. Property transfers to William Sharon, a millionaire banker and Ralston's close business partner, prompted rumors of the mammoth project. Toward the end of the year, other plans for another extension of Montgomery Street by Ralston's company surfaced, but by then the street project had been deemed by the press "a failure and an elephant of the most unprofitable kind." The hotel project would be Ralston's final attempt to make good on his expensive real estate investment that had already cost him two million dollars.[28]

"The Greatest Caravansary in the World"

By any standard, the new Palace Hotel was huge. It stood 120 feet high, its seven stories towering over the city like an enormous fortress. The hotel occupied an entire city block bounded by Market, New Montgomery, Jessie, and Annie streets, just across New Montgomery from the now-dwarfed Grand Hotel. Its frontage on Market ran 275 feet, on New Montgomery 350 feet. An 1890 panoramic photograph taken from a perspective south of Market shows the hotel a good four stories taller than any other city building, its immense size looming in startling contrast over the smaller buildings lining the streets in all directions. Taking note of the hotel's effect on the city's landscape, London's premier architectural journal, *The Builder,* remarked, "This architectural monster lifts its colossal bulk above the business and social centres of the city." In square footage, the Palace eclipsed by far that of the nation's newest hotel construction that followed Chicago's disastrous 1871 fire that destroyed the city and its hotels. Its ground floor covered two and a half acres, while the subterranean service areas reached three.[29]

John P. Gaynor, the hotel's architect, was born in Ireland in 1826. He first practiced architecture in New York City, where he won national acclaim for his land-mark Haughwout Building (1856) at 488 Broadway. Known architecturally for its remarkable cast-iron facades of repeating arched and colonnaded windows, the Haughwout Building acquired additional significance as the site of Elisha Graves Otis's first commercial safety elevator installation. After moving to San Francisco in the early 1860s, Gaynor built one of the city's first cast-iron buildings for the Savings and Loan Society. Ralston hired him to design the Grand Hotel as well as the financier's pretentiously elaborate estate, Belmont. According to hotel lore, Gaynor's preparation for the commission included visiting all of the best American and European hotels so as to be able to incorporate and surpass existing technological, architectural, and decorative elements.[30]

The hotel's construction began in mid-1874, and San Francisco watched as first the excavation and then the erection of the vast walls proceeded. Well aware that a series of six fires had repeatedly destroyed the city in its short history and having experienced the recent 1868 earthquake, Ralston determined to construct a building that would prove impervious to those threats. He assumed a certain technological audacity in his plans to make the Palace earthquake-proof. *Frank Leslie's Illustrated Newspaper* described the hotel as having the "strength of a fortress." Its foundation walls, built of stone, iron, and brick, measured twelve feet thick at their bases, narrowing to a still substantial fifty-four-inch thickness at ground level. All the exterior, interior, and partition walls were banded together with iron bars, forming an iron basketwork lattice throughout the entire building. At the lower elevations, the iron bars were five feet apart. As they ascended the height of the building, the bands became increasingly closer together so that on the uppermost seventh story, they were two feet apart. The stone and brick walls were built "around, within, and upon" this huge skeleton of wrought iron, creating a citadel against natural disaster. The bands weighed three thousand tons and were fastened together with huge bolts that extended through to the outside of the building in a regular pattern. Decorated with gold paint, the bolt ends glittered against the background of the building's white painted brick. In this way, the structural components merged with the hotel's flamboyance, contributing to the facade's astonishing impression.[31]

Despite the boundless size and expense, the building's exterior architecture did not win much praise. Bay windows extended from each of the 348 exterior rooms, establishing the building's most predominant architectural feature. The six stories of bay windows rising above the ground floor prompted one observer to describe the hotel as a "huge building 'broken out into bird-cages.'" The *Sunday Chronicle* noted that those who failed to appreciate the advantages that bay windows offered

A widely reproduced image of the exterior of the Palace Hotel featuring its bay windows and the busy street. *Courtesy, The Bancroft Library, University of California, Berkeley.*

in catching additional sunlight and the bay breezes might find that the repetition created "a slight appearance of monotony." The *Daily Alta California* described the hotel as "plain," with none of the usual features such as great columns, domes, or steeples that might normally augment buildings of such size. Even the hotel's own literature characterized the architectural style as being "severely simple," acknowledging that the "myriads" of bays served to mitigate its "oppressive massiveness." However, the bay window was *the* distinguishing feature of San Franciscan architecture. One architectural historian noted that 95 percent of all San Francisco dwellings had at least one bay window. The fresh air and additional sunlight that the hotel's windows showered on its guests in this fog-bound city contributed to the luxurious amenities of the hotel. Despite the reserved reviews that the windows generated, Ralston's insistence on them introduced the San Francisco style to a worldwide audience and once again reworked the generic hotel standard to showcase provincial distinctiveness.[32] Ralston sheathed his version of the monster hotel with San Francisco's one distinguishing architectural feature to situate it firmly in its city, regardless of where around the world its image appeared.

"Promenades amidst Tropical Verdure"

Only the building's immensity provided clues to the opulence that awaited guests inside. On New Montgomery Street, a forty-four-foot-wide colonnaded entrance-way passed through cast-iron arches and led into the magnificent grand court. Here, a circular driveway allowed carriages to discharge passengers in a seven-story-high atrium crowned by a glass roof. The entire court, including the marble-paved promenade and orchestra pavilion, measured an inconceivable 144 by 84 feet. In her 1877 travel memoir, newspaper editor Frank Leslie's wife described how driving through the iron gates and stone archway of the hotel's entrance reminded her of Paris's Grand Hotel, but on a larger and more regal scale. Her description no doubt captured the effect that Ralston was hoping to achieve:

> Seven tiers of balconies surround the four sides, ornamented with frequent tubs of flowering plants, cages of singing birds, and sofas and chairs, where groups of guests sit to chat, or promenade up and down. A glass dome covers the whole, giving a soft and tempered light during the day, while at night the place is brilliantly illuminated by gas. On the ground floor the court is faced with white marble, and a circular carriage-drive sweeps around the centre, where stand groups of palm and banana trees, and vases of beautiful flowers. Chairs and settees are dotted around the pavement, where sit the *flaneurs* who like to watch the constant arrivals of guests, and visitors' carriages standing in waiting complete the lively and picturesque scene.[33]

The six floors of guest rooms above opened to the atrium; twelve-foot-wide plant-filled verandahs separated the guest room doors from the railing that looked down on the scene below. On the seventh floor, a rooftop garden with broad halls and observatories allowed guests to partake of panoramic views of the city and the bay. Many visitors in addition to Mrs. Leslie acknowledged a similarity to Paris's Grand Hotel, but the size of Ralston's creation swept away any comparisons.[34]

Vases of flowers, exotic tropical trees, the expanse of space, and singing birds combined to fashion a "natural" world where, unlike nature, everything was not only both clean and controlled but also made possible through expensive technological systems that created the environment. The iron and glass skylight with its

This drawing of the entrance to the Palace's Grand Court helps the viewer understand just how large the hotel was. The center arch accommodated the passage of two carriages. *California Spirit of the Times,* March 11, 1876, 5. *Courtesy of the California History Section, California State Library, Sacramento, California.*

ingenious mechanical window-washing apparatus ensured filtered natural light by day, while after dark, 516 gas jets flickered in burnished gilt fixtures, creating a starry-night atmosphere. The guests' carriages left the chaotic urban world behind at the sidewalk as they crossed the hotel's threshold, sweeping into its carefully constructed and sanitized urban environment along the indoor roadway that bridged the two cities. For there was no question that this immense structure housed a "complete metropolis within itself"; as one writer observed, if not for the exceptions of a theatre and chapel, a guest might never need to leave until "his final necessary pilgrimage to Lone Mountain."[35]

While the central court was certainly the most visually arresting of the hotel's features, the other public rooms were stunningly elaborate as well. With the court in the center of the hotel's rectangular layout, the office, private dining rooms, and large breakfast room flanked the court to the left, while the enormous formal dining room and ballroom did the same on the right. Two lesser light courts, opened to the building's full height, ran parallel to each of the center court's long sides and separated the aforementioned rooms from still other outer hotel areas. These courts had driveways leading from the basement service areas to Annie Street at the building's rear. They allowed natural light into all the guest rooms. Stores lined the hotel's main frontage on New Montgomery and Market streets to the right of the light court beyond the formal dining and ballrooms. Following custom, the shops that occupied the frontage on Jesse Street all catered to men: barber shop, billiard room, bar, and committee meeting rooms. Thus, as in earlier hotel plans, the main floor divided into gendered regions, with the ladies' reception room, formal eating and social areas, and shopping to the right; and the offices, "gents" reception room, private meeting areas, and masculine activities to the left. All of these rooms, except the stores, had twenty-seven-foot ceilings, and were of enormous proportions. The "smaller" of the two dining rooms, for example, measured 110 by 55 feet—an area large enough in which to fit a small hotel.[36]

The ground floor provided space for eighteen retail shops. Ralston had hoped to draw other retail businesses to the New Montgomery Street area, creating a lucrative new shopping district. The Palace's enormous bulk, however, standing next to the Grand along Market and just south of the Lick House, compounded the effect of the large buildings and seemed to discourage traffic from penetrating beyond the large hotels. By this time, the *San Francisco Real Estate Circular* had turned somewhat hostile; in commenting on the erection of the hotel's seventh story, it remarked that the building "stands like a dark menace to Montgomery street, as though it said, 'Thus far shalt thou go and no further.'"[37] In addition, the Palace's stores were just thirty feet deep, not large enough to satisfy San Francisco retailers who needed larger spaces. In yet another derisive snit, the *Real Estate Circular* blamed this faulty decision on New Yorker Warren Leland, whom Ralston hired as the Palace's proprietor. "The shallow depth named might do on Broadway, New York, where land is worth two to four times what it is on the south side of Market street, between Second and Third," the paper charged, "but it will hardly do in the Palace Hotel, since the location is not a first-class one for either wholesale or retail business." This represented a complete reversal from the paper's attitude of the previous year. It also revealed bitterness at Ralston's hiring of a New York hotelier as his advisor. Leland, one of a three-generation dynasty of hotelkeepers, enjoyed a national reputation as the longtime proprietor, with his brothers Charles, Simeon, and Aaron, of New York's Metropolitan Hotel.[38]

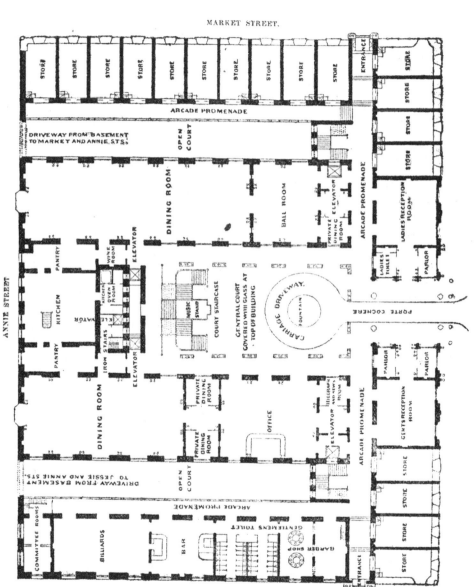

MARKET STREET.

STORE · STORE · STORE · STORE · STORE · STORE · STORE · STORE · STORE · STORE

ENTRANCE

STORE · STORE

STORE

STORE

ARCADE PROMENADE

DRIVEWAY FROM BASEMENT TO MARKET AND ANNIE STS.

OPEN COURT

DINING ROOM

BALL ROOM

ARCADE PROMENADE

LADIES RECEPTION ROOM

PRIVATE DINING ROOM

PRIVATE ELEVATOR

LADIES TOILET

ANNIE STREET

PANTRY

WINE ROOM

ELEVATOR

KITCHEN

OVEN ROOM

ELEVATOR

IRON STAIRS

MUSIC STAND

COURT STAIRCASE

CENTRAL COURT COVERED WITH GLASS AT TOP OF BUILDING

CARRIAGE DRIVEWAY.

FOUNTAIN

PORTE COCHERE

PARLOR

NEW MONTGOMERY STREET,

KITCHEN

PANTRY

ELEVATOR

DINING ROOM

PRIVATE DINING ROOM

PRIVATE DINING ROOM

OFFICE

ELEVATOR

TELEGRAPH ANNEX ROOM

PARLOR

PARLOR

GENTS RECEPTION ROOM

ARCADE PROMENADE

STORE

STORE

DRIVEWAY FROM BASEMENT TO JESSIE AND ANNIE STS.

OPEN COURT

ARCADE PROMENADE

STORE

STORE

COMMITTEE ROOMS

BILLIARDS

BAR

GENTLEMENS TOILET

BARBER SHOP

ENTRANCE

STORE

JESSIE STREET.

The stores, contiguous under an interior arcade, formed a "bazaar," where, in typically gendered language, women guests could indulge in "shopping to their hearts' content" without worrying about the weather. The *California Spirit of the Times* noted that "ladies, who are chronic shoppers and snappers up of unconsidered trifles, may take their afternoon constitutional and find everything they wish without braving the breeze . . . or wetting their soles."[39] The arcade or promenade ran along the smaller light court. Each of the stores had a large glass "show-window" that faced the promenade, thus bringing the usual retail street scene indoors. According to the hotel's promotional literature, the stores were elegant. One sold Remington sewing machines. In keeping with the hotel's high-tech profile, the *San Francisco-Newsletter* observed, "The proprietors of the valuable patent have determined to take premises worthy of such a grand innovation upon old time sewing machines, and have, with a discretion which does them credit, chose to locate under the great center of attraction, The Palace Hotel, No. 629." This little ad found itself in the middle of a long descriptive tabloid dedicated to describing the new Palace. The tabloid itself focused almost entirely on the hotel's infrastructure, such as the pipes, wiring, and other such mechanical matters, and the Remington shop was the only store described or mentioned. Somehow, for the writer, the sewing machine shop, by virtue of its selling modern machinery and that with a particularly female audience, bridged the hotel's functional and fanciful characteristics.[40]

"Richly Furnished Apartments"

The hotel's second through seventh floors accommodated the guest rooms, suites, and parlors. Each had similar floor plans, with the exception of the second floor, where some space was dedicated to communal amenities such as parlors, private dining rooms, a children's dining room, and a ladies' billiard room. The children's dining room appears to be a new development, perhaps first adopted at Chicago's 1873 Grand Pacific Hotel. There, as at the Palace, children dined in this newly dedicated space with their family servants and were only allowed in the drawing rooms if accompanied by their parents, thus establishing restrictions governing children's behavior. These regulations removed children from the family setting, relegating them to lonely meals with the nursemaid, while, at the same time, reinforcing the

Architectural plan of the Palace Hotel's ground floor. William Laird MacGregor, *Hotels and Hotel Life, San Francisco, California in 1876* (San Francisco: S. F. News Company, 1877), 14. *Courtesy, The Bancroft Library, University of California, Berkeley.*

formality of adult hotel dining. In private homes and hotels, dining had become an elaborate ritual that children easily disturbed. A new emphasis on dining *à la Russe* resulted in more elaborate food, specialized and slower service, and elegantly decorated tables. Charming, gay conversation and prescribed manners required a studied attentiveness that precluded noisy and restless children.[41]

The ladies' billiard room also represented a dramatic departure. Hotels had regularly devoted an enormous amount of space to men's billiards rooms, universally in conjunction with the bar. The billiard room was a male enclave for smoking, drinking, playing pool, and passing the time. Billiards had always enjoyed an upper-class following, in addition to its less respectable adherents in the "sporting life," the working-class male subculture that patronized cheap boardinghouses and their prostitutes. In the same way that the nineteenth-century upper middle class emulated other aspects of refined upper-class life, billiards, too, became a fashionable pastime. In 1859, *Frank Leslie's Illustrated Newspaper* began to publish a billiards column, indicating the game's widespread acceptance and popularity. By 1914, Chicago's Brunswick-Balke Company, who supplied the Palace's tables, was marketing billiards as a healthy avenue for "manly training"—interestingly enough, for both boys and girls—and offering endorsements of the game's benefits by ministers, doctors, educators, and sociologists. The Palace offered its women guests the opportunity to play a popular sport in a respectable gender-segregated parlor that not only protected them from male strangers but also allowed them to bend over the tables and otherwise move freely without compromising their dignity.[42] At the same time, it protected the men's freedom to smoke, spit, drink, and engage in friendly wagers in their own traditional sanctuary.

In the usual fashion, guest rooms could be combined in almost any number to form suites. These were good-sized accommodations with at least half the rooms being twenty feet square, with none smaller than sixteen feet square. Each had its own washbasin and clothes closet. The configuration was such that every two bedchambers shared a toilet and bathtub between them. Each had a fireplace, marble mantle, plate-glass framed mirror, four-burner gas chandelier, electric annunciator, fire alarm, and window to either the street or one of the three interior light courts. Carpets covered the floor, and furniture constructed of California woods filled the rooms. Ralston insisted that all the bedrooms be furnished equally so that no one would encounter less than a first-class experience.[43] In addition, a complex ventilation system ensured fresh air to every room and bathroom. Two thousand and forty-two ventilating tubes led outward to the roof from every room, closet, and bathroom, giving the appearance of a giant pipe organ. Pneumatic tubes whisked letters, messages, and packages to and from the various floor stations, while a mail tube carried guests' postal letters from the sleeping floors to the government letter-box in the main office.[44]

Needless to say, the numerous publications that showcased the new hotel described the interior decorations as being neat and elegant, modest, and in the finest taste. These comments were *de rigueur*, and not only reiterated those heard about every other luxury hotel built and reported on since the 1820s, but were always stated in the context of being in opposition to those "other" hotels whose decors were gaudy and of a "worn-out" style. At the Palace, the carpets in particular received special notice. Not only was the supply house, W. J. Sloane of New York City, the largest carpet supplier in the world, but the contract was the largest ever undertaken by any firm in the country.

Continuing to emphasize the grandeur of the carpet contract, articles assured the reader that the Bigelow Carpet Company, the carpeting manufacturer, was the largest manufacturer of Brussels carpeting. Moreover, the Axminster carpets were the most expensive known, and the Axminster parlor rug for the ladies' parlor was the "largest ever woven in one piece for any customer in the United States—larger than even those in the White House in Washington." Promotional literature stated that if laid end to end in three-quarter yard strips, including the runners and small rugs, the hotel's carpeting would measure forty-five miles. No superlative proved too great when describing this hotel's floor coverings. Twenty looms worked steadily for five months to weave the carpeting. Ralston imported the Axminster parlor rugs from Glasgow, and smaller Axminster rugs accessorizing the bedchambers required the handiwork of two hundred sewing women. Indeed, the building's vast scope inspired gushing commentary about virtually every aspect, including the glassware (11,700 pieces of "fine cut crystal"), the china ("half the thickness [of that] in use in other hotels"), the table knives ("the finest ivory handles"), and the gas globes (12,500—"without doubt the best in the market"). Hyperbole ruled every line of description.

"Electric Bells Everywhere"

Unquestionably, the reporting reserved its most lavish and unrestrained praise for the hotel's technical departments. Paragraph after paragraph in local, national, and international periodicals, newspapers, and promotional literature reviewed the various systems and features of the hotel's infrastructure. Certainly the extraordinary proportions of everything that went into the building distinguished the Palace Hotel's technology. For example, the plumbing required twenty-three miles of gas pipe, six miles of sewer pipe, eight miles of iron pipe for steam, twenty-eight miles of vulcanized rubber pipe for hot and cold water, and five miles of cast-iron and rubber-coated pipe for the firefighting system. The San Francisco firm of Bush and Milne installed the plumbing, gas, and ventilation systems and did so, they said, "without one single leak in the building." Steam heated the corridors and public rooms. There were 437 bathtubs, 492 water closets, and 891 wash basins. Even in a world of hyperbole, it seems perfectly believable that this was the "most extensive job ever done in the world in this line."[45]

The electrical system also earned noteworthy attention. 1875 predates electrical lighting by a few years, so the papers described a battery-powered signaling system used for communication between guests and the office and also for management to keep track of their employees. One thousand thermostats distributed in the rooms, closets, passageways, and service areas served as fire detectors, not temperature regulators, and signaled the office should the temperature in any room reach 120 degrees Fahrenheit. Using a newer, more complex invention, the annunciator, guests could signal floor clerks for service as well as signal the elevator operator to bring the car to their floor. The system required bellboys to check in by signaling from their stations so that the office could keep track of where they were. Watchmen, too, had a route along which they touched signals at appointed times to verify that they were indeed on watch. These devices were called "tell-tales."[46] With a twist of irony, the hotel described the fire alarm thermostat as a "sleepless watchman," which, of course, was the point of the tell-tale, to ensure that the human watchmen were not asleep. Finally, sixteen electric clocks completed the electrical workings, all synchronized by a patented regulator clock. Travelers had complained consistently over the preceding fifty years about the regimentation of hotel meal times. Hotels had already established time patterns divorced from the natural cycle of the day. Throughout the long nineteenth century, the agents of commerce, particularly railroads, adhered increasingly to standardized times as a necessary and logical component of economic progress and growth. Time became a commodity, and accurate time, purchased from Western Union, became a source of prestige.[47] In a place like the Palace Hotel, accurate time became one more symbol of commerce, luxury, progress, and the hotel's place within the international commercial world. And, too, when envisioning the hotel as a self-contained city, the hotel's clocks served both to regulate "the well-regulated machine" and to connect to the city working outside its walls.

The hotel had five elevators, more than any one hotel had ever had—or needed—

A view of the Palace Hotel's front desk after the 1891 renovation that replaced the marble floor with a mosaic one and the gas globes with electric lighting. The desk lined one wall of the immense office. On the far wall, bellboys await instruction. Behind the desk, the large electric clock received its signals from Western Union and regulated the workings of the hotel. *Palace Hotel (San Francisco: Palace Hotel Company, 1897), Courtesy, Wolfsonian Museum.*

before. These were hydraulic elevators, a new development in elevator technology just coming into use. G. W. Dickie of San Francisco's Risdon Iron Works, described as one of the West Coast's most prolific inventors, designed the elevators. The servants' elevator was located near the kitchen behind the Grand Court. The others were in the Grand Court's four corners and served the hotel's guests. One of these went all the way to the roof, taking resident-adventurers out onto an observation deck. The largest of the elevators could carry forty passengers, while each of the other three could hold thirty. It took one minute to travel from the ground floor to the seventh, the motion being described as "almost as imperceptible as that of a balloon." Light wood paneled the elevator walls, while decorative mirrors, upholstered seats, and a gas chandelier transformed the machines into small parlors.

Fireproof brick walls encased the elevator shafts from the basement to the roof, a lesson learned from the 1867 fire that destroyed St. Louis's great Lindell Hotel, which had enjoyed a reputation as the West's largest and finest hotel. There, a relatively controllable blaze destroyed the four-year old building in a matter of hours after spreading through the elevator shaft to every floor. Not only did the fire spread rapidly, but water from the city's fire engines also could not reach the building's sixth and seventh floors, so the fire hose had to be hand-carried up the stairs. By then, nothing could stop the fire's force, and it completely devastated the building. With San Francisco's history of fires, Ralston took these lessons to heart.[48]

Ralston involved himself in every detail of the hotel's construction, and his concerns about fire evidenced themselves in the elaborate systems installed for preventing, detecting, and suppressing them. The threat of fire was significant not only from the usual sources such as the kitchen but also from the hundreds of fireplaces that heated the rooms. In addition to the detection system that included both electrical and human surveillance, Ralston specified construction materials and methods that would deter a fire's spread, including a separate water piping and pumping system. Four artesian wells, two in each outer court, together with the city's water supply, supplied a main reservoir located under the Grand Court. The reservoir held 630,000 gallons of water and had five-foot-thick brick and cement walls supported by buttresses and banded together with iron rods. Seven rooftop iron tanks held an additional 128,000 gallons of water pumped from the basement reservoir by Knowles steam pumps. Two more steam fire pumps stood alongside the main boilers in a brick and iron vault under Annie Street. These began to work automatically as soon as anyone opened any one of the fireplugs. Water mains running throughout the building had 392 outlets, to which attached fifty- and one-hundred-foot lengths of fire hose. The Goodyear Rubber Company supplied a total of twenty thousand feet of carbolized hosing. Eight fire mains under the hotel's sidewalks connected to the city water supply so that city engines could assist in fighting a major conflagration, although the city supply, due to the long, dry summers, was often unpredictable.

The working departments' walls, floors, and ceilings were constructed of materials considered "absolutely fireproof." In the basement, the floors were either marble or asphaltum (asphalt). Iron beams and brick arches formed the ceilings of all the storerooms, kitchen areas, coal vaults, heavy equipment rooms, and servants' quarters in the basement and on the first floor. The marble floors of the first-floor kitchen and work areas, main corridors, pantries, and wine vault were laid on iron beams, while all other marble flooring lay atop foundations of brick and cement. The upper floors were wood, which were set on a composition of wood strips, mortised with a mixture of sand and plaster so that, when dried, it became as solid as stone. Brick fire walls extended the building's entire height and through the tin roof to prevent a fire from jumping from one section of the hotel to another. In the hotel's decorative areas, liberal use of marble and teak, a highly fire-resistant wood, augmented the protection as much as possible.[49] Thus, Ralston incorporated every known precaution to protect his investment, guests, and employees against the very real and serious threat of fire. These measures added a half-million dollars to the building's cost. And, while the system worked for thirty-one years, the great San Francisco earthquake of 1906 provided a challenge even Ralston could not have foreseen.

"The Genius of W. C. Ralston Illustrated"

Ralston envisioned his project not only as the largest hotel in the world but also as one that would speak for California's achievements in industry and craftsmanship. Part of the hotel's folkloric history recounts the extremes to which Ralston went to secure materials and furnishings that came from California and its manufacturers. One such story had Ralston purchasing a ranch in the Sierra foothills near Grass Valley so that he could use its stand of magnificent oak trees for the hotel's floors. Only after the sale went through did Ralston discover that the wood was inappropriate for that use. On a more positive note, Ralston set a local locksmith up in business to supply the thousands of locks needed. He bought a foundry to produce nails. He established a furniture manufactory to fashion the hotel's furniture from California woods. Ralston's silk company in San Jose produced the silk hangings and upholstery, while some of the blankets came from his San Francisco woolen mills. Stories claimed that even the horses that pulled the hotel carriages were the product of the mares and stallions from Ralston's Belmont estate. In this way, the hotel served as a showroom for California enterprise. As San Francisco's *News-Letter* observed, "Soon we will have the visitors from every quarter of the globe coming to stop at our Palace Hotel . . . and the fame of our furniture will be spread throughout not only America but Europe." Ralston sought to place the Palace, San Francisco, and California on a world stage.[50]

The fable of the California furnishings and the degree to which the story was disseminated belied Ralston's worldwide search for exotic materials and luxurious appointments. A series of letters reveals an intense quest for oriental woods to be used for paneling. Ralston had men in Hong Kong, Japan, China, Bangkok, Singapore, and Manila looking for suitable wood. One agent lamented the inability to secure timber from old Japanese temples, saying, "There is some first class timber but it cannot be gotten or any like it except by a fluke." While San Francisco work-shops produced most of the heavy materials such as the iron work, marble side-walks, hardware, asphalt, and elevators, the large majority of decorative accessories found their way to the Palace from the eastern United States and Europe. Many of the San Francisco firms credited with supplying the hotel did so with goods from the East. O. Lawton and Co. furnished the Haviland china and English grooming sets. Edward McGrath contracted for the marble through fifteen New York suppli-ers. Thomas Day and Co. bought the gas fixtures and chandeliers from Archer and Pancross Manufacturing Company, a well-known New York manufacturer.[51]

Frank Leslie, the New York newspaper publisher, was fascinated with luxury hotels and published descriptions of many as they came to be built. His article on the Palace played on this fascination but also paid homage to New York City's manufacturing interests. He pointed out that Ralston was determined to outfit the hotel with the best of everything, that his procurers were "thorough businessmen" with access to the world's goods, and that they had the experience to "make the best bargains possible." And Leslie noted with satisfaction, they did much of their shopping in New York. "The fact that so many of the supplies were obtained in this city," he stated, "is a high compliment to our resources and the enterprise of our businessmen." While Leslie congratulated San Francisco for having "the finest hotel building in the country" and "probably" the world, he also basked in civic pride by enumerating the contributions of New York businesses. A. T. Stewart, the celebrated department store, furnished upholstery, lace hangings, bedding, blan-kets, and towels. Morris, Delano and Co. obtained the plate glass and enormous mirrors from Europe. W. and J. Sloane furnished all the carpeting, and Hayden Gere and Co. supplied the washbasins and brass fittings for the plumbing and steam layouts. The mantels came from yet another New York firm, as did the Gor-ham silverware. Of course, Leslie's list did not include firms such as Philadelphia's Morris and Tasker, who supplied the radiators, or equipment such as a French mangle installed in the laundry to press fancy linens, or the billiard tables that came out of Chicago.[52]

In effect, Ralston exploited two images for the hotel. One featured the Palace as a product of an upstart western city that in the short span of twenty-seven years had grown into a world-class metropolis capable of constructing the world's largest, strongest, and safest hotel building. The structural elements—bricks, iron, concrete, bolts, screws, washers—and the skilled mechanical expertise required to forge these materials into a durable edifice came from local sources. Yet even these mechanics and merchants, by virtue of the city's youth, were themselves necessarily imports. Published descriptions invariably began with the news that architect Gaynor, an ex–New Yorker himself, had been sent all over the country and Europe to learn the ways of modern hotels. Instant cities like San Francisco, because they grew so quickly, had to imitate eastern and European models. The furniture of California woods was perhaps the hotel's only original California feature. Even while being touted as a regional product, the hotel's immense size, cost, and its modernity were elaborations on a proven national concept.[53]

Nonetheless, state and city pride imbued Ralston with a mission, and he was not alone in his attachment. Writing in 1879, John Hittell, author of the widely read book *The Resources of California* and editor of the *Daily Alta California*, noted that "the Californians who have been here from fifteen to twenty-five years are proud of their State and carry their pride so far that it is observed as something exceptional in the United States." Hittell attributed this fierce bearing to the self-satisfaction city pioneers enjoyed from the rapid and impressive growth to which they contributed. Over the span of a mere quarter-century, the town had grown into a major city that San Franciscans believed was "the best place in the world for the enjoyment of life." Hittell boasted that the city was "destined to have a prosperous and glorious future" and that it would be "a center of the highest civilization." Ralston intended for his hotel to herald the city's maturation. Not only did San Francisco's industry and master craftsmen produce the hotel, but its wealth and cultural achievements also indicated a need for one.[54]

While local skill and materials built the foundation and structure, New York, Europe, and the Orient contributed the hotel's decoration, its fancy dress. The drapes, upholstery, linens, china, silver, large pier mirrors, light fixtures, woods, carpets, marble, and mantels all came from places far away from California. These were the materials through which the guests experienced the hotel. This sophisticated facade that represented San Francisco's coming of age spoke the language of world-class luxury. Nearly every commentary compared the Palace to the hotels of Europe, particularly those of Paris, with the Palace emerging as the victor in terms of size, cost, and elegance. Of course, this is ironic in itself. By this time, European hotels were patterning themselves after American palace hotels. However, the seductive aura of aristocratic gentility continued to exert influence over American culture through standards set by material consumer goods. Furthermore, there existed a large cosmopolitan population of travelers who were discriminating enough to recognize these hallmarks of sophisticated taste. Ralston overlaid his celebration of San Francisco's skill with a veneer of aristocratic pretension—and drove home the point with the hotel's name. Like other large city hotels, the Palace strove to distinguish itself as a local production, a representation of civic pride and place, all while adhering to a standardized model that adopted European luxury goods as its cultural measure.[55]

"Warren Leland as the Prince of Hotel Managers"

The mantle of royalty conferred by the hotel's name fell upon Warren Leland, the man to whom Ralston entrusted the hotel's management. Alternately referred to as "the prince of landlords" or the "king of Hotelmen," Leland came to the Palace with a reputation as one who "Knows How to Keep a Hotel." Leland hailed from the country's foremost family of hotelkeepers. One historian claimed that every city in the nation had, at one time or another, a hotel managed by one of the three generations of Lelands in the business. The *News-Letter* noted, "You can hardly name a place where some member of the family does not conduct some large business in the hotel line. Wherever civilization is known or appreciated, the name of Leland stands pre-eminent." The Lelands were well known among hotelmen, so much so that an 1878 issue of *Hotel World* published their aggregate weight, the ten Leland men of the "kindred fat men's association" averaged 246.5 pounds each.[56]

Leland's father, Simeon Sr., began as a hotelkeeper, stage proprietor, and part-time lawyer in Landgrove, Vermont. Four of the elder Leland's five sons learned the hotel business. Five more Leland men from the next generation also became hotelmen. Warren earned his reputation serving as proprietor with his brother Charles for twenty years at New York City's Metropolitan. There, all family members served their apprenticeships. Evidently, the family bred the qualities of good hotelmen that the *California Spirit of the Times* described as "illimitable patience, boundless *bonhomie* and actual knowledge when a thing is well or ill done." By 1876, the two generations of Leland brothers, uncles, and cousins had managed or were managing hotels in New York City; Philadelphia; Albany; Saratoga Springs; Springfield, Illinois; Long Branch; Baltimore; and of course, San Francisco. Despite suffering a major financial disaster at his Union Hotel at Saratoga Springs, Leland recovered through a subsequent hotel venture and moved to San Francisco at Ralston's request. Ralston and Leland worked closely during the planning and construction stages.[57]

Part of Leland's responsibility was to hire the hotel staff. A full year prior to the hotel's opening, Leland established himself in Chicago to recruit craftsmen and managers. Chicago's post-fire rebuilding produced many of the nation's foremost first-class hotels as well as an active group of hotelmen. During the late 1870s, the hotel industry began to organize itself there. Hotelmen established the nation's first hotel professional society, the Chicago Hotel Association in 1877. The country's first hotel weekly, *The Hotel World*, began publication in Chicago in 1878. Hotel professionals organized the national Hotel Men's Mutual Benefit Association in Chicago in 1879. Hotel news editors indulged in personal news about hotelmen and guests and often resorted to biting gossip and fierce accusations, revealing the informal yet solidifying nature of Chicago's fledgling hotel industry. This newly established communication network served as a context for Leland as he entertained potential workers, hired them, and shipped them out to California. They were, as Ralston's friend W. F. Coolbaugh quipped, "Sheridan's boys going to help out Custer at the Black Hills."[58] The key personnel, then, those responsible for ensuring the new hotel's expert management, would hail from Chicago.

Of the hires, the most important person next to Leland himself was the French chef who was to preside over the kitchens. Leland hired Jules Hardes and a number of sous-chefs, also French. At the Palace, in concert with the royal theme, Signor Raffa, the Italian pastry chef, ruled his "little kingdom of pastry and confectionery," while Herr George Shafer, a German sommelier, governed as "monarch of the wine room," thus completing the hotel ruling dynasty.[59] Dishes on the menu reified this ongoing tension between American spirit and aspirations to royalty. "Lyonaise Potatoes," "Consomme Fleury," "Roulade of Veal à la Bourgeoise," "Salad à la Mayonnaise," and "Gateaux à la Royale" commingled uneasily with American and regional specialties such as pancakes with Vermont maple syrup, clam chowder, and boiled codfish. The exigencies of responding to American taste required a certain mitigation of the hotel's commitment to French dining standards.

In 1876, William Laird MacGregor, a Scottish visitor to San Francisco, published a guide to San Francisco hotels that included a lengthy description of the Palace as well as the other city hotels. MacGregor also provided some data on hotel employment. In his book, MacGregor recorded that first-class hotels required between 120 and 140 servants to complete the daily work, including forty waiters, eight bellboys, eighteen kitchen workers, eight porters, and one or two each of engineers, elevator boys, watchmen, yardmen, carpenters, plumbers, and upholsterers. Women workers included thirteen chambermaids, eight laundry maids, and seven scrub girls. A new class of middle management personnel consisted of head clerks, night clerks, a housekeeper, a caterer, assistant clerks, bookkeepers, storekeepers, a head porter, a laundry superintendent, and two assistant managers.[60] Presumably, the Palace required many more servants because of its larger than average size.

Leland also stunned San Francisco society by employing only black workers, whom he brought from the East in the tradition of eastern resort hotels. The daily newspapers pointed out this practice as a "vast improvement upon the impudent 'white trash' who have exchanged the hod for the napkin, but still retain the manners of their native hovels." Nativist disdain for immigrants who typically filled hotel service positions was a national pastime that found its early roots during the Irish immigration before the Civil War. The *News-Letter* went on to note that they had seen "some of the prettiest kind of colored girls flitting around the bedrooms in their nice, clean dresses, and felt very much inclined to take a suite of rooms in consequence."[61] The black waiters, experienced at serving affluent urbanites, contributed to the hotel's atmosphere of gentility through their erect bearing, graciousness, and military-like efficiency.[62] This tradition of employing African Americans lasted at the Palace over twenty years. In 1896, manager John Kirkpatrick, pinched by a soft economy, wrote to his boss, F. W. Sharon, that he "sent away all the colored Bellmen" and replaced them with less costly white men. He continued, "By this means I get very much quicker and more intelligent service, and also save money; as I paid the colored bellmen $30.00 per month, and pay the white $20.00." Attitudes had changed, as ideas about race hardened toward the end of the nineteenth century. Still, few other industries offered blacks jobs that paid more than whites.[63]

Bank Failure

Beginning in May 1875, five months before the hotel's official opening, Ralston suffered a series of economic setbacks exacerbated by the national economic depression that, while hitting the East Coast nearly two years before, finally bore down on California. California businessmen transacted their business with gold and depended on its easy availability. A number of factors, such as the competing demands of both the wheat harvest and the eastern banks, combined to drain the gold reserves from Ralston's Bank of California. During the summer, Ralston sold William Sharon half-interest in the Palace Hotel to raise cash. Ralston also resorted to illegal means to prevent the bank's collapse. Taking gold bullion that did not belong to him, Ralston struck it into coin and charged it to his personal account. He borrowed money on stock he did not own. He over-issued stock in the Bank of California, raising over $1.3 million in falsified financial documents. On August 25, 1875, the San Francisco stock market began to plummet. Sharon sold all his mining stock, causing panic and inciting a run on the bank based on rumors attributing Sharon's sale to the bank's precarious state. The bank's dangerously low reserves could not withstand the run. With the vaults emptied, Ralston closed the bank at 2:35 p.m., twenty-five minutes shy of the normal end of business.

By the next day, the bank's directors found out that Ralston was over $9 million in debt. Of that, he owed the bank over $4.5 million, precisely the amount the bank needed to become solvent. The directors forced Ralston to rewrite his will, naming Sharon as beneficiary to all his property, including the hotel and his Belmont estate. That way, Sharon, as one of Ralston's principal creditors, would take care of straightening out the mess. The directors then forced Ralston's resignation from the bank's presidency. Ralston left the bank and went to North Beach for a swim, something he often did for exercise. Observers from a nearby boat watched as Ralston swam a distance and then began to flail, either from the currents or some physical cause. By the time the boat's men brought Ralston back to shore, he was dead.[64]

Rumors flew that Ralston had committed suicide. However, a coroner's jury decided that his death was accidental. John S. Hittell's 1878 history conjectured that Ralston's death had come at an opportune moment: "His frauds were numerous, and the proof, though then known to only a few individuals, conclusive; the punishment would inevitably be severe." Citing the bank's failure and the panic that ensued as having destroyed many of the region's fortunes, Hittell surmised that Ralston's "great pride would not permit him to submit to the degradation from the highest social honor to the state prison. There was no other secure refuge, save death." Even though Ralston's involvement in other schemes contributed to his sizable debt, the Palace's $5 million cost, owned in the clear by Ralston until the sale of the half-interest to Sharon, created a burden not easily overcome. Rumors also circulated that Sharon, who had won election as senator from Nevada, had manipulated the stock market so as to ensure Ralston's fall, suggesting an almost unfathomable ruthlessness. Other public comments asserted that Ralston's financial problems could have been solved had his "friends" combined to help him. Sharon, in the meantime, inherited Ralston's property, including the hotel, Belmont estate, and Ralston's share in the bank. Sharon moved to Belmont and "allowed" Ralston's widow to live in the servant's house. Over the next six weeks, Sharon reorganized the bank. It reopened on October 2, 1875. Sharon marked the event with the illumination of the Palace's Grand Court the following night.[65]

"Ralston Has Just Begun to Live"

In an editorial published just after the Palace's official opening in mid-October, the *Daily Alta California* offered a tribute to Ralston's manly leadership role in developing the state, contrasting it to the lack of integrity displayed by the "prostitutes of Wall Street." Through a flurry of sentiment, the paper celebrated Ralston's energy and idealism and urged the city's business leaders to emulate his spirit in planning for San Francisco's great future. With enormous irony, the *Alta California* concluded that California had within its power the ability to transform what it characterized as the East's speculative business practices through new ideas—born of the West—that prized integrity and sound business practices. It urged San Franciscans to "cherish Ralston's idealism and have no more of the gross materialism" that disgraced the current generation. In death, Ralston had achieved a quick and complete rehabilitation. Additionally, the *Alta California* seemed able to distinguish between the speculative practices that had fueled San Francisco's growth and those associated with Wall Street, and between the hotel's acclaimed extravagance that inspired the editorial and some other offensive notion of "gross materialism."[66]

The hotel opened the night of October 14, 1875, with a banquet honoring Philip Sheridan, the Civil War general and Union hero. Hundreds of spectators packed the central court, hoping to catch a peek of the decorated banquet hall. Ladies fortunate enough to have escorts found themselves favored with the chance to walk inside for a brief glimpse of the splendor. Leland stationed policemen in the court to ensure a clear path for the 216 invited guests. The crowds jostled around the parlors' glass doors with "ill-mannered curiosity," while those outside clambered to the windowsills from the sidewalk craning to see the spectacle inside. The combination of the famous guests, the flowers, the decorations, the masterful compositions of cakes and fruits, and the brilliance of the hundreds of gas lights created a display of wealth and power unequalled in the city's history. Leland, according to the *San Francisco Chronicle*, had hit the "happy mean between ostentatious and plain display." Amid patriotic bunting and the lavishly appointed banquet tables, the city's aristocracy gathered for self-indulgent adulation. The *San Francisco Chronicle* described a scene resplendent with pyramids of flowers and "fantastic castles built of tuberoses, camellias, lilies, and other patrician plants, with the trailing smilax clambering over all, and extending from one to another in tasteful figures upon the cloth." Three immense tables accommodated more than two hundred place settings. The light from the enormous chandeliers cast a glow on the decorations. Following the reception, guests indulged in a thirteen-course French dinner. Speeches and toasts began at eleven and continued long into the night.

The twin themes of economic development and civilization's progress permeated the speeches. Called upon to honor the nation's centennial, the Honorable Thomas Fitch, a prominent mining lawyer, recounted his recent experience traveling by train through the Mont Cenis Tunnel to France after visiting Italy's Renaissance-era art galleries. He likened it to a trip through time, from the sixteenth century to the nineteenth, "from allegorical representations of the Alps seven centuries old to a double-track railroad under the Alps seven miles long." Invoking the technological feats of Franklin, Morse, and Fulton, Fitch articulated the shared sensibility of living in an age of wonders—and nothing so clearly demonstrated that age than the Palace itself. The speech culminated with a reference to the evening's venue that served as a symbol of American civilization. "It is indeed true that across the gulf of twenty-five centuries the spirit of Pericles whispers though every architrave and pilaster and angle of the beautiful structure in which we assemble tonight," he remarked. "But Pericles," Fitch continued, "could never have ascended on an elevator, or banqueted by gaslight." Applause and laughter ensued. The legacy of the ancient republic of Greece that once spoke to American political revolutionaries resided solely in the room's artistic embellishments. The technological achievements represented the great accomplishments of civilization and progress; they not only revolutionized life experience but also found a salutary sponsor in the modern American republic. In conclusion, Fitch declared, "We are a very great people, and when that is said, all is said—except, perhaps, that in the near future we are destined to be still greater." Certainly the men gathered in the Palace Hotel's banquet hall that night shared his vision of national, regional, and personal destiny.[67]

The Palace opened forty-six years almost to the day after Boston's Tremont House. A comparison of the two opening dinners provides a window on continuity and change over the nearly fifty years. Local and national commercial and political leaders attended both of the men-only dinners. The format, too, remained the same: guests seated themselves at long, sumptuously decorated tables; feasted on magnificently prepared delicacies representing the nation's natural and commercial bounty; and then congratulated themselves through a series of long, self-glorifying toasts and speeches. The speeches, however, point to changes over the intervening half-century. At the Tremont, renowned political figures such as Edward Everett and Daniel Webster invoked a strong tradition of patrician leadership and benevolence that characterized the hotel project as an example of responsible development undertaken for the public good. They spoke gravely and optimistically about the balanced interdependence of farmers, manufacturers, tradesmen, and merchants but lauded the wealthy merchants in particular as men whose risk-taking enterprises would benefit all of society. The songs and poems emphasized values such as thrift, graciousness, sobriety, sense, and decency and never failed to shower appreciation on the artisans who built the hotel. There existed a general sense of appreciation for every class and for the communal participation the event embodied. For example, the papers gave special notice to the baskets of "fine grapes" donated by a gentleman farmer to honor the occasion. As the *American Traveller* noted when praising Boston over New York, "The Bostonians build a tremendous great house, eat a dinner in it, and say nothing more about it."[68]

The Palace's opening occurred on the eve of the nation's centennial, and rather than offering cautious prescriptions for continued national political success, the speeches exulted in the nation's one hundred years of progress and material gain. They celebrated military victories, technological advances, and the accumulation of wealth, all hallmarks of a strong, proud nation. Indeed, the hotel served as a backdrop, its opening merely an occasion for mutual adoration, its magnificence a symbol of wealth rather than of communal endeavor. The Palace was an individual commercial undertaking that represented a shift from holistic ideas that interwove political, cultural, and economic ideologies to ones that focused primarily on the acquisition and display of wealth. The Tremont's "fine grapes" may have served as a metaphor for the change from producer to consumer values, but the cascades of forced flowers and elaborate confectionery display that adorned the Palace's table represented the unlimited possibilities of luxury's infinite sweep. Certainly, having onlookers scrambling at the windowsills to catch a glimpse of the event would have horrified the proper Bostonians. The dinner had become a commodity in its own right, as hoards of unbidden voyeurs sought to snag a piece of it by any means.

While the *San Francisco News-Letter* may have declared the hotel, "The Greatest Caravansery in the World," the hoopla that attended its construction, opening, and continued publicity invited the attention of satirists, who found a wealth of material in the city's newspapers. One reporter, writing to the Hoboken (New Jersey) *Democrat* from his room somewhere at "an elevation above Mont Blanc," targeted the hotel's size and technological attractions. Recognizing the ironic interplay between the much-hyped technological magic and more traditional ideas about personal service, he situated these somewhat oppositional features within the overblown rhetoric on size through a description of the bell service. "There are 25,000 bellboys," he wrote, "one for each room, and numbered." The thousands of boys were corraled in a large basement room that communicated with a series of trap-doors. Once a lodger rang the bell signaling that he or she required service, "down goes the clerk's foot on a corresponding pedal and up shoots a bellboy. Sometimes a dozen rise so at once. He is put in a box, shut up in a pneumatic tube and whisked right into the room designated by the bell-dial. A door in the wall opens to receive him, an automatic clamp catches him by the coat-collar, and he is quietly dropped to the floor." The bellboy's soft landing tempered the drama and violence of his call to action, so that he could still present the calm and reassuring face of service to the impatient guest.

The article also described dress-makers and milliners "in direct telegraphic communication with Paris" who stationed themselves in the miles of parlors, rooms notable for their solid gold spittoons. Three hundred elevators carried guests a half-mile up. The ladies' elevators had "toilet tables and accessories, refreshment counters, full-length mirrors and sofas to recline on." The men's elevators, representing a perfectly gendered elevator world, had a bar and restaurant attached and also a barber-shop "run by the motive power of the elevator." The elevator's stops and starts caused an occasional slicing off of an ear or nose, but new, better-looking ones were available upon application at the office. Faucets fed a supply of wines and brandies into every room, while gauges enabled the office clerks to monitor the guests' sobriety. Waiters dashed about the immense dining halls on roller skates, and circular railways on each floor facilitated ladies' visiting hours. According to this New Jerseyan, only about a dozen people a day died in the hotel, an astonishingly low number for its size, and the hotel's staff included its own undertaker, doctors, and druggists. Other than the Palace Hotel, San Francisco was otherwise remarkable "only for earthquakes, wind and dust, Chinese hoodlums, bar-rooms, pretty women and fast men," all meant to combine into a pathetic soup when compared to the aforementioned extravaganza. By focusing on an amusing mix of technological and human resources, the writer both captured the hotel's public image and parodied its bombastic self-promotion with precision.[69]

"It Seemed to Dance a Jig"

The Palace immediately became a San Francisco institution. Articles describing the building continued to appear in national publications for several years. Hotels began to call themselves the "Palace Hotel" of their city. An 1878 article in the *Hotel World* took umbrage at a piece in the Philadelphia *Times* that irresponsibly passed the title of the world's most magnificent hotel to a new Paris one. Pulling out all the statistics yet again, the *Hotel World* matched those of the Palace to Paris's Continental Hotel with complete confidence in the Palace's superiority. In number of rooms, elevators, and gas burners (the Palace had ten thousand compared to the Continental's paltry six hundred), and every other quantifiable standard (including cost), the Palace continued to reign. The article conceded that the Paris hotel might have fancier furnishings, but even then, if one included the Belmont estate as part of the Palace's amenities, the San Francisco house held its own. It concluded, "When one wants to see the best thing in hotels that the world has ever produced he must visit San Francisco."[70]

Over the years, the Sharon family continued to update the hotel. In 1878, Sharon obtained a system of Brush electric arc lights for the hotel. Ten electric lights displaced nearly eleven hundred gas jets, with two three-thousand-candle lamps replacing the 510 gas jets in the Grand Court alone.[71] In 1879, *Harper's Weekly* published an illustration documenting former president Ulysses S. Grant's dramatic entrance into the hotel's Grand Court. A sixth-story arc lamp showers a canopy of light rays down upon his carriage and team of elegant white horses as he passes through the court's arches. Periodicals framed Grant's visit with inspired descriptions of the various electric light installations whose modern brilliance reinforced those ideas of a progressive destiny spoken of just a few years before at the hotel's opening. The public interest in electricity kindled by Grant's visit coincided nicely with the startup of San Francisco's first central generating station.[72]

In 1891, the hotel underwent extensive renovations, including a reconfiguration of the entire first floor. Management added two new entrances and replaced the various dining rooms with several restaurants, each in a different style. The small interior light courts were covered with glass roofs, creating additional usable space. A new mosaic floor replaced the more traditional black and white marble diamonds characteristic of nineteenth-century hotels. Large electroliers with multiple incandescent bulbs lit the new dining areas, and new porcelain tubs and water closets replaced all the old plumbing fixtures. Additionally, new furniture, upholstery, and wall hangings refreshed the public rooms. These improvements came at the direction of William Sharon's sons, who rather unhappily inherited the family business. In an 1892 letter, Frederick W. Sharon admitted to his brother after being chastised for his hapless efforts during the remodeling, "It has never been my good fortune to be obliged to earn my own livelihood." By the end of the century, other improvements included heated bedrooms; more new carpeting, furniture, and bathtubs; and faster elevators.[73]

The hotel enjoyed the patronage of the city's famous visitors, including the Emperor Dom Pedro of Brazil, the Chinese Embassy, Sarah Bernhardt, Oscar Wilde, Rudyard Kipling, Oliver Wendell Holmes, Julia Ward Howe, King Kalakaua of Hawaii, Theodore Roosevelt, and countless other dignitaries and celebrities. On April 18, 1906, while Enrico Caruso's opera company was sleeping at the hotel, the great

Former president Ulysses S. Grant arriving at the Palace Hotel with Brush arc lights illuminating the Grand Court. *Harper's Weekly,* vol. 23, October 25, 1879, 849. *Courtesy, Kelvin Smith Library of Case Western Reserve University.*

San Francisco earthquake struck. This was, of course, the ultimate test of the protective technical systems that Ralston had engineered into the building thirty-one years before. William J. Dutton, president of the Fireman's Fund Insurance Company, experienced the earthquake at the Palace and, not too surprisingly, suffered a heart attack there as the long-feared fires began to sweep the city. First-person accounts from those at the Palace during the quake described some of the most fearsome shaking experienced anywhere in the city.[74]

More than one observer reported feeling as though the building had twisted on its axis. Ernest Goerlitz, the manager of the Caruso's opera company, wrote one of several narratives that attempted to chronicle the incredible event. He described being awakened by a loud noise and the building's trembling, which he first thought was an explosion in the hotel's cellar or a nearby building. However, he soon realized that it was an earthquake as the shaking intensified over the next few seconds: "Furniture, bric-a-brac from the mantelpiece, especially around the bay window, fell topsy-turvy around us. The noise from the rumbling of the earth, and as I afterwards, saw, from the falling of buildings, was deafening. It is impossible for me, and perhaps for anyone else to accurately describe our experience. I looked out the window and saw the Grand Hotel, situated opposite our room being shaken as by a terrific explosion. I cannot better describe the motion of the Palace . . . it seemed to dance a jig."[75] Bedlam struck the hotel as guests poured out of their rooms in their bedclothes. Some calmly settled their bills and checked out, but most simply ran into the streets alongside the guests of the Grand Hotel, joining the crowds of frantic, directionless residents.

Fires broke out all over the city. By noon the fire department had dynamited the Hearst and Monadnock buildings in an attempt to keep the fire from jumping Third Street and threatening the hotel. Hotel employees successfully battled the flames approaching from the south and west with the hotel's handheld hoses. It seemed that the Palace might be saved. Soon, however, a fire department hose cart made its way up Market Street from the east and, in the words of an eyewitness, "stopped at the Palace Hotel hydrant, screwed on a hose and went off with the water into Sansome Street." The fire department drained the Palace's water supply to fight the fire raging at Battery and Market. When the fire finally found the hotel, there was no water left, and the defenseless structure burned with the rest of the city.[76]

The hotel suffered a tremendous and complete loss. Insurance paid just over $1.5 million, nowhere near its $6 million value. However, a bizarre tribute to Ralston lies in the fact that the hotel's skeletal structure suffered almost no damage. Despite the violent twisting reported during the earthquake, the iron-reinforced brick walls remained standing, almost unscathed, even though the fire had utterly consumed the hotel's interior. It took nearly eight months for three hundred men working night and day to raze the building. The lime and cement mortar defied demolition, and the bricks had to be scaled by hand. The dynamited walls fell in strips because the iron bands held the bricks together. The bricks failed to separate even upon hitting the ground. It cost $90,000 to raze the hole and cart away its thirty-one million bricks. So Ralston had indeed succeeded in building an earthquake-proof hotel, but the disaster's magnitude left the building vulnerable to unimaginable forces of a completely unpredictable nature.[77]

Conclusion

The Palace Hotel's story evokes classical warnings about luxury's evil toll on humanity. Ralston's commitment to constructing the largest, most expensive hotel in the world contributed to his death. His hotel surpassed all existing hotels in size and cost, and it did so with such excess that it maintained its dominance for many years. The 755 rooms and $5 million price tag were more than any city needed or reasonably could afford. The Grand Pacific, Chicago's 1873 landmark hotel, cost under $2.5 million and had five hundred guest rooms. Even Chicago's 1889 Adler and Sullivan Auditorium, a multi-use masterpiece containing a seventeen-story office tower, a four hundred-room hotel, thirteen elevators, and a four-thousand-seat opera house, cost $3.2 million.[78] Five million dollars was a staggering sum to spend on a hotel in 1875. Even without the extra half million that the earthquake and fire protection added, the Palace was still an expensive project. It was, however, a project conceived in a context completely at odds with the parameters of moderation or business conservatism.

San Francisco society embraced hotel life enthusiastically. Because of the city's youthfulness, luxury hotels, perhaps more ardently there than elsewhere, symbolized the city's growth and coming of age. The economy's speculative nature—with its wild booms and busts—created a climate of risk-taking in which a man like Ralston flourished. Empowered by success, wealth, energy, and dominion over his many huge enterprises, Ralston epitomized the ideology of the self-made man. He entered adulthood as an Ohio farmer's son and emerged as one of California's most powerful men. His own growth and success coincided with that of California. The hotel project became a statement of both his personal achievements as well as the rising star of San Francisco. Despite Ralston's increasing economic desperation, his pride and his confidence in his ability to succeed prevented any scaling back of the enterprise that would speak to the world in a way that only a conspicuous, pretentious hotel could.

At the time of the hotel's construction, both the *Sunday Chronicle* and the *California Spirit of the Times* described its best characteristics. They noted, "The qualities of the building are safety, durability, convenience and elegance. The beautiful is not sacrificed to the practical, but made everywhere subservient to it." Despite the hotel's size and expensive decorations, its strength and utility were the qualities that Ralston, and Sharon after him, stressed as foremost in importance. The technological systems contributed to the hotel's functionalism as well as to the guests' comfort and safety. The structure's immensity and solidity faced off not just against nature but also against fashion and the rapid obsolescence and transience characteristic of modernity. This hotel, despite its wholehearted adoption of the modern ethos that dictated seizing upon the latest and newest, was built for permanence. The twelve-foot-thick brick walls banded together by iron strips were intended to withstand both earthquakes and time. Ralston meant for his hotel to be a lasting monument even though his own life had witnessed changes more dramatic than any other age.

Despite the Palace's complete destruction, Ralston's vision for a permanent monument succeeded in a very interesting way. Once the smoke cleared and the rubble was removed, the city immediately began to rebuild. By 1906, steel-framed construction had revolutionized urban buildings. Pioneered by Chicago architects, the steel frame enabled the construction of tall skyscrapers that not only made efficient use of expensive urban property but also responded successfully to functional requirements such as durability, fire resistance, economy, and versatility. New hotels in San Francisco such as the St. Francis, built at Union Square shortly before (and then rebuilt after) the earthquake, were skyscrapers. The St. Francis's twelve-story triplicate wings towered over its neighboring buildings. One could imagine, given the task of rebuilding a hotel with this brilliant new technology at one's disposal, that plans would call for a new, modern skyscraper similar in style and certainly taller than those growing up elsewhere in the nation. These included Washington's thirteen-story (new) Willard, Philadelphia's eighteen-story Bellevue-Stratford, and New York City's spectacular eighteen-story St. Regis.[79]

However, over the years the Palace had endeared itself to San Francisco. Writing in 1930, its proprietor explained the hotel's enduring role as the center of the city's social and professional life. "Everyone met there," he recalled, "everyone of importance stayed there, and many prominent San Franciscans lived there permanently. It was more than an hotel: it was a public institution . . . a source of pride and satisfaction not only to the people of San Francisco, but to thousands of travelers from all parts of the world." In an extraordinary tribute, the new $9 million Palace, built in 1909, duplicated exactly its predecessor's footprint. The Sharon family's descendents recreated the hotel's shape and incorporated its most beloved feature, the Grand Court. The new Garden Court, rather than opening the building's full height, instead boasted a three-story vaulted ceiling, crowned by a breathtaking leaded glass skylight. After its $150 million restoration in 1991, the Garden Court remains San Francisco's only landmarked interior space. Once again, the new building claimed the latest technology, the finest kitchen, the most modern elevators, and every other imaginable convenience necessary to ensure its position as a hotel without peer. As a symbol of the city, the Palace Hotel endured beyond anyone's expectations, except perhaps those of William C. Ralston.[80]

....

CHAPTER SEVEN

THE "NEW" MODERN HOTEL, 1880–1920

....

"It Is Part of the Hotel Business to Hide All These Things from View"

W HILE THE STORY of the Palace Hotel might seem idiosyncratic, driven as
it was by an overriding commitment to excess by its fanatical promoter,
economic and political leaders in American cities large and small shared an understanding about the way grand hotels, their architecture and interiors, served as
material testimony to a city's economic and cultural ranking in the world. After
Chicago's disastrous fire in 1871, that city rebounded in part by building four of the
nation's most important hotel buildings, the Grand Pacific Hotel, and the Tremont,
Palmer, and Sherman Houses. In addition to the understandable claims for their
superior fireproofing, these buildings boasted the quintessential urbane signifiers:
grand lobbies, monumental staircases, elegant parlors, cafes, barber shops, bridal
suites, dining rooms, ballrooms, promenades, hundreds of private bedrooms and
baths, and all the latest luxuries. These four Chicago landmarks not only spoke to
the city's remarkable recuperative power but also served as benchmarks for successor hotels, including the Palace. An 1873 headline in Chicago's real estate newspaper *Land Owner* proclaimed "Chicago Ahead of the World," a headline that was
tied to a description of the Sherman House.[1] Large luxury hotels like these continued to play a critical role for cities in their ability to attract visitors, business, and
ongoing development.

San Francisco's Palace Hotel capped the age of enormous masonry-constructed
hotels. Although the particular circumstances of San Francisco's economy allowed
the construction of the Palace, the nationwide economic depression dampened
building construction elsewhere in the country, with building starts reaching a
low in 1878 not equaled again until World War I.[2] In the 1880s, the hotel industry

seemed to mark time. Some monumental pre–Civil War hotels such as the St. Nicholas and the New York began to close, making way for new development.[3]

In his published memoir of his return 1879–80 tour of the United States, George Augustus Sala dwelled once again on American hotels, and his descriptions hardly varied from his earlier observations, except to note with relief the demise of the hotel gong that called guests to meals.[4] The industry, taking on a more cohesive organization, began to publish not only national trade journals and city-specific hotel registers directed at hotelmen and traveling salesmen but also comprehensive national hotel directories to aid travelers and merchants.[5] In addition, several hotelmen published exposés of hotel life. The 1884 *Horrors of Hotel Life by a Reformed Landlord* was a particularly vivid and gruesome book that seemed capable of singlehandedly destroying the industry.[6] These activities were prelude, however, to a new building boom that seized on modern advances in construction technologies and promised a new era in hotel accommodations.

By the time the economy recovered, revolutionary construction methods relegated "monster hotels" to a seemingly premodern past. The Equitable Life Assurance Building's 130-foot height (New York, 1868–70) quickly inspired building technologies that enabled skyward growth such as was seen in New York's 260-foot Tribune Building (1875). By the late 1880s, twelve- to sixteen-story buildings had been built in considerable numbers in both Chicago and New York. These technologies included the development of fast, safe, and dependable elevators; fireproofed structural iron; steel-frame construction; electric lighting; the telephone; elaborate plumbing systems; temperature controls; and healthful ventilation.[7] Nearly all skyscraper histories focus exclusively on the corporate office building, but urban luxury hotels participated wholeheartedly in this structural revolution, as they had with other architectural shifts. One architectural critic noted that the turn of the century's opulent hotels came to represent the full expression of New York as "the rich man's city."[8]

Seeking to capture the spirit of the decade of growth that began in the early 1890s, *The World's Work* published a 1903 article called "The Workings of a Modern Hotel." The author, novelist and writer Albert Bigelow-Paine, introduced the article with words meant to impress: "Among all our institutions of progress, there is none more amazing than the modern hotel in immensity, in complex activities, in social significance." Thus began a lengthy description of a modern hotel's ingenuity, buttressed by stunning statistics of every kind that reinforced the sheer magnitude of the building and the services offered within. Subtitles such as "A Story of Organized Luxury" and "A Vast Machine of Well-Regulated Activity" promised to reveal the inner workings of "a human system on a mighty scale." This was a system of luxury that transformed each fee-paying guest into "a living embodiment of human irresponsibilities." In other words, once checked in, a guest only needed to make known (with as little effort as possible) his or her needs. It became incumbent on the hotel to take care of anything and everything with no further effort or thought required or expected on the guest's part.[9]

Among the many statistics Bigelow-Paine employed to astonish his readers was the eighteen thousand eggs a large hotel was reputed to use in the course of a single day. As just one of hundreds of ingredients and varieties of foodstuffs procured, stored, prepared, and served to guests, the egg vividly represented the complexity of the hotel and its immersion in systems of production, transportation, communication, energy, waste disposal, and machine technology, to list only a few. Bigelow-Paine dwelled on the daily feat of soft-boiling several thousand eggs, a task accomplished by a mechanical egg-cooker whose perforated dipper cradled and dropped the egg into boiling water and jumped out at the correct moment, determined by an automatic timing device. Bigelow-Paine observed, "There are men who do nothing else but fill and watch and empty these dancing dippers, and it seemed to me great fun."[10] Setting aside the issue of "fun," the staggering number of eggs raises many questions. Where did those eighteen thousand eggs come from every day? Where were they stored? Who purchased them? Who delivered them, and how often? What kind of energy powered the delivery vehicles, the egg-cooker, and the other cooking equipment? Who served the eggs? Who supervised the eggs' preparation and delivery? Where did these employees eat? Where did all the eggshells go? While Bigelow-Paine's comments invoke the typical hotel boosterism of the long nineteenth century, the number of eggs and the questions it raises suggest a whole new level of technological and managerial complexity that indeed separates the newer hotels from their predecessors. Even as these newer hotels drew on and embellished established principles of luxury, size, service, and technological innovation, the scale and complexity of these buildings ratcheted dramatically upwards, a direct result of the introduction of a new urban institution, the tall building.

Housed in these new tall structures, urban luxury hotels exhibited elements of both continuity and change. Size, cost, technological innovation, and opulent decoration continued to define the buildings. The two metaphors used to describe luxury hotels in the mid-nineteenth century, the hotel as a self-regulating machine and as a city within a city, endured and flourished well into the twentieth century. Despite significant changes in shape and the introduction of new functional areas such as convention facilities, the building's general design remained recognizable with stores, services, lobby, gendered public spaces, and dining rooms on several lower stories both below and above ground, and private bedrooms and suites on the floors above. These "new" hotels continued to serve both transients and permanent residents, occupying prime real estate in the hub of a city's business and retail districts. Competition persisted as a compelling force driving construction, as the hotel industry attracted attention for its profitability. Moreover, hotels maintained their symbolic role as signifiers of American cultural and economic achievement, even as they lost their iconographic supremacy to the new corporate skyscrapers.

Despite these continuities, the skyscraper hotels also seemed to be stunningly different structures. The overwhelming impression from reading the architectural literature is one of breathtaking complexity. As the buildings rapidly grew in size to include one thousand, fifteen hundred, two thousand, and eventually three thousand bedrooms, the facilities required to support the needs of both patrons and employees expanded accordingly. With many interdependent and complicated technological systems integral to its design, the hotel became the province of the architectural and engineering professions. The architect needed to engage structural, mechanical, electrical, gas, plumbing, and kitchen engineers in order to ensure a safe, healthy environment for the thousands of people who visited, stayed, and worked in the building each day. In addition, despite all the "labor-saving" devices, these hotels employed one to two people *per guest,* requiring a management organization that rivaled the most complicated corporations. At the same time that hotels promoted their expanded menu of technological luxuries, they also began to emphasize personal service in an attempt to combat the impersonality of these enormous structures. Like his contemporary Henry Ford, hotelman E. M. Statler capitalized on the cultural movements of standardization, efficiency, and mass consumption but also promoted the idea of perfect service as part of a holistic system that included architecture, engineering, and human service. Based on these concepts, his hotel empire brought the luxury hotel idea to the middle class and demonstrated that a branded design idea could be replicated in city after city. Thus, someone like Sinclair Lewis's character George Babbitt could "seek out the best hotel in town, no matter in what town," and still feel at home.[11]

The "new" modern hotel that evolved between 1880 and 1930 continued to represent and shape cultural changes in American society. As corporations gained economic power and the site of a city's face-to-face business interactions moved from the public forum of the hotel exchange to corporate headquarters, architects responded by redefining space to accommodate the new trends. Voluntary and professional organizations began to hold regular meetings and conventions in hotels, and, as a result, architects designed huge new spaces such as separate banquet rooms and exhibition halls. Hotels persisted as sites of informal commercial activity and branch offices for drummers, but these interactions became far more privatized than was common even twenty years before. Lobbies became more like public living rooms and waiting areas, while private parlors and other meeting facilities, both large and small, swept private associational gatherings deep into the recesses of the hotel.

Not coincidentally, architects began to domesticate the hotel through its architecture and function. Hotel proprietors began to promote their buildings as venues for private domestic entertainments. They courted society doyennes to hold their debutante balls, charity events, and bridge parties there. Other changes included the shift in hotel dining from the American Plan, which included all meals in the room rate, to the pay-as-you-go European Plan. Architects continued to carve out gendered spaces such as the barber shop, men's grill, bar, ladies' parlor, tea room, and writing room, but these spaces became increasingly marginalized spatially as heterosocial activities became more acceptable. Grand new lobbies, glass-walled restaurants, and long corridors not only heightened opportunities for personal display but also encouraged a voyeuristic public to participate. In a final irony, as the complexity of the technological systems intensified and the skyscraper acquired a distinctively American character, the interior decorations kept slavishly appropriating European models, with Versailles serving as the decorative standard until it was replaced in the 1910s by the English Adams style. Seemingly unlimited financial resources helped produce hotel after hotel that broke records for size, cost, and engineering challenges. These four decades witnessed a shift from older conceptions of luxury that emphasized elitist individuality to models that promoted mass-produced standardization accessible to a broad-based consumer market.

The Waldorf-Astoria

Almost as if there were not sixty-plus years of hotel history preceding it, architects and architectural observers regarded the Waldorf-Astoria in New York City as the first "modern" hotel. As Frank Crowninshield, editor of *Vanity Fair* magazine, rather myopically observed in his book about the Waldorf, "The best hotels . . . were never thought of as social centers at all. They were merely convenient and orderly places in which to eat and find lodgings. There was never a question of giving fashionable dinner parties in any of them; nor were suppers ever ordered, dances arranged, or receptions held in them."[12] His obvious bias was shaped by the Waldorf's well-known social scene, but the building trade, too, recognized the Waldorf-Astoria as a real demarcation between the older block hotels and the newer steel-framed towers of luxury. For example, in a 1905 issue of the *Architectural Record*, Herbert Croly, the journal's editor, using his pseudonym Arthur C. David, wrote that "the design of the contemporary American hotel must be traced in its origin to ... the Waldorf in New York, which with its bigger brother, the Astoria, indicated the main lines of the design of a hotel 'sky-scraper.'"[13] Others continued to acknowledge the Waldorf's place in history even as late as 1924, when an article in *Arts and Decoration* called the Waldorf "the real forerunner of the hotel of today."[14]

The Waldorf-Astoria was actually two hotels. The Waldorf opened in 1893, the Astoria in 1897. Joined by a large elegant corridor, together these hotels had one thousand rooms, making the hotel the largest in the world at the time. William Waldorf Astor, following in the footsteps of his grandfather John Jacob Astor, who had built the Astor House, constructed the Waldorf hotel on the site of his inherited family home at the corner of Thirty-Third Street and Fifth Avenue. His cousin, John Jacob Astor III, lived in a similar mansion on the adjoining lot to the north at Thirty-Fourth Street on which he eventually built the Astoria Hotel. (In 1929, this Waldorf-Astoria closed when the land was sold as the site for the Empire State Building.) William Waldorf Astor left the United States to live in England in 1890; hotel folklore claims that Astor built the hotel as an act of vengeance toward both the "rubbernecking" sightseers who so disturbed his life and the district voters who had defeated his election bid to Congress. By erecting a towering commercial building in a formerly sacrosanct wealthy residential district, Astor reduced the nearby three- and four-story mansions to what one historian characterized as "squat hovels."[15] Strained relations between the Astor family's two branches seemed to preclude any future cooperative ventures, yet delicate negotiations finally resulted in an agreement to construct the second hotel, the Astoria. The two hotels were to be managed as a single entity under the direction of George C. Boldt, the Waldorf's legendary proprietor. However, the contract called for the contingency of a division wall that, should the combined hotels falter, would enable the latter Astor estate to separate the two buildings completely and distinctly—this despite plans for a grand unified ground floor.[16]

Architect Henry Janeway Hardenbergh designed both hotels. Born in New Jersey, Hardenbergh studied architecture under Detlef Lienau, a pupil of Henri Labrouste, a famous Neo-Grec architect at the École des Beaux-Arts in Paris. In 1870,

Hardenbergh opened his own practice, but his first decade in business served more or less as a continuation of his studies. The Waldorf was the first important building for which he is known and the first of his many significant commissions, which included the Dakota apartment house (New York, 1884), the second Plaza Hotel (New York, 1906), and the new Willard Hotel (Washington, D.C., 1906).[17] The Waldorf rose thirteen floors, while the Astoria eclipsed it with its sixteen stories. These were tall buildings for their time. The sixteen-story World Building (1890) at City Hall Park remained the tallest building in the city for nearly a decade. Still, by 1897, there was a fourfold increase in the number of tall buildings in New York to ninety-six, with the tallest, the American Surety Building on lower Broadway, reaching twenty stories. The great majority of the new tall buildings were between nine and fourteen stories. The Waldorf-Astoria, sitting far north of the new business clusters, was a contending participant in the emerging Manhattan skyline.[18]

Hardenbergh's vision for a skyscraper hotel differed from that of the tall office building. In his entry on hotels in Russell Sturgis's 1901 *Dictionary of Architecture and Building,* Hardenbergh outlined his belief that a hotel should serve as a temporary or permanent place of residence, in other words a home. He wrote, "The modern hotel should not only afford ample means of furnishing lodging and food to those seeking those necessities, but such privacy, comfort, luxury, or means of entertainment as may be secured in a private domicile, and in addition every means of carrying out the domestic, public, or social functions of life."[19] The exterior of the Waldorf-Astoria represented this domestication of the tall building. The turreted and dormered roofline expressed the hominess of a German Renaissance dwelling, while Hardenbergh punctuated various other levels with balconies overflowing with lush greenery, presenting the appearance of a (very tall) German castle.

Hardenbergh's concept for softening the skyscraper's impact proved influential. In 1905, the *Architectural Record* observed that the Waldorf-Astoria conformed to the "regular 'skyscraper' convention" by dividing the facade into a solid base, a monotonous treatment for the main shaft, and an elaborately treated crown. This design scheme responded to how the buildings were viewed from the ground. However, Hardenbergh differentiated his hotels from office buildings by using "warmer and more attractive materials" and by treating the roofline with dormers to evoke a domestic appearance for the building. The article noted that with only one or two exceptions, architects adopted Hardenbergh's concept for all subsequent important metropolitan hotels, an important break in terms of architectural representation. For example, the early-nineteenth-century Greek Revival hotels conformed

The Waldorf-Astoria Hotel. The Astoria Hotel towered over the Waldorf an additional three stories to dominate its younger half, which ran parallel and behind the massive Astoria frontage on 34th Street. *Postcard, collection of the author.*

to civic and commercial architectural conventions, distinguishing themselves as hotels only through size. The mid-nineteenth century's "*palazzo*" hotels also out-sized other commercial block buildings such as department stores to emphasize their position in the commercial world. However, the growth of corporate organizations, with their bureaucracies, specialized office buildings, and advances in communication—specifically the telephone—threatened to make obsolete the hotel's once primary importance as a site of face-to-face commercial exchange.

Hardenbergh's architectural vision recognized this change and further marginalized the hotel as a city's commercial center while at the same time established the luxury hotel as a profitable industry in its own right. The concurrent evolution of apartment or family hotels contributed to a sense of architectural confusion that occasionally surfaced in the architectural journals in discussions among architects concerned with the skyscraper's treatment as either a commercial or domestic building in those varying circumstances.[20]

In elaborating his vision of luxury hotels of the future, Hardenbergh further wrote that the hotel should "afford means of offering diversion or amusement to those abiding under its roof."[21] The idea that the hotel should be an "amusing" place was one that hotel architects took to heart. Bigelow-Paine's article in the *World's Work* captured this enthusiasm. He described the hotel as being so energized with distractions that guests could happily amuse themselves for a week without ever going outside. "Entertainments are always in progress," he wrote. "Two orchestras supply music; objects of art and interest are on every hand, businesses of almost every sort are represented on the ground floor, and when at a loss for other amusement the visitor may ascend to the fifteenth floor and sit for his photograph, or spend an hour in a gay roof-garden."[22] A 1907 *Architectural Record* article asserted that the greatest challenge a hotel architect faced was somehow to convey to patrons that "their surroundings are novel and amusing."[23]

As an example of this design philosophy, the new Hotel Astor (New York, 1904) created a series of rooms in different styles. Arthur C. David's sardonic description of the Astor critiqued what was almost a manic desperation in this regard: "There are French rooms, which are extremely French, German rooms which are desperately German, Dutch rooms which are fearfully Dutch, a Pompeian room which makes one think of Vesuvius, Chinese and Japanese rooms which are as Oriental as a Buddhist god, a yachting cabin and a hunting lodge for convivial sailors and shooters, and finally, several rooms in the 'new art' style, which are about the most extraordinary things in the whole extraordinary collection."[24] The Knickerbocker Hotel installed the Maxfield Parrish mural "Old King Cole" in its bar room, complete with John Jacob Astor's face as King Cole. The *Architectural Record* article describing the Knickerbocker claimed that the hotel's function was to "help people to live temporarily in a more amusing way." It added, "The patrons of a hotel have as much right to demand amusement as the patrons of a theater, and everything about the building should be designed to keep them diverted and gay."[25] This new intensified focus on amusement represented an enormous shift in public and professional attitudes about the hotel's civic role. Where once the hotel's business exchange and bar room had served as a center of gravity for the business and political communities, architects now were situating the hotel as a social and entertainment institution, casting it almost as the city's public face of ostentatious frivolity.

The origin of all this amusement can be traced to the Palm Garden of Hardenbergh's Waldorf-Astoria. The Palm Garden, or Palm Room, stood exposed by virtue of its glass walls and doors that opened directly opposite the Waldorf's main entrance. One of the hotel's chroniclers claimed that it was the most fashionable restaurant in the United States, the exclusive province of New York elite society. The glass walls allowed observers a clear view of the privileged diners, and oglers accordingly jammed the corridors each night to watch the fashion parade. The corridor itself became known as Peacock Alley for all the strutting of fancy finery that went on there each day. Walter T. Stephenson, writing in London's *Pall Mall Magazine,* claimed that "it was perfectly fair to affirm that nearly every adult resident of Manhattan, male or female, excluding the labouring and poorer classes, visited the Waldorf once a week."[26] Another of the hotel's historians insisted that as many as twenty-five thousand people strolled through the corridor each day, with that number climbing to perhaps thirty-six thousand if a president or other celebrity was a hotel guest.[27] When the Astoria opened in 1897, the activity transferred itself to a new three-hundred-foot-long corridor that ran nearly the hotel's length parallel to Thirty-Fourth Street. From that time on, the Palm Room, a room "which pretends to be out of doors," sited between the hotel's entrance and the main dining room, became an absolute necessity for luxury hotels nationwide.[28] In descriptions of new hotels, references to the "usual palm room" or "the inevitable palm room" littered the pages of the architectural journals.[29] Architects assembled various combinations of Grecian columns, palms, lattice, greenery, running water, stained glass ceilings, and brilliant lighting to produce a garden room that they hoped would provoke the same interest as that of the Waldorf's and would serve as a popular restaurant and destination for afternoon tea.

Most of the Waldorf-Astoria's earlier histories focus on the fantastic doings of New York's Gilded Age society, and this activity took place within a building decorated with unprecedented expense. One magazine estimated the price tag of the furnishing alone to be nearly $2 million, in addition to the building's assessed value of nearly $12 million.[30] The expenditures for hotels that succeeded the Waldorf-Astoria recall the *Daily Tribune*'s 1852 comment regarding the St. Nicholas: that "henceforth [luxury hotels] must be furnished without regard to cost."[31] Architects and developers continued to heed Boldt's and Hardenbergh's interpretation

Peacock Alley, the three-hundred-foot-long corridor that connected the Waldorf and Astoria Hotels. This illustration from a 1900 issue of *Harper's Weekly* illustrates the nightly gathering of the city's fashionable elites in the hotel's "amusing" spaces. *Harper's Weekly*, vol. 44, no. 2256, March 17, 1900, 248–49. *Provided courtesy HarpWeek.*

of luxury for years. In 1905, in writing about the Philadelphia's new Bellevue-Stratford, Arthur C. David noted, "The interior ... constantly suggests, as it was doubtless intended to suggest, the Waldorf-Astoria. It was evidently the intention of the owner to produce the same impression of overpowering and spacious magnificence upon his patrons as that produced by the famous and popular New York hotel."[32] Again, in 1924—and only a few years before the hotel's close—*Arts and Decoration* noted the Waldorf-Astoria's intention to "offer an environment transcending in its glamor the ordinary environment of the guest." The Waldorf-Astoria served as a model in terms of "marble, gold and plate mirrors."[33] As easy as it is to trace all the various elements of the Waldorf-Astoria and its contemporary imitators to their mid-nineteenth-century predecessors, the scale of expense and size inherent in the tall building made it all seem extraordinarily new. Moreover, as Hardenbergh's professional influence combined with the consolidation of business activities in corporate headquarters to drain the hotel of one of its core purposes, the luxury hotel transformed into a playground that masked its continuing commercial role in the city.

The Back of the House

In 1892, a writer for the *Atlantic Monthly* questioned whether or not modern technology had increased one's comfort at a hotel. Lamenting a decline in personal service, he wrote, "Hotels are not what they were in the essentials of comfort." Despite the increase in technologically produced "conveniences," such as gas, electric bells and lights, steam heat, elevators, and hot and cold running water, he asked a key question: "Do they increase one's comfort?" The writer levied his complaint against the lack of attentive service he experienced in the "dreary and oppressive and costly luxury" found in the large hotels, ones he characterized as hotel-keeping ventures for "raising millionaires." He blamed the new hotels' impersonality directly on their technology. Electric lights in one's rooms enabled nighttime solitary activity and thus replaced the cozy latenight drawing room fires around which fellow travelers had forged lasting friendships. The elevator imposed a "leveling tyranny," eliminating the camaraderie of young bachelors' "attic cells." Table delicacies served out of season, a feat made possible by modern transportation, communication, and refrigeration, dampened appreciation for them during their "real" season. Admitting himself to be an "old fogy," the writer nonetheless lamented the passing of the personal attention he had associated with midcentury hotels, assigning his disquietude to the intensification of the new hotel machine.[34]

That machine was one that seemed to have taken on a life of its own. Analogies covered a wide field, some describing these technological buildings as huge complex machine-systems in and of themselves, some as living organisms, comparable in complexity to a human body. Two decades before Le Corbusier introduced the house as "a machine for living," the *World's Work* referred to the hotel as a "huge living machine." The newspaper *Everybody's*, in describing the enormous engine room of a large hotel, explained that "to find room for it all, the machinery must veritably be packed and interpacked like the organs in one's body." At the St. Regis, promotional material described the kitchen as the "soul of the hotel," as though the building were a spiritual being whose sustenance depended on the nourishment and creativity the kitchen provided. And, like the body with its unknowable internal organs, *Scribner's* noted that "it is part of the hotel business to hide all these things from view."[35] These metaphors recognized the energy that flowed from and between the technological systems and the people who lived and worked within.[36]

The machine metaphor pervaded the architectural literature. In a 1923 hotel planning issue, the *Architectural Forum* argued that the modern hotel's essence was that of a "service machine" and that all architectural considerations needed to be secondary to that idea. The Waldorf-Astoria, again, set the precedent. William Hutchins, writing in 1902 for the *Architectural Record,* captured the Waldorf's complexity as well as the interplay between the hotel's machinery and its workers: "There are in this colossal building room for fifteen hundred guests who require almost as many servants. The enormous electric plant, 3,000 horsepower, 16 elevators, a plant making fifty tons of ice a day, boilers using 100 tons of coal a day, 1,300 bedrooms, 150 hall-boys, 400 waiters, 250 chambermaids, 40 public rooms . . . gives only an inadequate idea of its scale." This followed an 1897 twenty-four-page illustrated *Scribner's Magazine* article that delved into what it called a "many cog-wheeled machine." Although purportedly based on a loose amalgam of New York's different luxury hotels (without naming any one in particular), the article's publication in the same year as the Astoria's opening meant that every reader would instantly think of the grand Waldorf-Astoria. The article gave a rich description of the late-nineteenth-century technical labyrinth that gave substance to the building and the workers who inhabited that world. The dark, almost brooding, hand-drawn illustrations of the "back of the house" work sites contrasted starkly with the glittering descriptions of the Waldorf-Astoria's other half.[37]

As earlier in the century, the two places that received the most attention for their mechanization and output were the laundry and the kitchen. A series of illustrations in the *Scribner's* piece began by picturing long-gowned and capped laundresses hanging linens out to dry on the roof, in reference to the "old-fashioned out-door process" that the new machines had outmoded. The caption informed readers that the hotel often processed over twenty thousand pieces of laundry in a day, so very little could actually be dried as shown. As many as two thousand pieces came from the dining room with orders that it be processed and ready for use within one hour. Architects usually designed one of the basement levels to hold the laundry so that, among other reasons, soiled linen could be pushed down chutes, saving the use of large dumbwaiters for the clean linens. The Belmont (1902) had five such underground stories to accommodate all of its power plants and machinery. There, turnaround time for dining room linen was purported to be as little as ten minutes, although this sounds as preposterous as it did when William Chambers made similar observations about the Metropolitan Hotel's laundry in 1854.[38]

In a large hotel such as the Waldorf-Astoria, as many as sixty persons worked in the laundry department. A series of enormous washing machines, extractors, and various ironing machines filled large rooms. A separate facility handled guest laundry. An all-electric laundry measuring twenty-thousand square feet installed in the

Hotel Pennsylvania (New York, 1919) included ten 42-by-72-inch cascade washers, twelve overdriven centrifugal extractors for damp drying, three steam-heated drying tumblers for towels, five ten-foot-long flat-work ironers (including one simply—ominously—called the "annihilator"), a curtain dryer, and an entirely separate facility to launder guest clothes. The Hotel Pennsylvania's laundry employed two hundred persons and could turn out one hundred thousand pieces a day.[39] In an answer to the *Scribner's* piece, an 1897 *Overland Monthly* article described the "bare-armed, white-aproned French women" who did the finer ironing at one of San Francisco's great hotels. This boosterish article asserted that California offered hotel accommodations of equal quality to that of New York City; while unnamed, the hotel that served as the author's example was unquestionably the Palace. The *Overland Monthly* suggested that French women were superior and more sophisticated ironers. *Scribner's,* however, depicted a similarly large room with as many as twenty "bare-armed, white-aproned" women bent over wide tables ironing, supervised by a head ironer who inspected every piece of finished work. Other kinds of laundry workers included the manager, washmen, wringermen, head washmen and wringermen, loaders, starchers, bosom press operators, lady clothes ironers, shakers, folders, feeders, bookkeepers, and the marker/sorter.[40]

As much as laundry rooms incorporated increasingly more sophisticated technological and managerial systems, their tasks were fairly straightforward. Dirty linens needed to be laundered and pressed and returned to the various departments and guest rooms in good condition, ready for reuse. Kitchens, by contrast, proved to be far more complicated, given the large number of different dishes prepared and served throughout the day. Mistakes in the kitchen were costly, and the potential for large mishaps grew as the number of meals and patrons served escalated. New skyscraper hotels, with many hundreds and even thousands of rooms, served meals to several thousand people each day and greatly amplified the challenges to the system. As hotels adopted the pay-as-you-go European Plan, each developed several public restaurants that served an even more transient population with individual demands. Some hotels continued to offer the American Plan and the European Plan side by side until the system shook itself out later in the twentieth century. Thus, kitchens had to service restaurants, bars, room service, private dining rooms, and even pet needs. Some hotels had as many as seven kitchens, including one for the employee cafeteria. Figures from the 1910s and 1920s show that nearly half of a hotel's total revenue derived from food service, making the kitchen's role as a profit generator a more significant one than in the mid-nineteenth century.[41]

Scribner's Magazine portrayed many of the hotel's workers in dark, brooding sketches, here showing women in the laundry department, folding the endless supply of sheets, towels, tablecloths, and napkins that had been ironed by machinery. *Scribner's Magazine,* vol. 21 (February 1897), 139. *Courtesy, Kelvin Smith Library of Case Western Reserve University.*

In addition, hotels added banqueting facilities and services in response to the great proliferation of professional societies that turned to hotels to accommodate their regular meetings and annual conventions.[42] The Waldorf-Astoria's 1906 "Entertainment Correspondence" book documents the wide range of organizations that patronized the hotel for their events. These included small dinners such as that for the twelve to fifteen directors of the New York State Society of Certified Public Accountants, a wedding dinner for the twenty-four guests of a suit manufacturer, and large banquets for organizations such as the National Association of Manufacturers, the Saint Andrew's Society of the State of New York, and the Beta Theta Pi fraternity's Silver Gray Reunion and Banquet. Other catered events included those for the Stationers Board of Trade, the American Association of Freight Traffic Officers, the Associated Press, the Amherst College Alumni, the National Metal-Fabric Company, the Kentucky Society, the Sons of the American Revolution, the Daughters of the Empire State, the Daughters of the Loyal Legion, the Daughters of the American Revolution, the Daughters of Ohio, the German Charity Ball, the Democratic Club, the Sphinx Club, and the YMCA, to name only a few. The entertainment rooms at the Waldorf-Astoria included the Grand Ballroom, which seated 744 on the main floor, with seating in tiered boxes that brought the capacity up to 1,054; the Astor Gallery, which accommodated 650; the Myrtle Room (280); the East Room (135), and the State Banquet Hall (100). These were in addition to the many smaller dining rooms that serviced small dinners, receptions, luncheons, and teas.[43]

The large kitchens depended, above all else, on "system." The chef ruled over the preparation of the food, while the steward managed the purchasing, the store rooms, and cost management. In order to maintain a seemingly effortless service for their patrons, the chef and steward worked under an imperative to cooperate well. *Scribner's* tried to capture the hectic ambiance of a busy kitchen by describing it during one of New York's premier social events.

> If you were to visit the kitchen of a large European-plan hotel at dinner-time some evening in Horse Show week, when every one of the seven hundred rooms is filled to overflowing, and the occupants all want dinner about the same hour, not to speak of a hundred additional diners from without, besides a dozen private dinner-parties in the small dinner-rooms, and a college alumni banquet in the ball-room, you would feel something of the same spirit that pervades a circus-tent five minutes after the evening performance is finished and the outfit has a fifty-mile run to make before daybreak, or in a newspaper office when the story of the day comes in ten minutes before the paper goes to press.[44]

The kitchen was nothing short of a factory that may have been invisible to guests but was key to their satisfaction.

In order to satisfy the competing demands of so many events, the kitchen was divided into many different departments. These included the enormous ranges, divided functionally into five or six sections, each with a separate cook staff. Other areas included the preparation areas, a station for cold meats and appetizers, butcher shops (with separate areas for fish, chicken, and meat), an oyster pantry, pastry and bake shops, kettle rooms for soups and hams, and the ice cream, salad, and coffee pantries. The dishwashing and silver polishing sections; banks of specialized refrigerators, broilers, and ovens; and the daily and longer-term storage rooms branched off the main cook areas. A strict hierarchy governing perhaps three hundred employees kept the kitchen functioning, if not always in an orderly fashion. The vast majority of kitchen workers were men. One *Scribner's* image showed one of the two women employed in a cook staff of sixty. The vegetable chef allowed her to serve vegetables but not prepare them, and she spent nearly all her time making toast. This was the one job in the kitchen that a woman could do better than a man because, as the article explained, "stewards find that a man cook considers this beneath his dignity, and will not take pains with it." The professional kitchen did not replicate the gendered work of the home kitchen, where women prevailed. Moreover, the head chef, who was almost always French, was said to earn as much as the state's governor.[45]

Architects usually placed the kitchen in the hotel's upper basement level, connected to the dining room floors by dumbwaiters, conveyors, and staircases. However, in 1910, the *Hotel Monthly* published a plan for Chicago's Blackstone Hotel kitchen that situated the kitchen adjacent to the main dining areas, a move considered to be a "daring departure." The plan's detail tells the story of the hotel kitchen's evolution toward a complicated technological production area. For example, the kitchen had at least eleven different sink areas, all specialized in shape and material to facilitate pot-washing, vegetable cleaning, different butchering tasks, coffee brewing, soaking, salad preparation, and cooking. There were eighteen custom-designed refrigerators for the different cook stations, cold meats, fish, lobsters, crushed ice, oysters, milk, ice cream, the pantry, and the bakery. Machinery included potato cutters and peelers, meat cutters, silver buffers, dishwashers, roll warmers, urns, kettles, steamers, dough mixers, cereal cookers, egg boilers, griddles, waffle irons, steam bowls, warming ovens, broilers, ranges, knife polishers, and silver washers. Service and work counters required specialized surfaces including silver, copper, wood, marble, glass, and metal to maximize the effectiveness and safety of different kinds of food preparation. Each of the mechanical tools needed to be powered from an array of sources, including gas, electricity, wood, charcoal, and steam. Heating, ventilation, noise concerns, and cleanliness influenced the layout and materials used. In addition, considerations were necessary for food, wine, dish, and silver storage; elevator service to the storage areas; banqueting facilities; guest floor pantries; dumbwaiters; staircases; checker stations; and all the waiters' service counters.[46]

Not only did the kitchen face the enormous task each day of preparing food, but a good many of the dishes were exotic and required stores of resourceful knowledge on the part of the steward. Certain months might require purchasing asparagus from Boston, while in other months, the asparagus might come from New Jersey. The steward and chef might court loyal customers by stocking sugar-cured hams from a particular farm in Virginia, Vermont maple syrup, hothouse tomatoes, and farm-fed geese.[47] In 1909, the chef of Chicago's Hotel LaSalle imported

The dishwashing department of the massive hotel kitchen, as portrayed in *Scribner's Magazine*, vol. 21 (February 1897), 153. *Courtesy, Kelvin Smith Library of Case Western Reserve University.*

sixteen green sea turtles, each weighing 250 pounds, which one *Chicago Post* reporter found sprawled on the sidewalk in front of the hotel. "The kitchen of the new hotel will be opened today," he wrote, "and for the next two or three days Charles Laperrique, the chef, formerly of the Plaza in New York and the Ritz in Paris, will be busy transferring the huge carcasses from their shells into convenient bottles, which will be tapped during the coming year whenever green sea turtle soup is ordered."[48] These creatures had been shipped live from Central America to the Fulton Fish Market in New York City and then expressed to Chicago, just one example of the hotel's immersion in world markets. Very little seemed beyond a luxury hotel's reach. The chef, trained in France, knew what he needed; it was obtained, paid for, and shipped, even by express. Chef Laperrique, working in a facility that could accommodate two tons of live turtles, transformed turtles into soup, live animals into bottled liquid, for a demanding clientele who themselves may have come from anywhere in the world.

In addition to the escalation in size of the kitchens and laundries, the hotel's me-

The woman who makes toast, surrounded by the nearly all-male kitchen and wait staff. *Scribner's Magazine*, vol. 21 (February 1897), 147. *Courtesy, Kelvin Smith Library of Case Western Reserve University.*

chanical plant grew to immense proportions as the buildings grew taller. Hutchins, in the *Architectural Record*, claimed that "the mechanism of a modern hotel is an expression of the most ingenious planning in the world. In that respect nothing could be greater except perhaps the planning of the modern steamship."[49] Arthur C. David, writing about the 1905 St. Regis, echoed this statement by saying, "A great modern American hotel is, among other things, probably the most complicated piece of mechanism which the invention and ingenuity of men have ever been called upon to devise."[50] David described the large amount of space necessary to house the boilers, engines, dynamos, and pumps, a design issue complicated by the small superficial area of many New York City building lots. Hotels often went three to five stories beneath the ground to accommodate the equipment, kitchens, and laundry. The square footage of the St. Regis lot was just over twelve thousand square feet, compared to the Palace's footprint of 113,900 square feet.

The peculiar demands of the hotel's daily rhythms made the mechanical requirements much more exacting and complicated than that of an office skyscraper. For example, because people returned to the hotel in the evening, the hotel drew far more additional power at night. The plumbing system—in particular the hot water supply—needed to "feed the several hundred bath-tubs between 8 and 9 o'clock in the morning." David took note of the requirements to store great stocks of food and wine, over and above the aforementioned situating of the kitchens and laundries. Then there were the elaborate communication systems of pneumatic tubes, telephones, and bell-services, as well as other specialized machinery such as icemakers, water filters, the "rubbish crematory," and a machine "for charging water with gas." In addition to accommodating guests, the hotel also had separate lodging, dining rooms, and bathrooms for the staff. "Even then," David concluded, "the network of pipes in the engine-room is utterly bewildering to a visitor, and would be so even to the engineer of the building were not the apparatus carefully mapped and numbered."[51]

The chief engineer of the St. Regis supervised a highly trained staff of thirty-six men whose duties entailed keeping the machinery functioning successfully. In the hotel's elaborate heating system, fresh air intakes were located at various floor levels rather than in the basement and required the engineers to monitor the weather constantly and adjust the system correspondingly. David cited one forty-eight-hour period when the heating system had to be readjusted seven times. Luxury hotels also began to install new devices in the guest rooms, such as electric clocks and telephones. The Belmont claimed to have a telephone exchange as large as that of a good-sized town. Hotels layered telephone systems on top of other communication systems such as electric bells, pneumatic tubes, and the telautograph, a relatively new device that used electrical impulses to produce reproductions of handwritten messages, a forerunner to the fax machine.[52]

The introduction of new technologies also dictated design changes. Perhaps the most significant was the disappearance of the ceremonial staircase. As elevators became the dominant means of vertical transportation, staircases became relegated to isolated corner fireproof towers, to be used only by employees and in the case of emergency. Sometimes architects designed a grand staircase to carry guests up to a second story banquet room, but architects soon isolated these more public facilities to keep transient partygoers separate from overnight guests.[53] With private-bathroom-to-guest ratios improving dramatically, architects eliminated public bathing and toileting facilities from the guest floors, limiting public baths to the barber shop and the Turkish baths, and toilets to the dining and parlor areas. Even though E. M. Statler introduced the idea of a private bath for each guest room during the twentieth century's first decade, luxury hotels still built bath-sharing rooms to accommodate guests who preferred a lower room rate.[54] As the hotel became structurally more complicated, so too did the business of running it. The business offices that once were located behind the lobby's large service counter disappeared to the hotel's hidden nether regions, in effect completing the distancing of the proprietor from his guests. Gone, too, was the lobby's huge gang of bellboys. With modern communication systems, the bellboys no longer needed to be within shouting range of the front desk. Following Hardenbergh's domestication ideas, the lobby became a public parlor, providing space for social encounters, rendezvous, and the pleasure of watching the world go by, completely losing its mid-nineteenth-century commercial identity.

The increasingly complex mechanical plant and enormous size of the new skyscrapers wrought another—almost counterintuitive—change, that of the huge growth in the number of employees necessary to keep the guests happy and the hotel functioning successfully. In the mid-nineteenth century, hotels first experienced the way in which burgeoning size and mechanization expanded the payroll. Then, the ratio was one employee for every two guests. The new skyscraper hotels needed at least one employee per guest, a figure that could climb to as high as two per guest depending on the level of services offered. A 1923 *Hotel Monthly* editorial observed that while ingenious labor-saving devices such as elevators, furniture casters, conveyors, and wheel carts had eliminated a fair portion of the physical drudgery from hotel work (at least for men), these machines needed skilled attention. Hotel employees had become "chauffeurs so to speak, of this or that contrivance, and the trained mind operating the trained hand [brought] into action the power behind the machine."[55] Later that same year, another editorial noted, "The more labor-saving devices introduced into hotels the greater the pay-roll, strange to say." It added, "Today the average first-class hotel will have approximately an employe [*sic*] to a guest, and some of them as many as two or more employes [*sic*] to a guest."[56] Large hotels catering to as many as fifteen hundred guest rooms might employ anywhere from fifteen hundred to three thousand workers. The hotel's management organization became as complex as its physical plant, transforming the proprietor from a host to a corporate manager. As one business writer observed in 1895, "There is no business more complex and exacting in details, or that requires greater ability in management." This was a continuing affirmation of the skills represented by the earlier nineteenth-century phrase, "to keep a hotel."[57]

The *Scribner's* article described the full range of employees in the modern hotel. En route to his lodgings, a guest encountered a room-clerk, key-clerk, bellboy, head-porter, trunk-porter, elevator boy, chambermaid, hall-maid, hall-boy, and possibly the head housekeeper. The guest's dinner represented work performed by the chef and his kitchen staff, supplemented by the steward's department, dining room staff, sommelier, hotel printer (who produced the daily menu cards), laundresses, silver cleaners and platers, specialty bakers, and the checker, who inspected every order received by the kitchen and every plate that left it. The firemen, watchmen, engineering staff, and house detectives kept the building safe. An army of craftsmen included painters, furniture upholsterers, carpenters, seamstresses, decorators, musicians, tinsmiths, blacksmiths, pillow-makers, and mattress refurbishers. Some men did nothing but move furniture; others washed windows, scrubbed floors, and wound clocks. The office employed cashiers, office clerks, bookkeepers, comptrollers, and stenographers, as well as timekeepers to keep track of employees' hours. Some of the larger hotels had a small surgery with doctors and nurses to attend to the employees' and guests' health needs. Finally, the Waldorf-Astoria had a paid staff of six guides who did nothing but lead tours of the hotel.[58]

The management organization needed to focus considerable attention on eliminating waste and maximizing profits in what proved to be a very costly business, where a good chunk of daily expense derived from payroll. Just as architects and the public came to view the hotel as a machine for living, its management embraced similarly technological values such as efficiency and functional regularity when it came to administering the hotel. Efficiency came to be an idea that, as one 1917 writer claimed, "overrides a generation."[59] Horace Leland Wiggins, vice president of United Hotels Company of America, which in 1926 owned twenty-eight small hotels, stated in an article on hotel administration that hotel management had become so overloaded with details that even he "confessed to a certain sense of bewilderment." When asked if the introduction of machines into the hotel had reduced the number of employees, he concurred with the other experts. He turned to ice cream, as an example of a product whose production technology had generated a complex system to support it. "In the old days," he noted, "one man could grind out enough ice cream to serve all the guests in the place." However, now ice cream was produced by machinery that required maintenance, and it had to be served on dishes, which in turn had to be washed. When guests purchased the ice cream, cashiers made change, bookkeepers kept track of accounts, bills had to be paid. Finally, he said, "about the only saving factor in the whole situation is that the modern hotel man does not have so much difficulty getting his profits to the bank, because, proportionately, the profits are about the only things which have not kept pace with increased business."[60] Wiggins was overplaying the situation a little—at least until the cone was invented, ice cream had to be served in dishes that required washing—however, he drove home that it was impossible to fine-tune the system enough to prevent duplication of services and overemployment.

With Frederick Winslow Taylor's ideas about efficiency and scientific management providing cultural context, the hotel was, indeed, an efficiency expert's dream—if he or she loved a challenge—where endless reorganizations were possible to achieve the "one best way" to do things. An organizational flowchart published in the *Architectural Forum* showed twenty-eight different departments, each with its hierarchy of employees to be managed.[61] Governed by a general manager rather than a proprietor, much less a host, the modern hotel had evolved to such a state that—one architect suggested—the complexity had reached its limit. The American hotel, he asserted, was an American invention, and, like other inventions, "if the hotel would avoid ultimate chaos as a result of this ever increasing complexity, it must, paradoxical as it seems, progress from this complexity to simplicity."[62] The man who would lead the industry out of its self-defined wilderness was to be Ellsworth Milton Statler.

E. M. Statler and the "Statler Idea"

Sinclair Lewis's 1934 novel *Work of Art* told the story of Myron Weagle, the son of a small-town Connecticut hotelkeeper who dreamed of owning his own luxury hotel. After a tour of Europe to learn what he could from the Continent's best hotels, he returned to the United States consumed with passion. But, as Lewis tells us, "Weagle privately jigged as he steamed toward his own native varieties of poetry—toward American elevators and circulating ice water and telephones that worked, all-night service, and corn pudding, and adjustable heating and free daily papers that *were* newspapers, and Alpine views from thirtieth-story windows, the American belief that quick-laundry service did not require special permits from the police, and even more surprising the belief that coffee should be served hot and

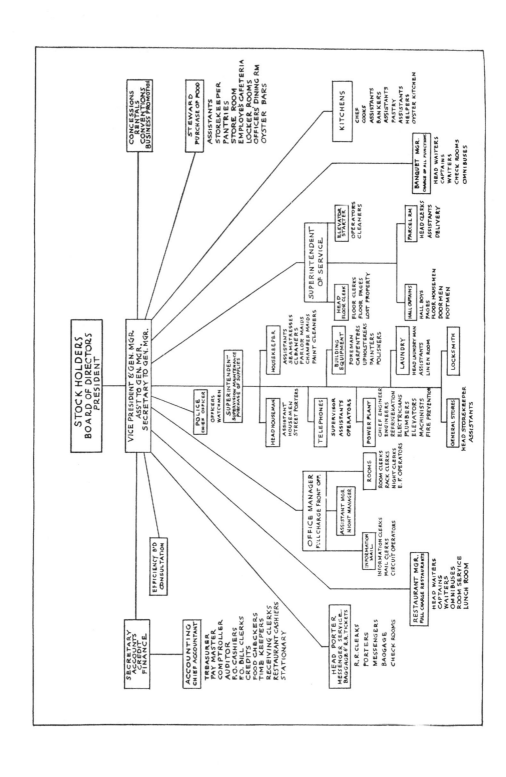

should, preferably, be made from coffee."[63] Myron Weagle's vision of the perfect hotel described flawlessly and completely that created by E. M. Statler in the first three decades of the twentieth century. Statler's own "rags to riches" story, his life of hard ambitions, and his success as the country's most renowned and respected hotelman served to construct an American hotel ideal packaged by modern advertising in an ideological message of stern producer ethics, salesmanship, and customer satisfaction. Statler's concept effectively blurred the boundaries between luxury and middle-class service in an age obsessed with the benefits of efficiency, standardization, and consumption.

Statler's death in April 1928 at the height of his career elicited national headlines: "Statler, Once $6 Bellhop, Studied Cranky Guests in Building His Hotels" (*New York World*) and "Life of Hotel Magnate Reads Like Movie Tale" (*Buffalo Evening Times*).[64] Statler's legend began with him growing up in the small town of Bridgeport, Ohio, across the Ohio River from Wheeling, West Virginia. The tenth of eleven children of a stern, unsuccessful, impoverished preacher and his wife, Statler went to work at age nine at La Belle Glass Works. For fifty cents per twelve-hour day, he carted coal to stoke the "glory holes," the ovens that heated and softened glass pieces in preparation for finishing. Despite the hellish and dangerous work, Statler and his two older brothers boasted of their endurance because, as Floyd Miller, one of Statler's biographers, explained, "they felt like men."[65] In 1876, at age thirteen, Statler found work (through obstinacy and refusing to accept no as an answer) as a bellboy in the McClure House in Wheeling. There, his unflagging energy and interest not only distinguished him from the "lazy, stupid, and insolent" bellboys but also set him on his ascendant path to building and owning a succession of "the world's largest hotels."[66]

Statler rose through the ranks of the hotel staff from bellboy to night clerk to day clerk. He kept track of his observations about hotel management in a little notebook. Eventually, Statler took over the owner's bookkeeping. He recognized and acted on business opportunities, taking a lease on and refurbishing the McClure House's billiard room, setting up a railway ticket sales office, investing in a nearby bowling alley, and adding a concession where he sold his sister's pies and the "best coffee he could get."[67] Every venture enjoyed success, with Statler's

This diagram illustrates the administrative reporting organization of a large hotel, with all departments reporting directly or indirectly to the general manager, who in turn reports to the president, directors, and shareholders of the corporation. *Architectural Forum* (November 1923), 242. *Courtesy, Kelvin Smith Library of Case Western Reserve University.*

appetite for risk always balanced by common sense, shrewd business acumen, and an unstinting capacity for hard work. Statler accumulated capital and built a reputation for honest business dealings. With credit from his Wheeling supporters, he invested in a large, unproven restaurant concept in a new downtown Buffalo, New York, office building. The restaurant's first year was a stunning failure, but under pressure from creditors, he instituted efficiency measures, began a series of advertising promotions, and finally, by offering twenty-five cent meals, turned the business around to become a surprise success. Buoyed by the restaurant's recovery and his love of the hotel business, Statler bid on and won the contract to provide a temporary hotel for Buffalo's 1901 Pan-American Exhibition.[68]

Statler proposed to build a temporary structure, the largest hotel in the world, one that would have 2,084 bedrooms, house up to five thousand people, and be easy to "junk" at the end of six months. Statler's ability to provide innovative solutions to expensive problems soon won him the admiration of the hotel community. By the turn of the century, the hotel dailies had been replaced by trade monthlies that reported on new construction, the professional movement of hotel men, management systems, and mechanical equipment. In August 1901, the *Hotel Gazette* acknowledged Statler's success by emphasizing his hotel's excellent service. "The system at Statler's is fine," the *Gazette* noted. "Manager Statler is a wonder. . . . In all the hustle and bustle the hotel has an air that is pleasing. They are never too busy but what a bellboy can be had at once. The cashier is never too busy to change a fivespot. The clerk, room, key or mail, finds time to answer courteously a thousand questions an hour. In the dining room you are promptly waited on and served with good food."[69] Built within one block of the exhibition, the hotel covered nine acres of ground, had three miles of hallways and a dining hall that could feed five thousand fair-goers at once.[70]

Unfortunately, the fair itself was something of a train wreck. Marred by relentlessly wet and chilling weather, the Pan American Exposition was a dreary event for visitors, despite an extraordinary light show that featured a sixty-foot model of Niagara Falls that celebrated the site's first hydroelectric power station.[71] As if things were not bad enough, the fair's showcased Eskimos caught pneumonia and died. Most of the visitors cut short their visits and headed to Niagara Falls. The one hope for recovery rested with President William McKinley's visit to the exhibition, an event that turned into an unprecedented disaster when he was assassinated. With the fair a failure, Statler's hotel never had more than fifteen hundred guests at a time; yet, even so, he managed to squeak out a few thousand dollars of profit. Already he had become known in hotels by his credo: "Small profits and big business."[72] His enthusiasm intact and even bolstered, Statler jumped at the chance to build a similar hotel for the 1904 St. Louis Pan-American Exhibition.

Statler called his St. Louis project the "Inside Inn." This $450,000 temporary hotel, his second "largest in the world," had 2,257 rooms, a staff of one thousand, two dining rooms that seated 2,700 people each, and, for the first time, a site within the fairgrounds which gave the hotel its name. Built of pine, Statler minimized the fire hazard as best he could by encasing the wood within a fireproof product called Sacket Patent Board.[73] Statler's policies and business ideology began to take shape with this venture. He paid $60,000 for advertising, but a shrewd manipulation of the nation's press gave him more exposure without additional expense. His appeal to prospective customers whose vacation time, too, was at a premium echoed his system of efficient management. One advertisement read:

THINK OF THE SAVING OF TIME
To one who must crowd all his enjoyment into a brief period. No time lost in the struggle for street cars,—no time lost in waiting in line to buy gate admissions,—no time lost by leaving too early in the afternoon because of over weariness and thus cutting one's self off from the Fair or pay a double gate admission. All your time for the objects for which you came.[74]

Statler advertised a "platform" for his hotel that promised consistent good service. He encouraged guests to make suggestions and register complaints. He promised to sack employees found guilty of giving poor service or having a bad attitude. Statler posted room rates so as to eliminate fudging by potentially unscrupulous room clerks. The Inside Inn proved to be as much interest to hotelkeepers as it was to visitors. The *Hotel Monthly* published several stories about the hotel, giving Statler national exposure among hotelmen.[75]

A serious accident nearly cost Statler his life and threatened the success of the Inside Inn. On opening day, a coffee boy summoned Statler to check a hissing fifty-gallon coffee urn. With Harry Watcham, the hotel manager, Statler and the coffee boy checked the urn, which exploded as they peered at it, drenching the three men with boiling coffee. The coffee boy died. Watcham suffered some burns, from which he recovered within a few weeks. Statler, however, hovered near death for close to a week. After a time, when it became clear that he would heal, his wife Mary moved him to the hotel where he could receive medical attention while managing the hotel, first from his room and later from a wheel chair. Although doctors predicted he would never walk, Statler's tenacity and will gave him the strength to persevere and make a complete recovery. While the accident was highly regrettable, it also generated considerable publicity. By the end of the fair, Statler had earned over $360,000 in profits from the hotel. Furthermore, he began to craft a public personality as a no-nonsense man of ability, courage, and irrepressible will. His staff at the Inside Inn presented him with fishing equipment inscribed with the message, "In recognition of splendid nerve, abundant horse sense, and executive ability." They also presented a plaque to his wife that read, "To the little woman who did her duty," presumably a reference to Mrs. Statler's care of Statler while he was incapacitated.[76]

Statler's next project was to build a permanent commercial hotel in Buffalo, New York. This hotel served as a laboratory for the ideas that Statler had developed through his experiences with the two exposition hotels and his research of the hotel industry. His major innovation called for every bedroom in the house to have its own private bath. While not the first hotel in the country to offer this, Statler's plan to feature it at a price that commercial traveling men could afford was revolutionary.[77] Already well known for his careful attention to finances, Statler calculated that the hotel would cost the same to build with logically arranged plumbing shafts as it would to plan the usual large public baths on each floor. The same shaft that carried water and waste pipes for the back-to-back, stacked bathrooms could also house heating ducts and electrical conduits. His slogan, "A Room and a Half for a Dollar and a Half," struck fear into the hearts of hotelmen everywhere. The days of the two-dollar hotel room had long since vanished, and it seemed impossible for anyone to make a profit selling rooms at such a low price. Statler was adapting the principles of mass merchandising to the hotel industry.

Statler turned to technological solutions to address both common management problems and guest comfort and convenience issues. For example, he wanted circulating ice water in each bathroom. This would eliminate the endless runs by bellboys delivering ice water to guests that Statler claimed constituted 90 percent of their job. Other specifications included electrically lit closets in every room, room light switches by the guest room door, full-length room mirrors, bathroom hooks for towels and nightclothes, and a host of other guest-oriented innovations that Statler believed would lead to higher occupancy rates and thus, higher profits. While these ideas seem commonsensical to us—and are things we take for granted—apparently no one had ever before thought to put them all together in a way that was consistent and affordable. The *Hotel Monthly* observed that Statler's new Buffalo hotel represented "more new ideas of the utility order than have heretofore been featured under one room, and many of these ideas are so clever they will doubtless be adopted by other hotel builders and furnishers."[78] Shrewd as ever,

Statler fed the details of the hotel's innovations to the press little by little, garnering free public attention as he did so.

According to Statler's biographer, the hotel structure itself was "graceless," and indeed, images of the thirteen-story structure confirm this as a kind judgment. The hotel brought such cosmopolitan élan to Buffalo, however, that the city embraced it wholeheartedly and made it a success. The three-hundred-room, three-hundred-bath hotel opened January 18, 1908. Within the first year, Statler built a 150-room addition to the hotel. In 1910, the *Hotel Monthly* described the Buffalo hotel as "one of the best known hotels of the world." In 1913, the same journal stated that due to its "many novel and practical features," the Hotel Statler-Buffalo had "become the Mecca of architects and hotelmen seeking new ideas."[79] Statler had successfully combined ideas about technological innovation, efficient management practices, guest comfort, and marketing into a system that brought hotel management and design practices to a new level.

Statler became something of a mythic ideologue in the hotel world as the result of his outspoken personality, singleminded ideas about hotel service, and nimble adoption of advertising and marketing techniques. When he opened his second commercial hotel (Cleveland, 1913), the *Hotel Monthly* described him as already having attained "first place in America as the builder of hotels on original lines."[80] This was quite an accolade for someone whose reputation rested essentially on a single hotel in a not-particularly-compelling midwestern city. Concepts such as "Statler-trained," "Statler service," and the "Statler Idea" had found their way into the lexicon of architects and hotelmen, who integrated them into their own management practices.[81] A well-planned marketing strategy disseminated these ideas and gave Statler's name and his hotels not just a brand but also a broad public forum. Over the years, right up to his death in 1928, Statler found various avenues through which to spread his message. These included general circulation periodicals such as *American Magazine* and *World's Work*, business periodicals such as *System* and *Nation's Business*, internal publications such as *Statler Salesmanship* and the *Hotel Pennsylvania Daily*, all the major hotel trade journals, and the architecture and building trade press.

The Cleveland hotel built on the lessons learned in Buffalo with a leap forward in sophistication and quality. Statler chose George B. Post as the architect of the new hotel. Post had earned his reputation as a consulting architect for the path-breaking Equitable Life Assurance Building (New York, 1868–70), the Manufactures and Liberal Arts Building at Chicago's 1893 World's Columbian Exposition, the Wisconsin State Capitol (Madison, 1904), and the New York Stock Exchange (1904). He enjoyed a national stature among architects, having served as president of the New York Architectural League, the National Arts Club, and the American Institute of Architects. Post died toward the end of 1913, but he had begun to pursue an interest in hotel design that culminated in his work with Statler. Post's biographer wrote, "Post's interest was chiefly structural: there is little artistic unity in his buildings. All are excellent machines, dressed in any decorative clothing that seemed expedient."[82] This fit in perfectly with Statler's ideas about efficiency and frivolity. One editor described Post's architectural firm as "being imbued with the Statler spirit." Choosing that firm lent credibility to the building within architectural circles and elevated the stature of Statler's organization.[83]

To balance Post's machine aesthetic, Statler hired Cleveland's foremost decorator, Louis Rorimer, president of Rorimer-Brooks Studios, to decorate the Cleveland hotel. Statler's choice of Rorimer also served to cultivate an image of discernment and elegance competitive with that of New York hotels. Statler's biographer, Floyd Miller, repeats this purported exchange between Statler and Rorimer: "'I like what you've got here,' [Statler] said, waving Rorimer's water color, 'but, my God, man, you want as much to do one room as a whole floor cost in my Buffalo house.' 'Yes, I've been in your Buffalo hotel,' Rorimer replied quietly, 'and it does give that impression.'"[84] Rorimer was trained in Europe and had achieved a national reputation. In 1925, he was a member of the Hoover Commission that represented the United States at the Paris Industrial Arts Exposition. Thus, his role in decorating the hotel not only created a more refined interior when compared to the Buffalo hotel but elevated Statler's stature among hotelmen as well.

A simpler, more graceful, less florid hotel style had recently become fashionable in New York, based on the work of late-eighteenth-century English architects and designers Robert and James Adam. As described in the *Hotel Monthly*, the Adam style was characterized by "classic designs, in particular low relief in mural and ceiling decoration; the introduction of Wedgewood cameos and porcelains, soft colorings and rounded treatment producing the essence of refinement and comfort."[85] Rorimer chose a color scheme for the entire hotel using soft tones of browns, grays, and blues. Not only did this unify all sections of the house, but it also played into Statler's notions of efficiency by limiting the number of different choices necessary for carpets, upholstery, draperies, and even dish services. This represented a huge departure from hotels of the previous decade that insisted on many wildly themed public rooms. The team of Statler, George Post and Sons, and Rorimer-Brooks proved so successful that Statler kept it in place for every hotel he built until his death. While the Buffalo hotel served as a proving ground for

Statler's ideas, the Cleveland house became the prototype for the rest of the properties, establishing a standardized presence in all "Statler" cities.

In all, Statler built seven hotels. In March 1914, the organization incorporated, combining the Buffalo Statler that Statler owned personally with the Cleveland corporation. The new company, Hotels Statler Company, Inc., would also build the hotel already planned for Detroit.[86] Following Cleveland in 1913, Statler opened the Hotel Statler-Detroit in February 1915, an 800-room hotel. By the end of 1915, the Statler corporation announced a 300-room addition to the 700-room Cleveland property and a 200-room addition to the Detroit building, bringing both hotels to 1,000 rooms each. The 650-room St. Louis hotel followed in November 1917. Then, Statler obtained the lease on the Hotel Pennsylvania opposite New York City's Pennsylvania Station. Owned by the railroad but leased and operated by Statler, the Hotel Pennsylvania was the only one of Statler's properties not designed by George B. Post and Sons. Instead, the prestigious architectural firm of McKim, Mead, and White designed the 2,200-room hotel, the "largest hotel in the world." It opened in January 1919 and gave Statler the presence in New York City that he had coveted ever since being snubbed by the financiers there. After the Hotel Pennsylvania, Statler built a new 1,100-room hotel in Buffalo, while retaining ownership but changing the name of the old Statler-Buffalo to the Hotel Buffalo. Finally, five months before he died, Statler opened the 1,300-room Hotel Statler-Boston, bringing the total number of rooms in the corporation to 7,700.[87]

As head of his corporation, Statler assumed the role of a strict omnipotent patriarch. He immersed himself deeply into every business detail, from architectural plans to kitchen recipes. He devised conduct rules and standards for all employees, publishing a service code and requiring employees to learn it and carry it with them at all times while on the job. He concluded one code instruction that warned against judging guests by their dress with a stern, "I said so."[88] Failure to follow the code resulted in dismissal, an act tantamount to being disowned by one's father. In his "Talk on Tipping," it is often unclear when using the subject "Statler" whether he is referring to himself in the third person or his company, signifying a complete oneness between himself and his organization. Even though Statler signed the article and wrote in the first person, he slipped easily into the third person, issuing authoritarian pronouncements like, "Statler can and does do this," or "Statler can run a tipless hotel if he wants to." Without question, the man exerted total control over every aspect of his business.[89]

The idea of service drove Statler and his corporate ideology. In his "Service Code," Statler stated, "A Hotel has just one thing to sell. That one thing is Service.

The hotel that sells poor service is a poor hotel. The hotel that sells good service is a good hotel. It is the object of Hotel Statler to sell its guests the very best service in the world."[90] Like his contemporary Henry Ford, who made "the greatest creation in automobiles" available to the masses, Statler sold standardized service on a mass-produced scale within a highly calibrated and reproducible environment. For Statler, service encompassed everything from direct employee-guest encounters to guest room conveniences to the placement of structural steel in the building. In this way, too, Statler emulated Ford, whose obsession with vertical integration—the need to control every aspect of production, from owning and extracting natural resources to final product delivery—was well known. Statler's methods drew on Fordist principles that featured complex production systems (in this case, the hotel as a machine for service), extensive division of labor, massive economies of scale, enlarged markets, vertical integration, and price management.[91] Yet unlike Ford, whose famously rigid dictum reassured the purchasing public that it could obtain any color automobile it wanted so long as it was black, Statler's enterprise offered a standardized product supported by standardized procedures that, nonetheless, needed to address individualized personal needs.

By linking service to a sales transaction, Statler removed himself from the traditional proprietary role of "host" and declared himself a captain of consumer capitalism. Statler regarded his guests as being completely passive, even resentful of the possibility of needing to take responsibility for their own comfort.[92] The way to please guests was through proper employee training and education and the simplification and standardization of all design, equipment, and processes within the hotel. Statler transformed each and every hotel employee into a salesman, selling personality and service. As Statler warned, "Statler Salesmanship is the art of satisfying Statler guests with Statler Service. Get that. It's important."[93] Statler's focus on service and salesmanship, rather than the hotel itself, was congruent with new thinking in advertising that by the late 1910s was undergoing an enormous shift from highlighting a product to emphasizing a product's benefits, such as soft skin, rather than the soap that produced it, or good lighting, rather than the lamp. Statler was not only a product of his time immersed in emerging cultural processes but also a cultural force with widespread appeal and influence.

Service broke down into two components, employee-guest relations and attitude, and physical design and efficiency. Statler instructed employees to treat everyone with courtesy. "Never be perky, pungent or fresh," the codes cautioned. Employees needed to understand and walk the fine line between servility and sincere interest. "The guest can tell,—so can I—whether you're interested in him or not," Statler advised, once again painting a portrait of the all-knowing father. Statler instructed managers to hire only people who were good-natured, cheerful, and pleasant and who smiled easily and often. Furthermore, managers shouldered a similar responsibility to "get rid of the grouches."[94] Statler encouraged guests to register written complaints. Managers of each hotel forwarded these complaints daily to Statler, who recorded and researched the sources of guest dissatisfaction. In 1913, Statler began publishing a monthly corporate magazine called *Statler Salesmanship* with the goal of unifying the service organization, increasing efficiency, and creating a feeling of family in employees among themselves and toward the corporation. Moreover, with *Statler Salesmanship,* Statler formalized a way to hammer home his ideology through suggestions for developing personality, tact, and the proper attitude.[95] Managers received weekly messages to distribute to all department heads, who then read them aloud to all employees. These directives covered everything from dirty fingernails to corporate waste.[96]

The other half of Statler's formula, that of the efficient physical plant, reveals Statler's conceptual talents and his ability, perhaps rare, to integrate both production and consumption and human and material capabilities into an organic whole, all the while streamlining it to generate the greatest profits at the least cost. In addition to the most publicized guest conveniences such as the bedroom/bathroom combination and the in-room running ice-water, Statler incorporated several others that, while simple, indicated an understanding of the consumer and what he or she needed. These included wooden transoms above the guest room doors that blocked the bright hallway lights, a light over the dresser mirror, an adjustable bedside reading lamp, portable telephones, outside windows that swiveled so as to permit cleaning both sides from indoors, bathroom radiators, individual room thermostats, sewing kits, an electric outlet convenient for curling irons, and an occupancy light that warned the maid that the guest was in. Statler also shaved a half-inch off the bottom of the guest room door to allow the daily paper to be slid noiselessly underneath.

With the opening of the Boston hotel, Statler installed radio receivers in his guest rooms system-wide. To celebrate and advertise, Statler paid for a national NBC radio broadcast featuring his hotel orchestras. The program originated from six cities with Statler himself serving as master of ceremonies from the Hotel Pennsylvania's main dining room. The radio broadcast enabled Statler to charm a nationwide audience with information about his hotels. The broadcast served to confirm that Statler Hotels offered cultured entertainment, were technologically progressive, and were administered by a "real" person who cared about his customers. Statler's production required a corresponding effort to convince his audience to be equally enthusiastic consumers.[97]

Statler hotels, however, distinguished themselves most impressively by the structural and design logic about which guests had no overt knowledge. In a series of 1917 articles written for the *Architectural Forum,* W. Sydney Wagner, Statler's point man at George Post and Sons, explained what made Statler hotels different from other great twentieth-century hotels. Wagner often found himself being asked whether Statler's success should be ascribed to efficient and farsighted management decisions or the buildings' design and equipment. Wagner, as the architect, diplomatically weighed in on the side of the buildings' plan and mechanical equipment, while crediting good management and service as a contingent element for success. The secret to Statler's success lay in the reduction in the number of the hotel's essential parts, meaning the bedroom, sample room, public, function, and service floors, and the simplification and standardization of those parts from their greatest to their most minute details. In concert with an abiding Fordist culture that espoused mass production, standardization, and efficiency, Wagner advised that this formula could and should be replicated, rather than trying to express individuality in different buildings.[98]

Wagner explored the advantages of Statler's midwestern cities and the particular needs and requirements of business travelers and local residents alike. He attributed hotel failure to promoters and financiers whose interests concerned only securing the deal. He cast blame upon the owners and architects unfamiliar with the specialized knowledge and equipment necessary to build a hotel and the managers unable to share the wisdom of their workaday experience with aesthetically driven architects. Other hazards included bad locations, unwieldy financing, oddly shaped building lots, and unforeseen competition. One of Statler's strengths lay in the integration of the promoter, financier, designer, owner, and manager in one highly competent driven man. With a commitment to simplicity, identification through replication, and service, Statler transposed his ideas into the material form of his hotel buildings.

A disciple of Statler's, Wagner framed the buildings' architectural solution to answer the same question of each of the hotel's parts, "Will it provide the best service?" With simplification as his guide toward this end, Wagner designed each of the parts as self-contained units. Floors contained, for example, only bedrooms with all bedroom floors grouped together. Sample room floors followed the same pattern, rather than mixing sample rooms and bedrooms or sample rooms and function facilities on the same floor. Wagner placed the women employees' dormitory floor between the public floors and the building's main bedroom shaft instead of at the building's top. By doing so, the dormitory floor acted as a buffer between the two sections and could carry the large, and now easily accessible, plumbing drainage, steam, and vacuum piping trunk lines, ventilation ducts, and electric wiring under the ceiling. Service elevators, rather than being separated from the passenger elevators, as was the custom, were placed back-to-back with them within the same bank, bringing all elevator machinery under one roof and also providing greater convenience to the front office, porters, and bellmen.

The placement of the structural steel was perhaps the most innovative change. Typically, an architect first arranged the elevators, corridors, and wings and then designated the location of the structural steel columns. The architect would proceed to fit the rooms into the spaces left, which often resulted in oddly shaped corners and wasted space. Wagner reversed the process by first planning the rooms according to their service requirements and arranging the steel columns and floor beams to support the standard design. This required more steel but resulted in the exploitation of standardized bedroom/bath units to the fullest extent possible. Furthermore, the architect no longer had to contend with the often impossible task of reconciling the exterior architectural expression to the interior use of space because, as Wagner stated, "the steel structure [that determined that expression] assumed its legitimate place in the scheme of architectural design."[99]

Finally, the arrangement of the individual baths followed the same principles of standardization and simplification. Improved ventilation technology eliminated the necessity for the bathroom to have an outside wall and window. Freed from that restriction, Wagner rearranged the back-to-back bathrooms so that rather than aligning with the length of the adjoining bedroom walls, they turned ninety degrees to run parallel to the hall corridor. This created a small vestibule into the room, with the closet on one side, and the bathroom on the other. More bedrooms could be thrown together *en suite,* instead of just two; bedrooms became wider adding width and greater stability to the building's wings; and the baths served as a buffer zone between corridor noise and the sleeping rooms. The service shaft rose between the two bathrooms, housing hot and cold water supplies, drainage, steam, ice water, other piping, and ventilation ducts. By placing the bedroom and bathroom radiators against the shaft wall, one steam line served four radiators instead of two, as when the bedroom radiator was placed under the outside window.

Statler embraced standardization for several reasons. First, it made business sense and lent order to a complex enterprise. Not only did Statler replicate the structural plant, but the consistent decorating schemes, even as they became more

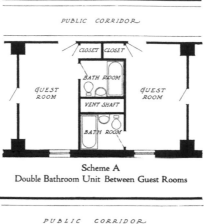

Scheme A
Double Bathroom Unit Between Guest Rooms

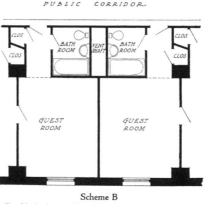

Scheme B
Double Bathroom Unit Between Guest Rooms and
Corridor

Scheme A shows Statler's first design for the back-to-back stacked bathrooms used at the Hotel Statler-Buffalo. The bathrooms separate the guest rooms but do not insulate the guests from hallway noise or light. Scheme B, used in Statler's second hotel, the Hotel Statler-Cleveland, buffers the guest rooms from the public corridors. Additional advantages included a greater number of connecting rooms and lower building costs due to the wider building wing that required less structural steel. *Architectural Forum,* vol. 27, no. 5 (November 1917), 168. *Courtesy, Kelvin Smith Library of Case Western Reserve University.*

elaborate and luxurious with each new hotel, also created a sense of familiarity that Statler hoped would make the guest feel as though he or she were returning home. Statler actively pursued the same reaction that Sinclair Lewis satirized in his 1922 novel *Babbitt.* On his way up the social ladder, George Babbitt, Lewis's symbol of cultural conformity, gives the annual address at the city of Zenith's Real Estate

Board dinner, held auspiciously in the Venetian Ball Room of the city's O'Hearn House. In a paean to the "Standardized American Citizen," Babbitt quotes his good friend, Chum Frink, who waxes on about the great state of midwestern civilization, symbolized by the standardized hotel. When on the road and feeling lonely, Frink would turn to the first-class hotel: "When I entered that hotel, I'd look around and say, 'Well, well!' For there would be the same news-stand, same magazines and candies grand, same smokes of famous standard brand, I'd find a home, I'll tell! And when I saw the jolly bunch come waltzing in for eats at lunch, and squaring up in natty duds to platters of French Fried spuds, why then I'd stand right up and bawl, 'I've never left my home at all!'"[100] Statler, of course, wanted customers to feel this way so that whenever they traveled to a Statler city, they would be sure to stay at the Statler hotel.

What began as an enterprise geared toward the "wants of the typical American" not only responded to those wants but also did so in a way that forced all other hotels to follow suit. Statler took the luxury hotel idea, simultaneously simplified it and expanded upon it, and made it available in large doses to an enthusiastic middle-class public. Standardization and simplification enabled Statler to satisfy his personal need for power and his public ambition to build the "largest hotel in the world." By the mid-1920s, one writer estimated there were 130 hotel chains in the nation, although over three-fourths of those were of the "two-link" variety. Yet some of the chains were sizeable. They usually specialized regionally or by hotel type, such as luxury, railroad, or locally owned hotels in small cities.[101] Statler's organization was different because he exploited the methods of mass production and consumer capitalism rather than conglomeration. Statler's product, like Ford's, fostered identification by merchandising a mythic personality and a big idea.

It may seem that the Waldorf-Astoria and the Statler hotels have little in common. The Waldorf epitomized luxury, extravagance, and cost without limits, while Statler's mass-produced interpretation of luxury took aim at the middle class. However, in 1923, Lucius Boomer, the Waldorf-Astoria's president, wrote an article for *System* magazine, "How We Fitted Ford's Principles to Our Business," that extolled the virtues of applying simplification and standardization practices to the hotel industry, indicating that, in fact, the history written in this chapter had come full circle. By 1923, Boomer controlled six big luxury hotels, four in New York, in addition to the new Willard in Washington, D.C., and the Bellevue-Stratford in Philadelphia. The examples that Boomer cited included the same measures that Statler had instituted early on, such as using a common dish service in all the hotels, reducing by two-thirds the styles of glassware and silverware used in meal service, and reducing by 75 percent the number of different carpet patterns. Offering guidelines such as "Find the best way to do things and make that method standard," Boomer wrote about standardizing and streamlining employee responsibilities, controlling the cost of supplies by manufacturing items such as mattresses and uniforms, and simplifying both products purchased and products sold. Boomer's article ostensibly drew on Henry Ford's ideology of mass production, but, in reality, he had adopted Statler's methods for running a profitable hotel empire.[102]

A Factory of Comfort

The Waldorf-Astoria introduced a new genre of luxury hotels, the hotel skyscraper. In the middle half of the nineteenth century, the hotel served an important commercial function as a business exchange. With the emergence of the iconic corporate skyscraper, the hotel found itself, even in its sophisticated new shape, domesticated by architects, proprietors, and patrons who indulged in new codes of behavior such as heterosocial restaurant dining and the use of hotels for private social functions. The lobby, once a scene of busy male commercial interchange, became a venue for display, relaxation, and rendezvous available to both men and women, much like a giant public living room. Statler carried the idea of the hotel as a home to an extreme, pushing to secure a loyal clientele by providing service and comfort. A 1928 photograph used for advertising his new radio hookup showed a dapper pajama-clad man, tucked into bed, his head propped up on two fluffy pillows, an open book on his lap, a radio headset on his head, the over-the-bed light on, and a telephone and glass of water on his bedstand. The caption read, "All the comforts of Home—A Typical Guest Room in Hotel Statler, Buffalo." One can hardly imagine this or any man, so fortunate to be at a Statler hotel, missing the wife and children with all those diversions as well as room service within arm's reach.[103]

The growth of the skyscraper resulted from and encouraged the development of new mechanical systems. In the hotel, these systems were particularly complex because of the unusual requirements of the guests, the management, and not least the thousands of employees. The increase in machinery resulted in a corresponding increase in the number and kinds of employees needed. Early in the twentieth century, hotels supplied much of their own power through boilers and dynamos. By the 1910s, they began to purchase power from private and public gas and electric companies. Some hotels located themselves next to or even on top of railroad

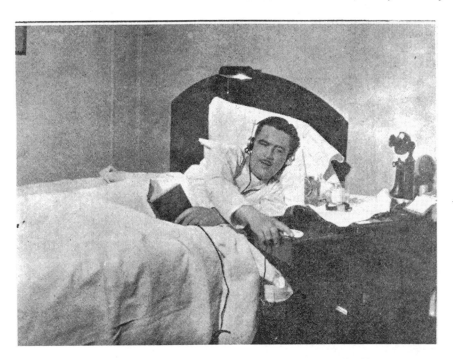

With the opening of his Boston hotel, Statler installed radios in all of his guest rooms systemwide, an addition to the list of conveniences each guest enjoyed, including a private bathroom, a telephone, circulating ice water, and the morning newspaper. *National Hotel Review,* February 11, 1928, 64. *Division of Rare and Manuscript Collections, Carl A. Kroch Library, Cornell University Library.*

and subway stations and connected to those stations, shops, and office buildings through underground tunnels. Theoretically, guests could arrive by train, transact business, enjoy an orchestra concert and dinner, and shop without ever going outside. It signified the complete connection to and dependence on transportation and communication systems, world markets, corporate financing, and advertising and the development of a highly competitive and potentially lucrative business. These hotels reached a stage of complexity certainly foreshadowed—but hardly envisioned—by their early nineteenth-century predecessors.

Despite the proliferation of luxury hotels, promoters still found themselves compelled to build "the largest hotel in the world." Statler did this three times; but even so, his 2,200-room Hotel Pennsylvania retained the title for only eight years. Because of the skyscraper's potential, there seemed no limit to hotel size. Three decades of profitable growth encouraged these kinds of ambitions. Even smaller cities like Cleveland and Buffalo supported two or more hotels of one thousand or more rooms. In 1913, one writer observed, "Today the McAlpin is 'New York's Newest Hotel,' but that is a distinction as fleeting as 'New York's Tallest Office-Building' for already there is talk of the new Hotel Biltmore."[104] The progressive yet destructive development that threatened Franklin in 1860 not only intensified but also had become an accepted characteristic of progress that hardly raised an eyebrow. The 1,500-hundred-room McAlpin was almost immediately outdone by 1914's 1,950-room Biltmore and 1919's Hotel Pennsylvania. The 3,000-room Stevens (Chicago, 1927) finally seemed to cap growth due only to the ensuing depression, but even the new Waldorf-Astoria, completed in 1931, claimed to be the largest—if not in number of rooms, then in height and cubic feet.[105]

The Waldorf-Astoria, even while introducing the "new" modern hotel, represented some earlier traditions through its individuality and expression of elitism and also as a standard bearer for technological luxury. However, it firmly anticipated the luxury hotel's role as a ready participant in the culture of consumer capitalism. Rather than serving its city by drawing business to its midst, the Waldorf-Astoria promoted ostentatious display and consumption, ironically limiting it to the very wealthy while encouraging mass voyeurism and imitation through society news and its own Peacock Alley. The new standards of technological luxury defined by the skyscraper hotel allowed for no personal discomfort. Insulated skyscraper microclimates created comprehensive entitlements for guests that stood in stark contrast to expectations of those who lived in the "go-ahead" age and marveled at and appreciated each innovation as though it were a miracle. Elevators not only controlled passage through time and space but also created increasingly greater assumptions about vertical travel that had formerly rested in one's own physical stamina. The luxury hotel had become a veritable factory of comfort. Statler's stress on exemplary but standardized service also generated a new measure of entitlement. Statler took the luxury hotel idea and shaped it through the imposition of mass production techniques and the employment of mass-marketing strategies. Despite, or perhaps because of, the industry's growth and its legitimization into a profession supported by trade journals, organizations, and professional schools, luxury hotels had become yet another expression of the vast producer/consumer matrix. As such, they ceased to command the sort of attention they had enjoyed before being bullied into submission by the economic world that had created them—and the tall buildings that stole their symbolic stature.

....

CHAPTER EIGHT

THE STEVENS HOTEL, CHICAGO, 1927

....

"Virtually a Multiple of Twenty-Five Small Hotels"

W HEN THE STEVENS HOTEL opened on May 2, 1927, nine thousand people came for dinner. During the more than two years that encompassed its planning and building, Ernest J. Stevens, the hotel's projector, majority shareholder, and manager, had thoroughly promoted his new hotel as the world's largest and greatest. The perfectly orchestrated opening festivities not only introduced the hotel to Stevens' guests but also demonstrated its enormously scaled capabilities and the fulfillment of promised expectations. In a number of ways, the Stevens serves as the perfect capstone to this century-long hotel history. Not only did the Stevens retain the "world's largest hotel" title for several decades, it represented many of the tensions that had built up over the century, including the drive for individual expression within a standardized form, the celebration of local enterprise within a global marketplace, the enormous challenge and cost of technological development, and a last great effort to corner symbolic capital against the new commercial urban icon, the corporate skyscraper.

Although the last chapter focused primarily on New York City hotels and Statler's strategy to standardize hotel luxury, Chicago, the nation's second-largest city, also emerged as a national center for competitive hotel development, particularly by the 1920s. Three of the major industry trade journals published from Chicago. Hence, every new plan for a large Chicago hotel received its share of attention. During the nineteenth century, Chicago had developed as a gateway to the West for eastern commercial districts. By the twentieth century, the city had become an industrial and financial center in its own right, represented by corporate giants such as Armour, Wrigley, Pullman, Swift, and Field. This growth created a need for important hotels; yet, during the century's first two decades, Chicago

The Stevens, Chicago. Opened in 1927, the hotel's 2,818 bedrooms earned the title as the world's largest hotel. *Postcard in collection of author.*

found itself consistently short on luxury facilities. As in New York, some of the city's early entrepreneurial elite—such as Francis C. Sherman, Potter Palmer, and John B. Drake—built and managed luxury hotels. In the 1890s, the Stevens family fashioned a financial empire, first in dry goods and real estate, then in life insurance, branching into the hotel business in the new century's first decade by building the very successful Hotel LaSalle. The Stevens Hotel continued the tradition, first begun in Boston one hundred years earlier, whereby upper-class businessmen sought to replicate, commercialize, and purvey an elite lifestyle in the name of civic enterprise and inter-urban competition.[1]

The Stevens Hotel story brings together several of our familiar threads. It represents the culmination of technological developments first set in motion by the Tremont House. The Stevens competed in the same sort of rivalry with New York City as had Philadelphia's Continental, and shouldered with San Francisco's hotels the burden of demonstrating cosmopolitan maturation and sophistication. Ernest Stevens, William Ralston, and E. M. Statler all revealed a common inner drive that compelled each to build the "largest hotel in the world." This drive, coupled with an incautious optimism encouraged by a thriving but temporary boom economy, brought financial ruin to both Stevens and Ralston, and death to Ralston and

Stevens's business partner and brother. Just as the depression of the 1870s destroyed Ralston's empire, the Great Depression of the 1930s did the same to the Stevens family. In trying to salvage the hotel, the Stevenses used money from their life insurance company, which resulted in charges being brought against them for embezzlement by the State of Illinois.

The promoters of each of the case studies championed their projects as being necessary to their city's commercial success and as representing their city's achievements and potential. Each drew on local artisans and industry to construct a local product that conformed to and embellished a national—by this time, international—hotel ideal. At the same time, as others had done before him, Stevens sought to individualize his hotel even as he drew on Statler's standardization and management methods. Chicago's Stevens Hotel was a product of and helped shape a reorganized society dominated by corporate and consumer capitalism. Its complete immersion into a national network of technology, communication, transportation, and marketing left it vulnerable, like most other things, to the debilitating effects of economic depression. It also delivered a hard lesson that effectively ended the construction of downtown luxury hotels for decades.

Ernest J. Stevens and the Hotel LaSalle

Ernest J. Stevens was born in 1884 in Colchester, Illinois, a small town thirty-five miles from central Illinois' western border. In 1890, his family, including his father's four brothers, moved to Chicago, where they engaged in various businesses, including dry goods and banking. Eventually, Ernest's father, James W., and James's brother, Edward, founded the Illinois Life Insurance Company, the business that ultimately generated the substantial wealth that financed other enterprises, including the two hotels that Ernest Stevens built and managed.[2] Ernest J. Stevens was a compulsive saver. His papers, deposited at the Chicago History Museum, fill seventy-six document boxes and include detailed records of the family businesses, scrapbooks of hundreds of newspaper clippings, personal papers, and correspondence. A considerable portion of the archive concerns the building and management of the Hotel LaSalle and the Stevens Hotel.

Ernest Stevens earned an undergraduate degree from the University of Chicago and a 1907 law degree from Northwestern University. While in law school, he started working at Illinois Life Insurance. After graduation, he went to work at the Hotel LaSalle while it was still under construction, serving as treasurer and vice president of the hotel company. He and his father conceived the project in order to make use of an idle piece of property on the northwest corner of LaSalle and Madison Streets in the city's financial district.[3] As they had no experience in the hotel industry, the two Stevens men hired George H. Gazley to manage the hotel. Gazley had been assistant manager at both the Waldorf-Astoria and the St. Regis in New York City. Despite those who predicted failure for a "Loop" hotel situated away from the Michigan Avenue hotel district, the shortage of hotel rooms in the city pointed toward a successful venture. A 1908 *Chicago Tribune* editorial compared Chicago's hotel facilities with those of New York. Claiming the same number of travelers for both cities, the *Tribune* observed that New York had ten times as many first-class hotels as Chicago. The editorial noted that during the last livestock shows, "Men and women with money to pay for first class service . . . were forced to go to boarding houses. In the hotels, cots, even, were at a premium." More good hotels would make Chicago a more attractive city. The lack of hotels, the paper asserted, kept visitors away.[4]

Stevens intended the LaSalle to be in the same class as the nation's other top-ranking hotels such as New York's St. Regis, Belmont, and Waldorf-Astoria, and Philadelphia's Bellevue-Stratford. William Holabird of the architectural firm Holabird and Roche was the architect. Holabird and Roche, a pioneer in steel-frame construction, was one of Chicago's leading architectural firms from 1883 to 1927.[5] The city watched fascinated as wreckers demolished the lot's old post-fire masonry buildings and excavators and foundation engineers poured 105 concrete caissons 110 feet down to bedrock. Although these techniques had been used in Chicago for over fifteen years, the *Chicago News* reported that "all classes of citizens appear still to marvel at the modern methods of skyscraper construction, though they have previously witnessed similar monuments of modern architectural skill going up toward the sky in various parts of the city."[6] With twenty-two floors rising above ground and over one thousand bedrooms, the LaSalle claimed its place as "the largest building of its kind ever built from original plans and opened as a hotel." All these qualifications were necessary to distinguish the LaSalle from the Waldorf-Astoria, which, despite its great size and one thousand rooms, opened in two stages.[7]

Critics questioned the LaSalle's location. The hotel's eventual financial success perhaps influenced Stevens when it came time to settle on the Stevens Hotel's site, which also had its detractors. The downtown office area surrounding the LaSalle emptied out at night, threatening the success of the hotel's restaurants that depended heavily on local patronage. As the *Hotel Monthly* observed, some thought "the large amount of space devoted to dining rooms was altogether out of proportion to what this section of the town would support." These fears appeared to be unfounded. Rather, the *Hotel Monthly* asserted that instead, the hotel brought new life to the area and in doing so influenced property values favorably and created new business.[8] The truth was that the hotel lost money for the first three years. Letters reveal certain strong-arm tactics that required suppliers such as those for coffee and meat to also invest as shareholders, their contracts contingent on their maintaining a specified financial position in the hotel.[9] The family resources and those of Illinois Life, plus a series of loans, kept the hotel afloat until 1912, when, as Stevens recalled in 1933, "We made the turn, and it became recognized as the most profitable hotel property in the world. We made over a million dollars a year for many years out of it."[10] Stevens fired his "New York management" in 1912 and took over the job himself, turning the hotel into the cash cow that it became. His successful experience at the LaSalle and his complete involvement in every detail of its day-to-day affairs, a management technique very much like Statler's, gave Stevens a foundation in the hotel business and nurtured his ensuing dream for another "largest hotel in the world."

Stevens developed a close working relationship with Statler. They shared similar autocratic management styles, and both believed that the most successful hotels were "one-man houses" whereby one person controlled both the corporation and the operation of the hotel.[11] Statler was a generous mentor. Stevens had, among his papers, profit and loss statements for the Statler hotels and restaurants.[12] By the time Statler opened the Hotel Pennsylvania in New York in 1919, Stevens was already thinking of something larger for Chicago. He spent time with Statler going over the Hotel Pennsylvania's plans and management system and came away convinced by Statler's ideas of economy of scale. Stevens later recalled, "Mr. Statler showed me how much more economical it was to operate a 2000 room hotel—his was 2200—than it was to operate the old hotel that had 1000 or 1100 rooms and that is what gave me the enthusiastic idea [for the Stevens.]"[13] Stevens referred to the one-thousand-room hotel as an "old" hotel, almost a dinosaur, implying that the two- or three-thousand-room house represented a new type, one that incorporated efficient modern business methods, perfected by Statler and based on the principles of mass production.

This business ideology dovetailed with the opportunity developing in Chicago in the 1910s and 1920s, which was fed by a dramatic increase in Chicago's population, the city's increasing attractiveness as a convention city, its corporate growth, its location as a terminus for all eastbound and westbound trains, and a continuing inadequacy in terms of hotel facilities. The LaSalle continually lost convention business because it had no exhibition space. In 1906, 201 conventions met in

Chicago. By 1924, that number had nearly quadrupled to 762 meetings, which were attended by 625,000 people.[14] The rise and expansion of professional, fraternal, and trade organizations—an outgrowth of the long nineteenth century's process of industrialization and professionalization—resulted in hundreds of local, regional, and national meetings that depended on hotels for places to gather. Statler's architect, W. Sydney Wagner, referred to these meetings as "that strange malady, ... the periodical invasion by the seven-day locusts of America,—the unnumbered trade conventions, clubs, and societies which fill the rooms and lobbies of the hotel to overflowing for a week and then disappear."[15] Stevens's astute business sense, his pride as a Chicagoan, success as a hotelman, and need to satisfy a personal quest led him to begin planning for the Stevens Hotel, the world's largest and greatest hotel.

Planning the Stevens Hotel

J. W. Stevens and E. J. Stevens began conceptualizing the new hotel as early as 1921. The architects, again Holabird and Roche, recorded a $100,000 commission check on January 1, 1922.[16] One partner, E. A. Renwick, remembered the earliest discussions as being shrouded in secrecy. He and J. W. Stevens would meet in a fourth floor room of the Hotel LaSalle. Stevens would take the elevator up to the fifth floor and walk down a flight of stairs to the meeting and leave by walking down to the third floor before taking the elevator down from there. Holabird and Roche kept their contract with Stevens sealed in a strongbox in the office to preserve secrecy. By the time Stevens announced the land purchase publicly, he was also able to give detailed information about the hotel's structure and design. E. J. Stevens later suggested that they kept the plans secret because of competition. Stevens was not the only Chicago hotelman with something up his sleeve. By the mid-1920s, Chicago witnessed the construction of over 6,300 more rooms at Chicago's best hotels, including a brand new 1,526-room Palmer House that opened just months before the Stevens. Chicago's March 4, 1922, issue of the *Economist* reported Stevens's purchase of the Michigan Avenue lot between 7th and 8th Streets and his intention to build a three-thousand-room hotel there costing as much as fifteen million dollars. By May 15, E. J. Stevens had an architectural model of the new hotel in his office to admire. The October 1922 issue of *Hotel Monthly* reported that Holabird and Roche had as many as sixty men working on the plans.[17]

The Michigan Avenue site covered eighty thousand square feet spanning the entire block of Michigan Avenue between 7th and 8th Streets. Michigan Avenue is Chicago's most prestigious thoroughfare. The section bounded by the Chicago

River at Wacker Drive to the north and Roosevelt Road to the south commands unobstructed eastward views of Lake Michigan and Grant Park. Over the preceding decade the city had developed Grant Park as part of the lakefront development scheme suggested by Daniel Burnham's 1909 Plan of Chicago, most of the rest of which would never be executed.[18] The Blackstone Hotel (1910) anchored the hotel strip's south end at 7th and Michigan. To the north, at Congress and Michigan, stood the Congress Hotel (1893), the LaSalle's strongest competitor, and Adler and Sullivan's celebrated Auditorium Hotel, finished in 1889 (now Roosevelt College).[19]

The site's advantages lay in the lower land cost due to its southern location out of the figurative and literal Loop and its proximity to the proposed Central Depot for the New York, Illinois Central, and Michigan Central railroads.[20] Unfortunately, the Central Depot was another aspect of the Chicago Plan that was never realized. Had the depot been built, the Stevens Hotel would have been the closest hotel, within just a few blocks of the station. E. J. Stevens estimated that a similarly sized lot in the Loop would have cost $14 million, compared to the $4.5 million that J. W. Stevens paid for the Michigan Avenue land.[21] Its location, however, was "in the opposite direction from the trend of development," as the hotel's 1934 Appraisal Report determined.[22] City images taken from the proverbial bird's-eye view show the Stevens anchoring the southern end of Michigan Avenue's skyline, with the tall office towers looming north near the Chicago River. Like the Palace, the building's sheer bulk seemed to prohibit any kind of development or foot traffic beyond it. E. J. Stevens purchased an additional piece of land behind the hotel, fronting on Wabash Avenue, upon which he constructed a fourteen-story service building to house the Stevens's mechanical plant. The Stevens family controlled nearly the whole city block.

Holabird and Roche completed the plans for the hotel by October 1922, but when construction bids came in, the Stevenses decided to wait until the financial climate seemed steadier. In September 1924, they issued a ninety-four-page prospectus that included highly detailed estimated building and equipment costs as well as estimated operating expenses. The hotel corporation secured an initial $5 million through a bond offering secured by the Hotel LaSalle. It raised the remaining money through loans and the sale of two million dollars' worth of common stock purchased in its entirety by James W., Ernest J., and Raymond W. Stevens, Ernest's brother and only sibling. The family used their experience with the Hotel LaSalle to project expenses and revenue. Based on the LaSalle's history that enjoyed an occupancy rate consistently above 82 percent from 1919 on, the Stevenses boasted that, "There is no question about our ability to sell 82% of the rooms in the

Stevens Hotel." And even though the title page of the prospectus proclaimed, "The World's Greatest Hotel, 3,000 Rooms, 3,000 Baths," in actuality the floor plans called for 2,818 sleeping rooms, and all figures referred to a 2,818-room hotel.[23]

Winning convention business drove the hotel design. The owners priced 2,254 rooms at five dollars or less based on two-thirds of them being occupied by a person traveling alone. Fifty-eight percent of the rooms measured between 136 and 169 square feet. These were small bedrooms designed for conventioneers of average means who typically would be spending little time in their rooms.[24] The plan included a thirty-five-thousand-square-foot lower level exhibition hall capable of accommodating heavy equipment such as might be displayed at truck and auto shows. The second story's south end housed a fourteen-thousand-square-foot Grand Ballroom with a capacity for two thousand diners at smaller tables or 3,500 people when set up for a convention. The ballroom's checkroom facility could handle the wraps, umbrellas, hats, and packages of 3,200 guests. Additional checking facilities boosted that number to five thousand. A separate 8th Street entrance with dedicated elevators kept the special events attendees from interfering with the everyday business of hotel guests.[25]

The Stevenses awarded the construction contract to George A. Fuller Co., a nationally prominent Chicago construction firm whose history with skyscraper construction dated back to Holabird and Roche's 1887 Tacoma Building (located in Chicago).[26] Work on the Stevens began July 23, 1925. The hotel opened eighteen months later, on May 2, 1927.[27] On May 5, 1926, steelworkers raised the flag on the completed steel frame, an event marked by a great deal of publicity, not all of it contrived. The building's design, with its enormous second story ballroom, presented an unprecedented engineering challenge. Most hotels put their large ballrooms on the building's top floors so that the structural frame above the ballroom needed to support only the roof overhead. The Stevens's ballroom's unobstructed open space, 87.5 by 169 feet, required a system of columns, beams, girders, and trusses that could carry the load of the twenty-two hotel stories above. The support structure began at the fifth story (the ballroom was thirty-five feet high) and extended three stories above that. Additionally, all structural members needed to be arranged so as to be concealed but not waste space.[28]

Detailed technical descriptions of the steel frame appeared in various trade journals in addition to the hotel journals. *Engineering News-Record, Western Society of Engineers,* and *Engineering and Contracting* all carried stories with diagrams and pictures.[29] Benjamin B. Shapiro, Holabird and Roche's structural engineer, penned an extensive summary for *Hotel Monthly* that conveyed the magnitude of the engineering that went into the building. He described the fifty-three columns above the ballroom that were carried by a system of auxiliary trusses and girders, which in turn were carried by four large main trusses. These auxiliary trusses were eleven feet deep, practically a full story high. The girders had large openings cut through them to allow corridors through. The loads from the four large trusses were carried on eight columns, each resting on a concrete pier that extended to bedrock one hundred feet below grade. He continued:

> The caissons supporting these large columns vary in size from 9 feet to 11 feet in diameter, filled with a very rich mixture of concrete. Upon these caissons large steel slabs are placed. These slabs are 96 inches in diameter and 11½ inches thick, weighing approximately fifteen thousand pounds each. The purpose of the slabs is to properly distribute the column load over the entire area of the caissons. After the slabs had been properly set and checked as to location, eight large columns were erected. These columns were 77 feet 7 inches in length from the base to the center of the pin hole on top; were built in one piece and varied in weight from seventy to ninety tons each.

Four large trusses with spans of eighty-six feet, each weighing two hundred forty tons, rose over the space of the ballroom. These were common trusses used for railroad bridges, but they had never been used to support such large loads. The "pins" that held the trusses together weighed about four thousand pounds each. In addition, a complete trestle had to be erected to assemble this part of the structural frame.[30]

This elaborate description demonstrates the project's almost incomprehensible immensity and the degree of technical knowledge and skill required to execute it. It also illustrates E. J. Stevens's personal influence. He knew how he wanted his hotel to work and encouraged the innovations required to execute his plan. At the flag-raising ceremony, Stevens gave a short speech congratulating the steelworkers. He also presented each of them with a reprint of the twelve-page *Hotel Monthly* article and a small monetary gift as a "token of friendship."[31] A *Chicago Tribune* editorial took the event out of its local economic context and placed it among historical American traditions that marked the country's emergence as an industrial world power. The editorial evoked "the technological sublime." Even though the "fast growing city" presented an ongoing panorama of architectural and engineering triumphs, raising the flag on the structural steel continued to elicit unabashed feelings of astonishment that glorified the works of man and his triumph over nature. Experiencing the sublime was both universal and democratic, not restricted to

The four giant trusses that support the twenty-two bedroom stories above the second floor banquet hall. *Engineering and Contracting,* vol. 66 (June 1927), 258. *Courtesy, Scranton Gillette Communications.*

those who financed, planned, and executed the project but extended to observers on the street, the steel men, and the community of citizens whose collective existence enabled and empowered it. The *Tribune* celebrated the hotel as one of many "great monuments to American labor and ingenuity." As part of a tradition that harkened back to those hardy men who scaled the square rigging of clipper ships and who built the great railroad bridges over canyons that dissolved "frontiers into a nation," the raising of the flag over the steel frame represented all that was good, superior, and powerful about "American mechanical genius."[32]

"The City of Stevens"

It would stretch the reader's patience to describe the Stevens Hotel in its entirety, but certain aspects of the building are significant because of the way in which Stevens combined older luxury hotel ideals that emphasized individuality and costly decorative floridness with early-twentieth-century ideas of efficiency, standardization, and economies of scale. The hotel, exclusive of the land, cost $19,792,014.25

to build. Stevens spent an additional three million dollars for furniture and equipment and to outfit the retail stores.[33] While sheer size contributed to the building's cost, certain design elements such as the second floor ballroom and the public rooms' expensive decoration added nearly $10 million to the early projections.

Stevens himself seemed torn when promoting his hotel as to whether he wanted it to be known as the largest hotel in the world or the greatest hotel in the world. One leading hotel critic concurred when he wrote, "There seems to be nothing about the Stevens that is not the 'largest in the world.' Then its magnificence dawns upon you." Even the architects exhibited this ambivalence. When asked to explain the hotel's guiding concept, John Holabird replied, "All of us merely tried to interpret Mr. E. J. Stevens's plans for a 'perfect' hotel.—Quite incidentally, it is the world's largest."[34] Stevens enjoyed comparing the hotel's cost to the mammoth battleship USS *Colorado* (commissioned 1923) that cost $27 million.[35] Perhaps the technological complexity, the ship's power, and the prestige of the United States military appealed to him in a way that by association cast him as the commander-in-chief of a large piece of complicated machinery.

The building's exterior stretched 405 feet on Michigan Avenue, nearly double that of Boston's Tremont House one hundred years before. Unlike skyscrapers in the Loop and north of the river whose architectural embellishments were lost among their neighbors, the Stevens's unobstructed front could assert its massiveness across the full height and breadth of the building. The exterior architectural elements integrated fully with the hotel's interior arrangement. The ground floor storefronts punctuated the sub-base of pink granite that supported three stories of Bedford limestone. An array of windows at evenly spaced intervals extended the full width of this base, allowing lake views from the important public spaces contained within, including the ballrooms, the main dining room, and the lounge. Nine pavilion bedroom wings rose twenty-one stories above the public rooms, arranged around nine interior light courts. Each of the bedroom shafts was sheathed in red brick for eighteen stories, with the top three stories reverting to limestone. A stone cornice capped the roofline, while an ornamented four-story tower rose above the center of the building.[36]

Entering through the main doors on Michigan Avenue, guests found themselves in the center of the Grand Stair Hall. Even though grand staircases had just about disappeared from the lobbies of the modern elevator-driven skyscraper hotels, Stevens showcased a two-story symmetrical double staircase that evoked the splendor of European royal palaces. The *Hotel Monthly,* after describing the hall's scale, its massive Travertine marble columns, and the way the staircases led to the grand public rooms on the floor above, continued, "Centering the stairways are bronze fountains, the subjects 'boys and fish,' the work of Frederick C. Hibbard, a famous sculptor, who took for his models the three sons of Mr. Ernest J. Stevens. The ceiling of this Grand Stair Hall is a cloud effect illusioning a vision, such as John Bunyan might have described, of angels in the sky."[37] What other kind of building could employ a traditional aristocratic parade route that encompassed features ranging from towering Grecian columns to the Stevens boys cast in bronze to the heavenly Celestial City—and this, no less, for the appreciation of middle-class conventioneers such as the National Society of Denture Prosthetists or the International Baby Chick Association? *Hotel World* went farther with its allusions to royalty: "I warrant you have never seen a lovelier room—not in Versailles the glorious, nor in the chateaus of France." Later, its writer claimed, "A coronation might be held here."[38]

The Grand Stair Hall evoked the ceremonial staircases typical of English aristocratic houses, whose masters withdrew from the common hall below to elaborate private chambers above, where they entertained visitors. The stair became a parade route that cast distinction on visitors as they ascended to a lordship's presence. Stevens applied this concept of climbing to royal chambers to his own great house. Certainly, he envisioned formally dressed guests ascending to elaborate events being held above, as observers watched the invitees flow gracefully up the great staircase that led to the Grand Ball Room, lounge, and main dining room. The lounge, as *Hotel World* gushed, "ought in reality be the drawing room of a great chateau." Its soaring windows looked out over the expanse of Lake Michigan, while its "embroidered silks" and tapestry, needlepoint, and brocade upholstery fabrics belied the furniture's sturdy commercial construction. *Hotel World* informed, "It is not 'hotel' furniture."[39]

The Grand Ballroom, however, was the hotel's centerpiece. Guests arrived at the second floor from either the staircase or the special 8th Street elevators and passed first through the Assembly Foyer, an immense room extending about two-thirds the ballroom's length designed to facilitate the flow of thousands of people in and out of the banquet hall. The grand hall was breathtakingly beautiful, according to *Hotel World*. In addition to being lighted by ten Parisian crystal chandeliers duplicated from patterns at Versailles, the vast room boasted two walls of windows that arched upward thirty-five feet to the ceiling. A balcony extended all the way around the room, providing additional space for unusually large crowds. Marble walls, ornamental plaster, exquisite draperies, a parquet floor, paintings in the style of Fragonard, and the expansive use of gilt refuted any assertion that artistic restraint

The Grand Stair Hall at the Stevens Hotel. From *Architectural Forum*, vol. 47 (August 1927), 109.

had been employed. *Hotel World,* completely overcome by the proportions and magnificence of the room, rhapsodized, "The splendor of the most noble courts in history is before you in a great sweep that has not its equal in the world. . . . You think at once of the masques and balls and promenades during the days of Louis XIV. And you vision gallants in powdered wigs and ladies in gay brocades moving with elegant decorum through the dances of their day. No court in Europe holds a lovelier room." Stevens, however, rather than visualizing bewigged gallants dancing quadrilles, had superimposed this courtly décor on a modern functional exhibition hall.[40]

The room was equipped with a concealed projection booth for showing motion pictures. Wall hangings suspended from hooks in the ceiling could enclose the center of the room, creating a perfect theater for watching films on a full-sized theatrical screen, also hidden from everyday view. Stevens designed the floors to be sturdy enough to accommodate truck, automobile, or heavy equipment exhibits. A service elevator large enough to accommodate a "threshing machine or a box car," a "printing press or an armored tank" enabled large machinery to be brought up to the second floor and into the room. This lift came up to the second floor right in the middle of a part of the banquet kitchen that extended into the service building. Stevens designed all the kitchen equipment in the area to be moveable to allow "the passage of a locomotive, if necessary." Additionally, the room could be divided into ninety-eight separate exhibit booths, each supplied with live steam, electricity, gas, and compressed air. Promotional material suggested that a circus could be held there, or an international tennis match. One can hardly reconcile the descriptions of royal grandeur with the utility and practicality of a truck show or the frivolity and mess of a circus, yet it is the perfect representation of the re- lationship between bourgeois luxury consumption and the capitalist commercial enterprise that produced it.[41]

The hotel guest rooms were similar to those of the Statler hotels. The means of obtaining variety among the furnishings, while at the same time limiting the chaos involved with decorating nearly three thousand rooms, was inventive and rational. Each of the twenty typical guest room floors had 136 rooms, each of the floors identical to every other. The *Hotel Monthly* described the Stevens as a "mul- tiple of twenty-five small hotels."[42] Dividing the typical floor in half produced two areas in mirror image to each other, with sixty-eight rooms each. The decorator, artist Norman Tolson, chose sixty-eight different ensembles, two per floor, that he replicated in vertical towers so that, for example, rooms 718, 718A, 818, 818A, 1218, and 1218A, and so on were all identical to one another while allowing for variety throughout the floor.[43] The guest experienced the impression of individuality that masked the logic of duplication in much the same way that the hotel asserted sin- gularity within the established hotel standard.

A floor clerk sat at a desk facing the elevator lobby and monitored the comings and goings of guests. He assumed all functions formerly provided by the business office, save registration and checkout. He managed duties such as holding room keys, distributing mail and telegrams, and arranging services. Fully wired by the standards of the early twentieth century, the floor clerk had the support of several communication systems including pneumatic tubes, annunciators, telephones, teletype, and the telautograph, plus the services of a full-time stenographer. These systems seem somewhat redundant, but their overlap suggests, as with other tech- nological systems, that their limits and potential had not yet succumbed to any obvious "one best way."

Typical guest floor plan, Stevens Hotel. Stevens's decorator, Norman Tolson, divided each floor into two mirror images of sixty-eight rooms each, which he decorated in individual ensembles, replicating the designs in vertical towers. *Engineering and Contracting*, vol. 66 (June 1927), 256. *Courtesy, Scranton Gillette Communications.*

Perhaps one of the most remarkable of the hotel's features was its fourteen-story service building. Stevens constructed it just behind the main building to house part of the mechanical plant and other services that did not necessarily have to be in the hotel proper, yet needed to be close by. The building fronted fifty-two feet on Wabash Avenue and ran 174 feet deep to the hotel, to which it was connected on several floors by corridors that ran over the intervening alleyway. The main hotel building housed the boilers, steam engines, and electrical plant in its four sub-basements. Unlike the trend in other large cities whereby new hotels purchased electricity from central power stations, the Stevens built its own electrical generating plant based on Stevens's previous experience at the Hotel LaSalle, which also had its own generating facility. Several journals ran lengthy justifications for Stevens's decision to build his own plant, suggesting some doubt in professional circles and perhaps some skepticism on Stevens's part regarding Samuel Insull's consolidation of Chicago's electrical service.[44] Walter Bird, the hotel's chief engineer, estimated the hotel would use ten million kilowatt hours per year, enough to generate about a half-million dollars of profit for a small public utility. The equipment's cost totaled about 1 percent of the hotel's total construction expenses, not enough to make a large difference in the capital investment. Stevens also based his decision on the fact that the hotel would be able to use the exhaust steam from the engines driving the electric plant. The exhaust would heat the building and hot water and run the refrigeration equipment.[45]

With the large power plant, boilers, and Corliss engines occupying the sub-basements, Stevens put most of the other bulky equipment, such as the laundry and auxiliary kitchens, in the service building. The basements held the exhaust and intake ducts and the refrigeration equipment. The loading platforms were on the first floor. Delivery trucks could drive right into the service building and onto the giant lift that could carry a loaded truck down to the basement or up to the second floor, where it could be unloaded. The service building's second and third floors housed the kitchen cold storage; bake shop; dish-washing service for the main and banquet kitchens; silver, glass, and dish storage; and ice cream and confectionery manufactories. The laundry machinery took over the fourth, fifth, sixth, and seventh floors, while the eighth through twelfth floors held the linen storage rooms and the many maintenance workshops—the electrical, paint, upholstery, carpet, drapery, furniture, sewing, carpentry, mattress, and print shops.

The uppermost floors, accessible by a bridge, held guest services. The thirteenth and fourteenth were called "club rooms," the lower floor for ladies and the top one for men. The ladies' floor had a lounge for "Kaffee Klatches," a gymnasium with exercise equipment, a snack room, a soda fountain, dressing rooms, and bathrooms. A "Fairyland" for children occupied part of the twelfth floor with a matron attendant, toys, and a soda fountain. The fourteenth floor held a lounge with a fireplace, a billiard room, a bowling alley with five lanes and mechanical pinsetters, a men's lavatory, and the ever-popular soda fountain. During nice weather, guests could play handball or tennis on the fenced-in roof.[46] Surely, the hotel provided an unlimited supply of balls.

The financial newspaper *Chicago Commerce* compared the service building to the "industrial portion of a city," distinct from the residential section.[47] The metaphor of the hotel as a city within a city proved to be unusually irresistible for those trying to describe the Stevens. The idea that a hotel could be described as a complete city unto itself had, as we have seen, originated in the earliest years of luxury hotel history. One hundred years later, the metaphor had merged with reality, with its industrial and residential neighborhoods, a complete electrical generating plant, and a population potentially approaching nine thousand. *Hotel World* went as far as to call it "The City of Stevens," recognizing that the building consolidated many aspects of an urban environment.[48]

The analogy's power lay for the most part in the sheer size and capacity of the hotel's technological systems. For example, the *Hotel Bulletin* described the telephone exchange as being appropriate for a "good-sized town." *Hotel World* noted that the exchange was large enough to serve a city of fifteen thousand people, for it was the world's largest private exchange, containing 340 city trunk lines.[49] *National Engineer* observed that the hotel's refrigeration equipment of five hundred tons capacity, "if used for making ice alone, [is] capable of producing 300 tons daily, more than many a city of 150,000 can boast of." These machines manufactured twenty-five tons of ice daily, froze 120 gallons of ice cream an hour, and cooled seventy refrigerators and storage rooms as well as the drinking water for the three thousand ice-water faucets in each of the private bathrooms. Additional compression machines maintained the summertime air conditioning for the large public rooms.[50] *Hotel World* compared the Stevens's electrical load to that required of a city of fourteen thousand homes, while the eighteen-inch-diameter water main connection to the Chicago city waterworks was the same as for a "fair-sized city."[51]

Stevens himself fostered the image of the hotel as being a private city. He named the long corridor that stretched the hotel's four hundred feet (between 7th and 8th Streets) Colchester Lane after his birthplace in western Illinois. *Hotel Monthly* bid its readers, "Let us take a stroll down 'Colchester Lane,' . . . sauntering on the way into the bypaths, meandering thru the Office Lobby, and leisurely viewing all the ground floor hospitalities."[52] The ground floor shops, designed as a main street service sector, were intended to supply the needs of the hotel's population, rather than to attract outside pedestrian traffic. Their resources included a pharmacy, clothing stores, a candy shop, a beauty parlor, a cigar stand, a newsstand, a theater ticket agency, a railway ticket office, telegraph offices, and a flower shop.[53] *Architectural Forum* referred to the fourth floor business office complex containing the manager's offices, the convention sales office, the bookkeeping department, the paymaster's office, and the employment offices, as the "City Hall of the Stevens City."[54] Nearby, a library with a fifteen-thousand-book capacity, staffed by Miss Gertrude Clark, a professional librarian from the University of Chicago, attended to the reading needs of the hotel's inhabitants.[55] The idea of the "city within the city" signified a comprehensive—if sanitized—version of the city that protected its inhabitants not only from the real city's unpredictability but also from the need to think about how to get things done.

Statler Lessons

Even while imprinting the hotel with his personal touch, Stevens clearly learned from Statler, whom he regarded with great professional respect. One of the most obvious Statler strategies he employed was propagating and indoctrinating a sentimental myth about himself and his family. While unable to foster a convincing "rags to riches" story in the Statler tradition, Stevens instead imagined an English family heritage to layer the hotel with old world aristocratic cachet. Stevens named his hotel's interior "street" Colchester Lane and gave the town's name to the main floor's Colchester Grill. Far from emphasizing the name's origins in small-town Illinois, Stevens made the leap back to Colchester, England, Britain's oldest recorded Roman town whose founding dated to 77 A.D. The contrivance of invoking Colchester's premodern character served to elicit heightened appreciation for Stevens's modernist creation. The contrast between an ancient British village and a rational, technological, self-contained urban simulation was stunning. It set up a contraposition between the "quaintness" of premodern life and Stevens's celebration of twentieth-century progress.

Stevens wrote to the mayor of Colchester, England, "for information concerning any of its special features." His Lordship the Mayor's widely reported response described an annual autumn oyster feast, which Stevens promptly integrated into the hotel's annual activities. Hotel promotion included a little blurb about Colchester being England's oldest town. Additional information got a little slippery, indicating that the Stevens family originated sometime in the distant past from Colchester, England, eventually settling somehow in its American namesake, Colchester, Illinois. The weekly hotel magazine *Stopping at the Stevens* described the hotel as being "fashioned from the childhood dreams of two little English boys who came to this country with their father to realize their aspirations."[56] It remains unclear as to who these two little English boys might have been. Among Stevens's papers was his acceptance into the Illinois State Society of the Sons of the American Revolution. He traced his ancestry directly to a paternal Stevens ancestor who served in the Revolution, more or less eliminating any of the Stevens hotel men as emigrating to the United States to fulfill their special hotel dreams.[57]

Stevens had discussed his hotel plans with Statler and incorporated many of Statler's standard conveniences into his own hotel. The bathroom's circulating ice water, full-length mirrors, headboard reading lamps, library, and the location of the sample room floors all drew on Statler's ideas of service and planning. Certainly, the way in which Stevens and countless other hotelmen described their enterprises in shorthand came from Statler. Stevens's promotional booklet declared, "Features of the Stevens, 3000 Outside Rooms, 3000 Baths." Stevens adopted Statler's design philosophy by planning his "perfect" hotel and expecting the architects and engineers to produce a structure that would execute the idea. Stevens also maintained a relationship with other hotelkeepers, often writing for their opinions on various matters, cooperatively exhibiting brochures for hotels in other cities, participating in the various hotel professional organizations, and keeping close tabs on the progress of other hotels.[58]

Stevens also made a special effort to accommodate women travelers at his hotel, something that Statler took extra pains with as well. For example, both Stevens and Statler had women clerks to "room" women traveling alone so that they did not feel uneasy being accompanied to their rooms by male clerks. Stevens designed a series of dressing rooms, each with a lavatory and dressing area, near the hotel's banquet entrance so that ballroom guests could dress at the hotel prior to an event without needing to engage an room overnight. He hoped that small touches such as a shelf of decorating books in the library and pretty prints on the bedroom walls would make women feel at home. Plus, he installed the "most modern and complete beauty parlor in Chicago." These extra efforts spoke volumes about the hotel's overall design. The commercial hotel was still a man's world, designed for men's work and comfort, an observation made clear by the need to "accommodate" women and reassure them by acknowledging their special needs for comfort and security.[59]

While Stevens's profile with the national public hardly approached that of Statler's, Stevens was often called to speak to various organizations, and his messages were ones that stressed professionalization and service, two of Statler's important themes. For example, when addressing a group from the American Home Economics Association, he spoke at length about the importance of specialized hotel management education and its place at the university, an idea Statler had promoted vigorously by helping to establish the training program at Cornell University. With the Foremen's Council of the Young Men's Christian Association, Stevens discussed the idea of service, emphasizing how personal service and relationships served as the backbone of the hotel business. He said, "Our duty is to keep our customers happy and we can only keep them happy by being happy ourselves and we can only serve them well by having an innate, natural subconscious attitude toward life which causes us to take delight in getting personal satisfaction out of rendering some service or doing some kindness to our fellowmen." While his style was a bit more sophisticated than Statler's, Stevens largely echoed Statler's edict to get rid of the grouches.[60]

Like Statler, Stevens also worshipped the gospel of efficiency. Stevens maintained excruciatingly detailed daily expense records for every department. He often found himself feeling victimized, despite his best efforts, by what J. O. Dahl, a 1920s hotel management expert, called the "human factor." According to Dahl, mechanization and design provided the means for eliminating inefficiency. Dahl stated unequivocally, "Carelessness on the part of employees will always be the cause of waste."[61] Stevens encountered this sort of frustration in several departments. For example, a few months after the hotel opened, he wrote a letter to Henry Hoppe, the hotel's laundry manager, who had previously run the laundry for Stevens at the LaSalle. Complaining about the laundry's operating cost, Stevens groused, "We have the most efficient laundry machinery in Chicago and we are the most inefficiently operated laundry. We cannot afford to pay a man a high salary who throws away $1400.00 a month in payroll." Stevens advised cutting payroll costs by lowering salaries.[62] Modern equipment alone did not ensure efficiency. It also required a human system.

Stevens followed a less aggressive strategy with the head chef, Ferdinand Karcher, who, as the reputation of many French chefs suggested, required coddling, even though he had worked for Stevens at the LaSalle for eighteen years. Stevens began a letter dripping obsequiously with compliments to the kitchen's food and service, adding that he "really hesitated to offer any suggestions or criticisms." Nonetheless, Stevens went in for the kill, observing, "The payrolls in your department total nearly $5000 a month more than we ever had any idea they would amount to." Stevens continued by suggesting a reduction in staff from ninety-four to sixty-six employees. A few months later, Stevens again wrote to Karcher, reminding him of the request. Even though Karcher had evidently reduced the payroll substantially, the number of employees had once again begun to climb. With patience wearing thin, Stevens wrote abruptly, "You have a total of 76 employees in your department. I want you to immediately reduce this to 70. Yours very truly."[63] Like Statler, Stevens knew every detail of the hotel's daily operation.

Despite the way that he appropriated and benefited from Statler's ideas and produced a building in line with nationally established hotel practices and expectations, Stevens also strove to differentiate his hotel from all others. The most obvious way that the Stevens differed from every other hotel in the world was through its size. *Hotel World* commended Stevens for having "placed the stamp of individuality" on the hotel. The journal indicated that some of Stevens's ideas, without specifying which, would "not meet with the entire approval of other hotel men." However, the journal predicted that these innovations would be ingenious for only a short time because hotelmen would copy them and make them standard. The only exception would prove to be the building's size. Here, the Stevens remained in a class by itself for decades to come. In Chicago, the Palmer House's 2,268 rooms and the Hotel Morrison's 2,500 came close to the Stevens's actual number of 2,818 bedrooms, but with the Depression of the 1930s bearing down, no developer could realistically continue to participate in the contest.[64] Indeed, the Stevens's size would prove to be a handicap that would spell disaster for the hotel and the entire Stevens family in just a few short years. Once again, subscription to the ideal of progressive development proved to have unanticipated consequences.

Success and Failure

The question that cries out for an answer is whether or not Stevens's predictions for the success of a three-thousand-room hotel were overly optimistic. The hotel opened in May 1927 and between that time and the end of 1930 the hotel's gross revenues were greater than their operating expenses, thus consistently showing an operating profit. However, after deducting capital expenses such as taxes, bond interest, and depreciation, the hotel never generated a profit. Stevens, when planning the hotel, anticipated that it would operate at a loss for at least three years, as the LaSalle had before establishing itself. However, the Stevens presented a far greater challenge to overcome because of the magnitude of its investment. The average daily room count hovered between a low in 1927 of 897 rented rooms per day and a high of 1,180 in 1929, far below the 70 percent occupancy rate Stevens projected as necessary for the hotel to be profitable.[65] But people were coming to the Stevens. In 1928, 1,238,501 people ate in one of the Stevens's many dining rooms. Nearly seventy thousand customers patronized the barber shop. Indeed, over two hundred thousand people registered at the hotel in the course of the year. All of these figures increased in 1929.[66]

Unfortunately, Stevens did not have three to five years to cultivate convention business or develop the dependable customer base essential to consumer capitalism's success. The Depression worsened in 1930 and continued to deepen with each passing year. By 1932, the daily average number of rooms rented was 543. In December 1932, the room count slumped to 317, barely exceeding a 10 percent occupancy rate, meaning, conversely, that twenty-five hundred rooms were sitting empty every day. The hotel showed a net loss of nearly two million dollars for the year.[67] Both the Stevens and the LaSalle went into receivership on June 4, 1932, with E. J. Stevens serving as the receiver. The Stevens was not the only hotel in the country with this problem. On June 7, 1932, the *Daily National Hotel Reporter* stated that "all hotels in the country show a greater average decline in revenue, both rooms and restaurants, than has been suffered by the Stevens or LaSalle in the past few months but the volume is so low as to make it impossible to operate at a profit during the present depression."[68] Still, the Stevens's size magnified the problem. The hotel's entire concept was based on ideas of mass consumption and standardization. Other luxury hotels facing receivership included Chicago's Drake, Blackstone, Sherman, and Morrison, and New York City's Plaza and the brand new Waldorf-Astoria.[69] In June 1932, the Hotel Management Institute published a list of suggestions for hotels in receivership, indicating a widespread problem. The answer as to whether or not the Stevens Hotel might have succeeded as the quintessence of design that integrated economy of scale with individual identity will never be known. While depending on an energetic market hospitable to mass production and mass consumption for success, even a savvy businessman like Stevens failed to consider that same market's historic potential for recession and depression and the obvious fact that it is impossible to scale back production far enough to save a three-thousand-room hotel.

With business going badly, Stevens was having trouble meeting the interest payments on the hotel's mortgage bonds. As part of a complicated story of interlocking interests, the family insurance company, Illinois Life Insurance, owned many millions of dollars' worth of these bonds, as did the Hotel LaSalle. E. J. Stevens, his brother Raymond, and his father J. W. Stevens owned 76 percent of Illinois Life stock and thus controlled the insurance company and the two hotel companies. Throughout 1931, Stevens borrowed money from banks and from Illinois Life to meet the quarterly interest payments, transferring stock between companies to serve as collateral. According to Stevens, the life insurance company had never before had to call upon its liquid reserves, nor had they expected it to. Because of the bank failures, policyholders had turned to their assets in the life insurance company, thereby creating an unprecedented situation in which the insurance company could not meet its obligations. It, too, eventually went into receivership.[70] Toward the end of 1931, the state's attorney general began investigating the Stevens Hotel and Illinois Life. He issued an indictment against the three Stevens men and arrested E. J. Stevens on January 28, 1932, for embezzlement and fraud in connection with loans improperly made by Illinois Life to rescue the insolvent hotel.[71]

The trial began in January 1933, producing ongoing front-page headlines in all the Chicago newspapers. The January 31, 1933, *Chicago Daily Tribune* gave equal exposure to Stevens's trial and Hitler's election as Germany's chancellor. On October 15, 1933, the jury found Stevens guilty of fraud in the amount of $1,200,000. Ernest Stevens remained alone to face the consequences. His father had suffered a debilitating stroke in March 1933, after which he was no longer able to manage his business affairs. That same month, tragically, his brother Raymond fatally shot himself in his Highland Park mansion, despondent over the problems with the insurance company, the trial, and his father's illness.[72] Ernest Stevens received a sentence of one to ten years in prison, though the Illinois Supreme Court later overturned the sentence in 1934 on appeal.

The public reaction was mixed. Stevens's papers are filled with letters of affectionate and sincere support from friends and business associates.[73] The *Chicago Herald* cast the men as victims of the depression yet contrasted them with the multitude who had "met their obligations squarely and honestly till they were wiped out," once again invoking contested ideologies of wealth production.[74] Still hostile after the reversal, the *Herald* listed the names of the Supreme Court judges, accusing them of siding with the Stevens family in finding a lack of criminal responsibility for having gutted the life insurance company to save the family's hotel enterprises. The paper stated, "The Supreme Court says that there is no fraud or embezzlement by Stevens because everything was done without concealment. Can it be the law that a man can do openly, without fear of criminality, that which would be a criminal offense if it were done secretly?"[75] The *Insurance Index* supported the Stevens family, arguing that the life insurance company and the two hotels could have been saved but for the arbitrary and unwarranted actions against Illinois Life that only enriched the "receivers, attorneys, and politicians." More pointedly, the *Insurance Index* objected to the jury that convicted Stevens, claiming that this "jury of peers," mostly artisans and tradesmen, could not comprehend the complicated transactions involved and that Stevens had been convicted of "once having been a wealthy man, of having built the world's largest hotel, and [of] the fact that Stevens was arrogant and vain."[76]

Because a federal judge had appointed Stevens receiver for the two hotels, he was able to continue managing them throughout the crisis.[77] Stevens hoped that the 1933 Century of Progress Exposition scheduled for that summer would pull the hotel out of its financial woes. Promotional literature showed that nearly one hundred conventions of two hundred people or more had booked their organizations into the Stevens Hotel during the months of the fair.[78] During that year, the hotel achieved a respectable 47.3 percent occupancy rate, falling to 35 percent in 1934. Prohibition also ended in 1933, and wine and liquor sales helped stimulate restaurant and bar revenues.[79] But things continued to go badly. In 1933, the *Herald* reported that E. J. Stevens paid $66.99 in personal property tax; his assets totaled three automobiles valued at $1,570, furniture at $900, jewelry, $300, personal effects, $100, and cash, $600. James W. Stevens died in June 1936. While the *Chicago Herald* estimated his fortune as once having been worth fifteen million dollars, a letter from E. J. Stevens to his nephews indicated that the estate's debts exceeded its assets by about two million dollars. The Hotel LaSalle Company was dissolved on June 12, 1936, and the Stevens Company followed suit on May 26, 1938. Thus had the Stevens empire fallen apart completely.[80]

On June 11, 1932, just after the Stevens Hotel and the Hotel LaSalle entered into receivership, E. J. Stevens received a letter of support from Burridge D. Butler, the publisher of a Chicago weekly, the *Prairie Farmer*. Stevens had written to cancel his advertising contract. Butler responded by saying he would continue the advertising in a show of solidarity in the face of hard times. "You are not going to jump out of any window," he wrote. "Because as I look at the picture, there are so many windows in the Stevens Hotel you couldn't decide which one to jump out of, and you would change your mind and buckle down to work and put the problem over."[81] Butler wrote with affection, but one cannot help but recall William Ralston's death after the collapse of his financial empire under similar public scrutiny. Other parallels exist. Both Ralston and Stevens shared a personal drive to build the world's largest hotel, both buildings were unprecedented in size and expense, both defined in part by their technological complexity and conceived in terms of civic boosterism, and both men suffered catastrophic failure brought on by a deep national economic depression. Even the reference to the numberless windows recalls the Palace's architecture with its endless bay windows meant to catch San Francisco breezes. It was as though history repeated itself—or that the lessons of history had gone unheeded.

Yet Stevens would have had little reason to think of himself in terms of William Ralston. Stevens was a man of the twentieth century. In his eyes, the modern luxury hotel had taken on a new character, grounded in the modern city's tall architecture and supported by mechanical systems of a complexity unimaginable fifty years before. The art of hotel management had been fine-tuned. Rather than turn to men trained informally through family dynasties, Stevens benefited from a professionalized industry supported by national trade associations and publications, trained in the ways of efficiency, and subjected to the most careful accounting and management systems yet devised. In addition to the ease and ubiquity of train travel, a new social order of organizations had grown up to feed the demand for and the demands of large downtown hotels. The luxury hotel industry profited from social and cultural changes rooted in the continuing expansion of the middle class, in mass culture and mass consumption. With its roots still clearly discernable in the patrician ideal of Boston's Tremont House, the Stevens Hotel represents the end of a century's worth of development that took basic principles and chased after technological progress, all the while combining ideas of technological luxury with technological possibilities to produce a building that would prove impossible to surpass for decades to come.

....

CONCLUSION

....

For Sinclair Lewis, George Babbitt represented America's everyman, and the downtown hotel—the material representation of Babbitt's world—was an integral part of his everyday life. It would be almost impossible to describe George's comings and goings without the luxury hotel as part of his *mise en scène*. While he regarded his hometown Zenith's thirty-five-story Second National Tower as "the temple-spire of the religion of business," the city's hotels supported the pattern of his daily life as a middle-class consumer and businessman. Zenith's various hotels offered shampoos and manicures, and their formal meeting spaces played host to Booster club luncheons, real estate conventions, and lectures by such organizations as the League of the Higher Illumination. Throughout Lewis's satirical portrayal of middle-class American life in the early twentieth century, the luxury hotel served as a powerful backdrop to Babbitt's attempts to secure status and social standing. The hotel concretized the creature comforts of the good life. Babbitt was a man who pictured Heaven as "an excellent hotel with a private garden."

In luxury hotels, Babbitt rubbed elbows with the upper crust and was—at least briefly—accepted as one of their own. In the ballroom of Monarch's Allen Hotel, Babbitt addressed the State Association of Real Estate Boards and found himself transformed from "a minor delegate to a personage almost as well known as that diplomat of business, Cecil Rountree." He made the acquaintance of a British peer, Sir Gerald Doak, in the lobby of the Regency, and he squired Mrs. Tanis Judique to lunch in the dining room of the Thornleigh. When part of the hotel world, either flowing through the golden doors of the ballroom or dining in the splendor of the Regency's Versailles Room, Babbitt, the "Standardized American Citizen," experienced life in unexpectedly intensified ways. While the hotel itself followed a standardized form clearly recognizable to Babbitt (and sharply skewered by Lewis), its luxury lifted Babbitt out of the everyday routine of his suburban Floral Heights life and reinforced his pride in being the type of man successful enough to afford it.[1]

The modern luxury hotel needed men like George Babbitt—men who regarded the hotel as a site for the pleasure of business and the business of pleasure. It is easy to envision Babbitt as a conventioneer at the Stevens Hotel, taking in Chicago's sights or, in an earlier era, as one of the "small-change rattling persons" loitering on the front steps of the Tremont or Astor House or perhaps, before settling down to family life, as one of the new white collar midlevel bureaucrats inhabiting the sky-parlors of the Fifth Avenue Hotel. Babbitt represents the tensions posed between luxury and elitism and the egalitarian expectations inherent in a "palace of the public." With rooms numbering in the hundreds and even thousands, large luxury hotels depended on men like Babbitt to fill their small but remunerative spaces. Yet, at the same time, the lavish designs, decorations, and advanced technology evoked an exclusive, even rarefied, atmosphere in order to draw the support of the more powerful political, social, and financial classes. While the buildings catered to the needs of a transient mobile population, they sought, too, their city's embrace, which was indispensable.

The American urban luxury hotel evolved under the pressures of the historical processes of the long nineteenth century that gave rise to the city as a place of business and consumption. Not merely a product of these forces, luxury hotels played a significant role in shaping urban development as well as social and cultural practices. Luxury, technology, and urban ambition are three themes that link the buildings—their materiality—to the social and cultural activities that bring them to life. Ideas associated with the gendering of space, the buildings' increasing size (which mandated a more inclusive clientele), and the class-based judgments levied on those who chose to live in and patronize luxury hotels were closely bound to the design of the buildings. Moreover, hotel experiences introduced both Americans and foreign visitors to urban culture. Because of the hotel's pivotal position in American cities, it served as a touchstone for analyzing those experiences, a vehicle through which urban dwellers and visitors alike came to understand American cities and urban life.

This history has focused on American luxury hotels, the class of hotels with the fewest numbers and the greatest visibility. Luxury can be defined in opposition to necessity, yet in the context of rising and more expansive consumerism and the growth of capitalism, the luxury market became, in the eyes of many producers and consumers alike, an engine for growth. As an institution at the center of the luxury debate, the American luxury hotel played an ironic role. Its outrageous extravagance, cost, and size conveyed exclusivity and differentiated itself from the smaller and more ubiquitous middle- or lower-class lodgings, yet the economic imperatives to find occupants for its many hundreds if not thousands of rooms required that it attract a large middle-class traveling public as well as permanent residents.

Equally interesting, as new wealth expanded the circle of financial elites, the hotel, with its elaborate social settings, enabled the inclusion of people at society events who existed beyond the boundaries of tight-knit social circles such as New York's infamous Four Hundred without compromising the sensibilities of private society.[2] An invitation to a charity event in a hotel ballroom, for example, carried far less weight and fewer concessions than an invitation to a private home. Newly wealthy families could use the palace hotel as a staging area to try to establish themselves in good society, but middle-class patrons did the same to improve on their middling place in the hierarchy of industrial society by appropriating the luxury hotel as a palatial extension of their own home. Thus, the hotel understandably came under attack from critics who sought to protect traditional values against the siren call of excessive consumerism.

The names on the top of the buildings clarified the differences between established elites, parvenus, and middle-class patrons. The Astor, Vanderbilt, Stevens, or even Statler name ablaze in lights above these structures reinforced the hierarchy of ownership that extended in a continuum from those who owned the land and the building to those who rented the one-, two-, or three-dollar-a-night rooms, whose names were etched far less impermeably in the guest book registers. Yet, the inclusion of the newly wealthy and middle-class served an important purpose. Enabling them to marry their children off in an Astor ballroom helped enlist their support for the industrial capitalist system that produced such extraordinary wealth for a limited few. Thus, an institution that had once been a cornerstone of republican life—the setting for political discourse in a burgeoning democracy—had evolved into a site of both cultural production and consumption dominated by the encoding and instantiation of social and cultural hierarchies. As early as 1844, the *New York Weekly Mirror* foretold the hotel's decline as a setting for the "political democratic freshet" and referred to the public table or *table d'hôte* as the "*tangible republic*—the only thing palpable and agreeable that we have to show, in common life, as republican."[3] The paper lamented the incipient privatization of sociability that it predicted would add to the hotel's exclusivity and, consequently, its decline as a characteristically republican institution.

Modern technology added new dimensions to developers' and consumers' understanding and experience of luxury. Structural advances, such as the steel frame, provided a means for hotels to attain their great size. Other key innovations, including those that heated and cooled interiors, powered lights and machines, and eased communication and transportation enhanced both hotel-keepers' ability to "keep a house" and guests' ability to experience comfort and convenience. Most technological applications, such as those in the laundry, kitchen, and housekeeping departments, remained virtually invisible to guests. They transformed the production side of the hotel, but indeed, their insulation from public view contributed to the guests' heightened sense of luxury. Guests interacted with other technologies more directly, particularly in their own rooms, such as lighting, thermostats, plumbing, radio, telephone, and other communication devices to once again maximize comfort and convenience.

Superimposed above it all was a complex system of management and organization that created an incomparable and characteristically American experience. Statler and others, including several luxury brands, applied Fordist principles of mass production to their organizations. However, unlike the production of hard goods, every unsold guest room remained unsold forever. Over the course of one hundred years—and certainly continuing to the present day—technological luxury became the moving force that underpinned competition among hotelmen and cities. Decorators could embellish on furnishings to a degree, but technologies have a history of improvement, innovation, and obsolescence that, without constant attention, puts the buildings at risk of "falling into the yellow leaf."

In 1829, Isaiah Rogers's concept for the Tremont House created a national design standard for luxury hotels. One hundred years later, E. J. Stevens pressed this standard to its limits. The Stevens Hotel conformed to the original structural configuration of basement service areas, lower floors of shops, parlors, and monumental public spaces topped by upper floors of private rooms. The Tremont's innovative characteristics—the consolidation of urban services, the appropriation of European models of decoration, the location on an important boulevard, and the extraordinary (and possibly irrational) extravagance in cost, size, food, and technological luxuries—continued through the long century with each successive wave of development.

While this system became standardized and nationalized, entrepreneurs such as Stevens, the Continental's promoters, and Ralston defined their buildings as local productions, celebrating their cities through the patronage of regional industry and artisanship, creating hotels that broadcast their cities' accomplishments to the traveling public and the world.[4] Even as hotel promoters engaged in a national competition, they also sought international recognition. It was never enough to be America's largest hotel; it always had to be the world's largest hotel, whether it was the Tremont, the Continental, the Palace, or the Stevens. Even in the early 1800s, the merchants and industrialists who sponsored the building of large luxury hotels did so within a cosmopolitan context, being familiar with and engaged in international trade and travel. As advances in communication and transportation mitigated the difficulties of continental and inter-oceanic travel and communication, hotel developers and backers kept abreast of luxury hotel projects around the world, either through their own travels or by sending representatives specifically to explore (and be ready to surpass) the competition.

Dramatic modern development also had consequences. The proprietors of the Tremont House embarked on their project fully aware that they were venturing into uncharted waters. The magnitude of their investment, while unprecedented for the time, nonetheless fell within acceptable margins of both personal and communal risk. By the time Philadelphia's businessmen had proposed the Continental, detractors were dubbing these large hotels "monsters" and this particular project "The Hotel Folly." The building's vast scale amplified the risk and threatened other hotelkeepers who, while supporting development, yearned for something more controlled. Still, civic leaders viewed the project as pivotal to Philadelphia's viability within an increasingly international economy. By midcentury the hotel standard solidified, and that ideal rejected parameters: all successive symbolic luxury hotels subscribed to the doctrine of excessive size, cost, and integration of modern technological luxuries. This became a standard that fostered overproduction, preyed on giant egos—like those of Ralston, Statler, and Stevens—and disseminated itself to small cities and towns in the form of the too large one-hundred-room hotel. Moreover, it ignored the increasingly more frequent economic cycles that subjected businesses and financial markets to periods of boom and bust and ultimately ground hotel development to a halt for decades after the Great Depression.

This study has played itself out in the major cities of the United States. However, smaller American cities also had hotel dreams that produced their own versions of the hotel ideal and were meant to address the same kinds of needs articulated by civic leaders in the larger cities. Three such hotels in different periods can serve as examples of this kind of provincial diffusion of the hotel idea. They are the Burnet House (Cincinnati, Ohio, 1850), the St. James Hotel (Red Wing, Minnesota, 1875), and the Hotel Cleveland (Cleveland, Ohio, 1918). Each of these cities was experiencing economic growth that city leaders believed generated a need for a hotel commensurate with its rising stature in the nation. Their stories not only demonstrate the diffusion of a common idea in smaller venues across the country but also highlight how that idea was implemented to represent a city in a particular time and place.

Emerging as one of several major regional wholesaling centers by the 1850s, Cincinnati, Ohio, experienced such population growth that in 1850, when the Burnet House opened, it was the nation's sixth-largest city, boasting a population over 116,000.[5] "Porkopolis," the nickname the city proudly adopted as the nation's largest pork-packing center, sat in a fortuitous location on the Ohio River and was an important stop for the vibrant riverboat trade as well. Over four thousand steamboats docked in Cincinnati in 1848, and that number doubled by 1852. In this context of dramatic growth, a "company of city gentlemen" commissioned Isaiah Rogers to design the Burnet House.[6] By 1850, Rogers's designs had become increasingly more elaborate, and the Burnet House evoked the nation's Capitol Building more than it resembled either the Tremont or the Astor House. A broad expansive ceremonial staircase led up to a portico similar to, yet far grander than, the Tremont's. Corinthian columns supported the portico, in contrast to the Tremont's and Astor's austere Doric columns. Symmetrical wings and recessed bays extended to the east and west, with the center section crowned by a forty-two-foot-diameter dome that sat one hundred feet above ground level and whose observatory commanded a magnificent view of the Ohio River and the blue hills of Kentucky.

The design evoked the nation's most eminent democratic symbol and represented the full expression of nineteenth-century classical architecture, yet in only three short years, the new New York hotels would adopt the *palazzo* style, thus rendering the Burnet House's design out-of-date. Moreover, the hotel's $250,000 cost paled in comparison to the 1853 St. Nicholas's $1,000,000 investment, which ushered in a new era of competition. Despite this, the Burnet House, named for Judge Jacob Burnet, a prominent early Cincinnatian and United States senator upon whose original land the hotel stood, was reputed to be "the largest, best constructed, and best managed hotel in the United States, and probably the world."[7] The hotel earned national and international press coverage, with *Gleason's Pictorial Drawing-Room Companion* giving it front-page coverage, and no less an authority than the *London Illustrated News* confirming its title as "the best hotel in the world."[8]

The Burnet House's first proprietor was Abraham B. Coleman, who had previously managed the Astor House. Francis Pedretti, the hotel's decorator, had decorated New York's Astor House as well as the residence of A. T. Stewart, the dry

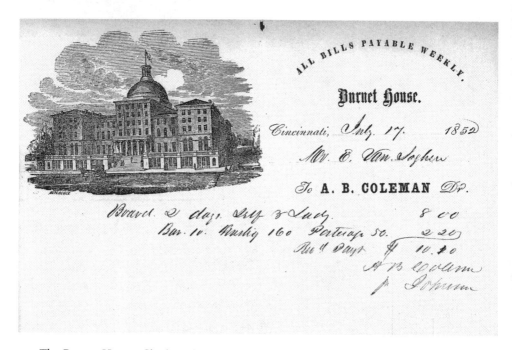

The Burnet House, Cincinnati, 1859–60. While continuing to draw on the Greek Revival style of architecture, the Burnet House was significantly more elaborate than the Tremont House, demonstrating both the increasing size of luxury hotels and Rogers's elaboration on the form. *Warshaw Collection of Business Americana—Hotels, Archives Center, National Museum of American History, Smithsonian Institution.*

goods impresario. Thus, the hotel became part of an established system that drew on architects, builders, and managers from the larger cities and imparted those cultural norms to the burgeoning midwestern city. Immediately anointed with a national reputation, the hotel enjoyed the usual roster of famous guests, including Henry Clay, Daniel Webster, Louis Kossuth, Abraham Lincoln, and, later, a host of Civil War generals. General William Tecumseh Sherman was said to have planned his march to the sea in one of the Burnet House parlors. In 1860, the eighteen-year-old Lord Renfrew stayed at the Burnet House during his tour of Canada and the United States. Lord Renfrew was Queen Victoria's eldest son, the Prince of Wales, who later would be crowned Edward VII. He was the first member of the Royal Family to visit the United States since the colonies had declared their independence from Britain.

The prince's order for rooms, food, carriages, and security for himself and his retinue of dukes, earls, lords, captains, and attendants is evidence that the Burnet

House was not only the finest hotel in Cincinnati but also one with the capacity and expertise to accommodate visiting royalty.[9] Yet in its report on the prince's Cincinnati visit, the *New York Times* observed, "At no other place has the Prince of Wales been so thoroughly initiated into the mysteries of Westernisms as here." Regarding a ball held in the prince's honor at the city's opera house, the correspondent noted, "There were, perhaps, one hundred well-dressed people in the room, while the rest were awkward, ill-mannered, ungainly, and dressed in the most marked wretched taste."[10] While national and international visitors recognized the Burnet House as a place that answered their requirements for quality accommodations, it would clearly take more than a fancy hotel to win the hearts and minds of New Yorkers. The Burnet House was an example of how a growing city could replicate the hotel ideal by hiring eastern experts familiar with national standards and integrate the hotel into an existing network of transportation and accommodations. At the same time, the city's distinctive western personality imbued the hotel with regional flavor—for better or worse.

The St. James Hotel in Red Wing, Minnesota, presents a quite different model than the Cincinnati hotel. In 1875, the year the St. James (and San Francisco's Palace) opened, most of the nation and the world was reeling from the onset of the difficult depression triggered by the 1873 collapse of financier Jay Cooke's financial firm. Yet Red Wing and surrounding Goodhue County were enjoying substantial economic growth due to bumper wheat crops and Red Wing's location as a busy port on the upper Mississippi River. In early 1874, Red Wing newspapers boasted that the city was "the largest market in the world for wheat brought in and delivered direct from the farmers' wagons."[11] Teams of wagons from regional farms carried unprecedented bounties of wheat to Red Wing. Both the farmers and the grain merchants needed hotel rooms and, in what is by now a familiar sequence of events, civic leaders mobilized to build a hotel that would represent the city's wealth and accommodate—in appropriate style—those successful businessmen now finding their way to Red Wing. The city's existing hotels were small and remarkable more for their ability to stable teams of horses than their ability to accommodate the farmers who drove them.

In early 1874, eleven Red Wing businessmen formed a hotel corporation that issued one thousand shares of stock at fifty dollars each, thus limiting the investment to fifty thousand dollars, one-hundredth the cost of the Palace Hotel. The incorporators represented the demographics of local Red Wing businessmen: successful middle-aged men who had immigrated to the region from either the American Northeast, the Germanic states, or Scandinavia. These were men who embodied heartland values such as "sturdy industry, strict economy, pluck, tenacity, and contentedness," which one writer used to describe Red Wing's population.[12] The corporation hired a local architect, Edwin P. Bassford, to design the hotel, a four-story 130-by-60-foot Italianate brick structure with a rear addition to house the kitchen, the laundry rooms, and the servants' quarters. The hotel could fit into the Palace's Grand Court (144 by 84 feet, eight stories high) with plenty of room to spare. Sited on a prime lot at the corner of Main and Bush Streets, just a block from the train depot, riverboat landings, and grain silos, the hotel incorporated many of the features that graced larger, nationally ranked hotels, such as street-level shops; a separate office; baggage and cloak rooms; reading, sample, and billiard rooms; a barber shop; bathing rooms; and the requisite fine dining room and separate ladies' parlor. Additionally, the hotel distinguished itself through its advanced technology and materials, such as through the use of French plate glass windows for the street-level shops, the installation of a Knowles steam pump as part of its fire protection system, advanced kitchen and laundry machinery, and a plumbing system requiring ten thousand feet of pipe. Everything was to be of "the very best material and latest improved apparatus" and "fitted up in the best manner."[13] Consistent with hotel development elsewhere, promoters acclaimed the way local contributions, products, and skills enabled them to construct a building that met and, they hoped, surpassed national standards. Eventually, the cost of the building ballooned to sixty thousand dollars, a situation that caused both consternation and self-approbation because of what the extra costs represented.

To commemorate the hotel's opening, a festive opening party held on Thanksgiving night engaged the entire city—no exclusive invitation lists here—with the building filled from top to bottom with an assemblage described as "the most intelligent and wealthy which could be gathered from our midst."[14] The St. James's amenities, its brilliant lights, the women's dresses, and the elaborate feast all spoke of the cultural success that economic prosperity fostered and that signified Red Wing's coming-of-age. Aglow with modern lights and filled with smart, wealthy, beautiful, well-fed people, the hotel gave Red Wingers reason to believe that they had successfully created an American symbol of urban sophistication that would stand them well in the larger national marketplace.

The example of Red Wing's St. James is important because it demonstrates the way the "hotel idea," as executed in America's large urban centers, became part of a national trope regarding capitalist development and perceptions of cosmopolitanism that diffused into the nation's heartland. Civic leaders in small nineteenth-

ST. JAMES HOTEL.

CORNER MAIN AND BUSH STS.,

RED WING, - MINN.

E. J. BLOOD, PROPRIETOR.

STRICTLY FIRST-CLASS ACCOMMODATIONS.

This advertisement for the St. James Hotel shows a lively street scene with many carriages filling the streets of Red Wing. At the rear, a Mississippi steamboat plies the waters as a train passes by, all signs of a prosperous city connected to the wider world. *Red Wing Pictorial Annual for 1884* (Red Wing: Red Wing Printing Co., 1884).

century American cities tried to appropriate the physical and functional elements of the metropolitan "monster" hotel, but the result was, not surprisingly, distinctly provincial. The rhetoric that drove development for accommodations representative of the community's success—that of community maturation, commerce, success, comfort, luxury, and service—echoed that of the biggest cities. Red Wing investors wanted a building that represented their little city's hard-won moment in the national consciousness, yet at the same time they understood that it was enough for their hotel to be the finest in a very circumscribed region, not the whole world. While very small in comparison to hotels such as the Palace, the Fifth Avenue, or the Continental, the St. James nonetheless was grand in Red Wing's context and grander still when compared to cities that had no first-class hotels. The St. James replicated a scaled-down version of luxury and service that larger cosmopolitan hotels offered and, as in other cities, celebrated the local industry and skill that produced a hotel consonant with a national model.

The third example is that of the Hotel Cleveland, which opened on December 16, 1918, just over a month after the armistice of World War I. The story of the Hotel Cleveland is interesting because it presents yet a third example of hotel diffusion. Instead of the hotel being an end in itself for its promoters, it marked the midpoint of a grand development scheme promoted by Oris P. and Mantis J. Van Sweringen, Ohio brothers who, beginning in 1905, developed the landholdings of the North Union Community of the United Society of Believers (more commonly known as the Shakers) into what is now Shaker Heights, a middle- and upper-class garden city suburb seven miles from Cleveland's downtown.[15] Their plans mushroomed from an already ambitious plan for suburban housing to one for a monumental modernist complex meant to emulate New York's Terminal City, an enormous mixed-use development that connected the spectacular Beaux-Arts Grand Central Terminal building to office buildings and several hotels, including the Belmont (1903), the Ritz-Carlton (1910), the Biltmore (1913), the Vanderbilt (1913), and the Commodore (1919). These New York hotels were highly complex technological buildings like the Waldorf-Astoria, Statlers, and Stevens, but their distinguishing feature was that the buildings connected to the railway terminal via systems of enclosed passageways. Thus, if they chose, patrons never had to confront the increasingly frenetic city outside their doors. As *Harper's Weekly* noted when describing New York City's newly built Belmont, "One can go from this hotel to San Francisco without leaving the cover of a roof."[16]

The Van Sweringen brothers encountered difficulties during the planning phase of their interurban light rail system meant to connect Shaker's suburbanites to the downtown. The rail system was key to the success of the suburb's development. The brothers were determined to build their own railway line after they were unable to secure the cooperation of the Cleveland Railway Company, a monopoly that controlled the city's streetcar lines and was the source of violent legal disputes with the city government during the first decade of the twentieth century. In 1916, in order to control the land for the railway, the Van Sweringens purchased New York Central's Nickel Plate Railroad, which held a right-of-way that every railroad entering Cleveland from the east or west had to cross. By 1929, they owned a $3 billion 30,000-mile railroad empire.[17] The Van Sweringens also purchased four acres of property on Cleveland's Public Square, the ten-acre park that defines Cleveland's center city, in anticipation of a need for a terminal.[18]

These negotiations and plans took place in the context of tremendous city growth as well as dramatic changes in city government under Cleveland's Progressive mayor Tom L. Johnson. Invigorated by the architectural success of Chicago's 1893 World's Columbian Exhibition, Chicago architect Daniel Burnham led a city-appointed planning commission that resulted in Cleveland's 1903 Group Plan, the nation's first major city plan outside of Washington, D.C., for the placement and design of government buildings.[19] The plan called for a grouping of monumental Beaux-arts public buildings including the city hall, courthouse, and library, sited around an open mall overlooking Lake Erie. The plan also envisioned a lakeside union station that would bring all railroads entering Cleveland into and through a central depot.

While the execution of the rest of the plan proceeded apace, disagreements among the railroads and then the disruption caused by World War I delayed the station's construction. In 1917, having purchased the land and the railroads, the Van Sweringens promoted an alternate plan for a union station facing Public Square. The station would accommodate both steam trains and electric interurbans in one multilevel structure. Supporters of the mall plan feared that the Van Sweringen's plan would drain stature from the civic center and also pointed to the great personal benefit the Van Sweringens stood to gain.[20] City Council stepped in to require the passage of a citywide referendum in order for the Van Sweringens to proceed. By the time the vote was held January 6, 1919, the plans included not just the station, but also a one-thousand-room hotel, three office buildings, a central office tower, an underground shopping arcade, and a United States Post Office.

The referendum was held just three weeks after the Hotel Cleveland opened in mid-December 1918. Of the multiple elements in the Van Sweringen's downtown plan, the hotel was the first to be built. Evidence suggests that the date of the referendum that would give the Van Sweringens their personal monument was carefully orchestrated to take advantage of the hotel's successful opening. The resulting building was representative of other hotels of the decade, and particularly resembled Statler's Hotel Pennsylvania, the new 2,200-room hotel connected to New York's Pennsylvania Station making national headlines as the largest hotel in the world.

The Hotel Pennsylvania opened only a few weeks after the Hotel Cleveland. The two hotels shared many of the same nationally renowned contractors, among them the W. G. Cornell Company (plumbing, heating, ventilation), the John Van Range Company (kitchens), Syracuse China (dishes), Berkey and Gay (furniture), the Oscar C. Rixson Company (door operating mechanisms), and the Servidor Company, which made hotel room doors with pass-through compartments to accommodate deliveries.[21] True to form, many Cleveland firms also filled important contracts, giving the city bragging rights for both local manufacturing as well as the ability to construct an important building that rose to national standards.

However, the Hotel Cleveland took on much more significance as it filled the role as linchpin in the Van Sweringens' campaign to win the support of Cleveland voters. On Sunday, December 22, 1918, the *Cleveland Plain Dealer* published a twenty-page special tabloid section that introduced the hotel to the city, describing just about every detail imaginable and portraying the building as the "first link of the chain that Cleveland is forging to create facilities to meet the demands of a future filled with the most optimistic spirit of prosperity." On the cover and then again in a full-page spread on page three, the editorial copy made clear that the brothers, with obvious support from the paper's publisher, intended the hotel to serve as material proof that they would make good on their promise to build a union station. The cover drawing displayed the official line drawing of the hotel, framed in a cornucopia of fruits meant to symbolize Cleveland's prosperity. The hotel itself spoke to the city in tortured first-person prose below the drawing. None too subtly, the hotel declared, "Conceived by the brain of a dreamer, who makes dreams come true, I was created by the combined genius and dollars of men who do things and do them well." Describing itself as a pioneer "in a new field of city development," the Hotel Cleveland went on, "If you wish, I shall soon have distinguished company and near me may stand that much talked of and long desired Union Station, a monument to Cleveland's patience, a product of civic pride."

As readers turned the page, they confronted another drawing of the hotel, highlighted by none other than Uncle Sam, whose image drew closely on the familiar James Montgomery Flagg "I Want You" figure used in World War I marketing campaigns. Uncle Sam pointed to a rendering of the proposed Union Depot, showing

Taking a full page in a special advertising section for the Hotel Cleveland's opening, the Van Sweringens and the *Cleveland Plain Dealer* aimed to convince voters that the hotel served as evidence for the success of the proposed union terminal's relocation. The hotel's tripartite wings anchor the right side of the terminal development in the lower image. *Cleveland Plain Dealer,* December 22, 1918, 3.

how the new hotel would anchor one side of the terminal/office/retail development. Under a headline, "A Union Depot Certainty," the paper proclaimed, "This is the big new Hotel Cleveland. A real improvement, an accomplishment that is a pledge of Union Depot *performance*. It is a monumental proof that your *Union Station will be built*." Uncle Sam underscored the message by advising Clevelanders, "A vote 'YES' on the ordinance providing for it, is a vote to bring this long needed and splendid improvement to Cleveland." What was a patriotic Clevelander to do?

The referendum passed, and the Van Sweringen's plans for a "bigger and better Cleveland" came to pass in the form of the Terminal Tower complex that, with its fifty-two-story office tower, was North America's tallest building outside of New York City until 1964, when it was eclipsed by Boston's Prudential Building.[22]

These examples—that of the Burnet House, the St. James, and the Hotel Cleveland—demonstrate several ways that the "hotel idea" was carried out in American cities across the nation. Another way to grasp the universal drive that cities had for new hotels is to look at the architectural journals of the early twentieth century. All of the major journals featured new hotels, often in dedicated hotel issues. For example, the *Architectural Review*'s April 1913 issue of nearly two hundred pages showcased new hotels in larger cities such as New York, Chicago, and San Francisco and smaller ones such as New Haven, Connecticut; Toledo, Ohio; Springfield, Illinois; Chattanooga, Tennessee; and El Paso, Texas. Civic leaders believed that a first-class hotel was an essential element of a prosperous city. It served as its public face and functioned as a cultural measure of a city's success. The degree to which a city's best hotel replicated the universally understood idea of "first-class" indicated how prosperous, sophisticated, and culturally adept a city was.

The *Hotel Monthly*'s July 1920 issue contained a short thought piece about "Hotel Life of the Future." The journal prophesied that "within a few years hotels will no longer be considered as mere eating and sleeping houses, but as the back bone of our civilization." The vision conceptualized hotels as "complete civilized communities" providing stores, shops, lectures, concerts, and even employment bureaus for transients.[23] Certainly, by the end of the 1920s, the "City of Stevens" was incorporating all that and more. Even taking into consideration the statement's rather parochial source, the idea that a complete community for transients with an average stay of 2.5 days would be regarded as the pinnacle of civilization is curious, to say the least. Social critics lamented the anonymity and impersonality of twentieth-century city life and examined the feelings of dislocation occasioned by technology's annihilation of natural concepts of space and time. The modern luxury hotel created an ironic version of civilization where people came and went, processed by a system of management and accounting, steeped in manufactured luxury, and yet completely disconnected from the "city" in which they temporarily lived. In addition, the viability of the hotel as "civilization" depended on external market forces rather than the hard work of the "citizenry" whose contribution to the artificial community's success lay in their collective role as consumer. There were no traditional "roots" planted here in the city of the hotel, only a superficially banal and temporary sense of belonging based on an ability to pay and dress the part.

One could and did leave without a moment's notice. In an hour's time the room could be "turned over" for the next transient urbanite. The hotel was more than the city's most arresting symbol; it was the apotheosis of the modern city itself.

Despite the Stevens Hotel's remarkable size and the hoopla attending its opening, the story failed to make front-page news. Unlike the era when a new hotel's opening captured prominent column space for days, the Stevens warranted only a small article buried in the *Daily Tribune*.[24] The days when the large luxury hotel stood as a representation of a city's entrepreneurial success and cultural sophistication had clearly given way to the more direct and narrow representation of the corporate skyscraper. The builders of these new corporate structures celebrated a notion of progress that focused on corporatization, centralization, and commercial development.[25] Stevens had trusted in the luxury hotel's historic image, that of a city's most important symbol, without recognizing that its day had passed. Even though the Stevens Hotel stood as the culmination of large-scale hotel development, it no longer stood as a symbol for the city it served, as its predecessors had.

This history ends neatly in 1929, with the juggernaut of the Depression bearing down on the newest of the nation's great hostelries exactly one hundred years after the opening of the nation's first modern hotel. The venerable Waldorf-Astoria closed in May 1929, its site acquired, tellingly, to build New York City's greatest icon, the Empire State Building. With financing in place before the onset of the Depression, the new Waldorf-Astoria opened in 1931 occupying an entire city block between Park and Lexington and East 49th and East 50th Streets. Its twin Art Deco towers laid claim to the hotel's ranking as the world's tallest. The new Waldorf-Astoria entered into bankruptcy immediately. The Depression and then World War II brought an end to major developments, thus ending this particular age of American hotel development. The growth of suburban living, the consequent decline of America's inner cities, and the development of the interstate highway system contributed to the ascendancy of the roadside motel and the corresponding decay of many of America's best hotels. Only the focus on urban renewal combined with significant tax incentives allowed many of these buildings to be restored in the 1980s.

The cult of technological luxury is one that continues to hold sway over contemporary hotel development. Current trends cater to the business traveler with connectivity for a range of personal technology devices, wireless internet, dual-line telephones with voicemail, automated check-in, and state-of-the-art work stations. Other technological luxuries include flat-panel televisions, sophisticated exercise rooms and health clubs, rapid-fill bathtubs, window blinds operated by a bedside switch, and video on demand. A recent CNN video highlighted Future-Hotel's "Hotel Room of the Future," an experimental center that tests guest room technologies. Reminiscent of *Jetsons*-styled décor with smooth, white, ultramodernistic moveable walls, the room includes a bed that rocks restless guests to sleep and floor sensors that measure heart rates and other biological markers to adjust lighting, heating, and even room color to maximize physical and emotional comfort. Infrared lighting in the bathroom promises to rejuvenate guests' skin so that they leave the hotel not only rested but also looking younger.[26] In an ironic twist recalling Statler's innovative addition of wall light switches, this hotel room eliminates them altogether, incorporating instead a multipurpose panel that responds to voice commands.

The modern luxury hotel remains a symbol of modernity with contested meanings. Between September 2008 and July 2009, terrorists attacked luxury hotels in Islamabad, Pakistan (Islamabad Marriott Hotel, September 20, 2008); Mumbai, India (Taj Mahal Palace and Tower Hotel and Oberoi Hotel, November 27, 2008); and Jakarta, Indonesia (J. W. Marriott, July 17, 2009). Each of the targeted hotels was located in the center of financial or government quarters of their respective cities and each had a reputation not simply for luxurious accommodations but also as a city landmark where wealthy citizens and foreign travelers congregated and stayed. Mumbai's Taj Mahal Palace Hotel's website summarizes its 1903 development as being the product of industrialist Jamsetji N. Tata's dream to build a "grand hotel, one that would enhance [its] reputation amongst the great cities of the world." In 1903, the hotel boasted French bathtubs, German elevators, and other western technology but showcased Indian architecture in the hopes of attracting visitors to India's richest city.[27] The Mumbai hotels were Indian, the others American, but all shared the now-familiar characteristics that defined the luxury hotel and created meaning for their cities. In an age of senseless terror, they became logical targets for groups bent on violence and enraged by the economic and cultural systems that these buildings symbolize. Luxury hotels represent the best and worst of industrial society. As symbols of great wealth, cosmopolitanism, and capitalist enterprise, they nonetheless flaunt the inequities characteristic of industrial and consumer capitalism. With the front of the house and the back of the house serving as an evocative metaphor for its history, the urban luxury hotel frames our own conflicted ambitions, to be created equal and yet be enormously and conspicuously special, all at the same time.

....

ACKNOWLEDGMENTS

....

As a project, *Hotel Dreams* has a very long history of its own, and, as a result, I am indebted to a great many people who have helped and encouraged me along the way. First and foremost, my mentor and adviser Carroll Pursell expressed enthusiasm for the project from its very first days, and his steadfast confidence in its significance and in me has helped me persevere whenever my own commitment to the task faltered. In equal measure, Michael Grossberg has been a continual source of inspiration. His willingness to read and comment on the manuscript at crucial times has also propelled me toward the proverbial finish line. Arthur Molella is a dear friend and colleague who helped me puzzle out more questions than is reasonable to ask of one person and always did so with great insight and humor. Cyrus Taylor, dean of the College of Arts and Sciences at Case Western Reserve University supported my work by recognizing the importance of scholarship despite the pressures of a busy administrative life and generously gave me time to write. I am also deeply grateful for the dean's subvention that supported the acquisition of images for this volume. Carroll, Mike, Art, and Cyrus have been much more than colleagues; they are wonderful friends as well. Words of thanks like these seem at best wholly inadequate.

My favorite part of writing this book was the research. This was helped significantly by two important fellowships. I was fortunate to be a Kate B. and Hall J. Peterson Fellow of the American Antiquarian Society. The librarians, archivists, and Fellows provided an unparalleled idyllic space for research and intellectual inquiry. In particular, I thank John Hench and Joanne Chaison for their continued friendship over the years and Ellen S. Dunlap, who long ago saw promise in the project. The Smithsonian Institution's National Museum of American History also awarded me a fellowship; I am grateful for the continued support of my colleagues there as well. In addition, I wish to thank the librarians, archivists, and staff at the many libraries at which I worked. These include the American Antiquarian Society, the Archives Center and Library at the National Museum of American History, the Avery Library at Columbia University, the Bancroft Library at the University of California at Berkeley, the Boston Athenaeum, the Bostonian Society, the California Historical Society, the Chicago History Museum, the

California State Library in Sacramento, the Carl A. Kroch Library at Cornell University, the Cincinnati Historical Society, the Cleveland Public Library, the Enoch Pratt Free Library, the Goodhue County Historical Society, the Hagley Museum and Library, the Historical Society of Pennsylvania, the Kelvin Smith Library at Case Western Reserve University, the Library Company of Philadelphia, the Library of Congress, the Maryland Historical Society, the Massachusetts Historical Society, the Minnesota Historical Society, the New-York Historical Society, the New York Public Library, the Ohio Historical Society, the San Francisco Public Library, the Western Reserve Historical Society, and the Wolfsonian Museum. The librarians at the Kelvin Smith Library—Joanne Eustis, Karen Oye, Mark Eddy, Bill Claspy, and Mike Yeager—have been heroic in helping me retrieve the books, articles, and documents I needed.

I am fortunate to be part of a community of scholars in the Society for the History of Technology. From my first days in graduate school, SHOT has provided an intellectual home that has supported and recognized my work. In particular, the members of Women in Technological History (WITH) have sustained me. I am grateful for the friendship of Debbie Douglas, Nina Lerman, Ruth Schwartz Cowan, Arwen Mohun, Gail Cooper, and all the many "Withies" who are my friends and colleagues and who have been an endless source of encouragement, intellectual vitality, and entertainment. Most importantly, I count my Case Western Reserve University cohort among my dearest friends. We have been through the best and worst of times together, and my life is immeasurably better because of Aaron Alcorn, Bernie Jim, Geoff Zylstra, and especially Susan Schmidt Horning. A special thank you goes to Aaron on two counts. Not only did he read the entire manuscript—some parts more than once—when he had far more important things to do, he also saved my life when I fell and broke my ankle while on a busman's holiday in the lobby of the Willard Hotel in Washington, D.C.

Many others have helped me. John Orlock is always quick to bolster any flagging spirits and provide useful comments. Sam Savin and Mark Turner also offered continuing support and encouragement. My department chair, Jonathan Sadowsky, implemented accountability measures at an important turning point. I thank Angela Woollacott, Jan Reiff, Catherine Kelly, Jim Edmonson, Gary Stonum, Phil Scranton, Bob Post, and James L. W. West III for their help during the early stages of the work. Professor West was particularly kind to me as a new scholar, and his generosity toward me—even though I was a complete stranger to him—has continued to guide my actions toward others. Robert J. Brugger, my editor at Johns Hopkins University Press, was stalwart while waiting for this manuscript, and I thank him for maintaining his interest over the years. I also wish to thank Josh Tong and Andre Barnett at the Press for their patience answering my many questions. I learned much from Leslie Sternlieb, who shares a passion for the subject and is an excellent editor and fun traveling companion. Jared Bendis was especially generous in helping me with the images as part of his work in the Samuel B. and Marian K. Freedman Center at the Kelvin Smith Library, Case Western Reserve University. I also acknowledge with gratitude my SAGES writing co-instructors at CWRU—Chalet Seidel, Danny Anderson, and Mary Assad—from whom I learned much about writing and who I hope will be pleased with my studious elimination of the passive voice and adherence to other academic writing conventions. In the dean's office, I thank Jennifer Dyke for helping me to keep track of the details of my life. Moreover, for more than ten years, Cynthia Stilwell, Denise Donahey, Ken Klika, Steve Haynesworth, Jill Korbin, Peter Whiting, and Marcia Camino have been my "work family," and I am grateful for their friendship and for the warm camaraderie of the office.

Finally, I wish to acknowledge my wonderful friends and family, who for all these years have had the good grace or good sense not to ask how the book was coming along and who, more importantly, bring such richness to my life. Laurel Hart, Hyla Winston, Margie Falk, Margie Reimer, Jane Berger, Jody Sather, Sue Horning, and Jill Korbin are my cherished friends, and I cannot imagine my life without them. As older brothers, Neil, Norm, and Jerry are good to have around so that I can forever be the baby. Above all, I have been blessed with the most spectacular children and grandchildren imaginable. Martha, Dave, Carrie, Brian, Angela, Andrew, Max, Alice, and Micah are the real measure of a life well lived. My love for them knows no bounds. In conclusion, I dedicate this book to the memory of my parents, Milton and Minnie Winger, whose own stories inspired me to become a historian and who I hope would have read and possibly enjoyed *Hotel Dreams* and maybe even mentioned it to their friends. More than anything, I wish they were here to do so.

NOTES

Introduction

1. Daniel J. Boorstin, *The Americans: The National Experience* (New York: Vintage Books, 1965).

2. I had first used the phrase "private bathroom," but recent trends in hotel design either remove the walls that separate the bathroom from the sleeping room or use glass walls, rendering the bathroom anything but private.

3. Works that employ the idea of the long nineteenth century include Michael Grossberg and Christopher Tomlins, eds., *Cambridge History of Law in America, vol. 2, The Long Nineteenth Century (1789–1920)* (New York: Cambridge University Press, 2008); Barbara Young Welke, *Law and the Borders of Belonging in the Long Nineteenth Century United States* (New York: Cambridge University Press, 2010), 213–15; and Edmund Burke III, "Modernity's Histories: Rethinking the Long Nineteenth Century, 1750–1950," University of California World History Workshop, Davis, California, May 19–21, 2000. Available at http://escholarship.org/uc/item/2k62f464, accessed March 30, 2010.

4. Jürgen Habermas, *The Structural Transformation of the Public Sphere: An Inquiry into a Category of Bourgeois Society,* trans. Thomas Burger with the assistance of Frederick Lawrence (Cambridge: MIT Press, 1989), 36.

5. Jefferson Williamson, *The American Hotel: An Anecdotal History* (New York: Alfred A. Knopf, 1930), 3.

6. Nicholas Pevsner, *A History of Building Types* (Princeton: Princeton University Press, 1976), 9.

7. Mary Ryan, "'A Laudable Pride in the Whole of Us': City Halls and Civic Materialism," *American Historical Review* 105 (October 2000): 1131–70; and Dell Upton, "Another City: The Urban Cultural Landscape in the Early Republic," in *Everyday Life in the Early Republic,* ed. Catherine E. Hutchins (Winterthur, DE: Henry Francis du Pont Winterthur Museum, 1994), 64–65.

8. Igor Kopytoff, "The Cultural Biography of Things: Commoditization as Process," in *The Social Life of Things: Commodities in Cultural Perspective,* ed. Arjun Appadurai (Cambridge: Cambridge University Press, 1986), 64–68.

9. Sandoval-Strausz constructs a typology of seven variants of hotels that includes luxury, commercial, middle-class, marginal, resort, railroad, and settlement hotels. This classification confuses class with function both among and within the various types, and he further adds residential hotels as a separate category. A. K. Sandoval-Strausz, *Hotel: An American History* (New Haven: Yale University Press, 2007), 81.

10. See, for example, Sandoval-Strausz, *Hotel*; Paul Groth, *Living Downtown: The History of Residential Hotels in the United States* (Berkeley: University of California Press, 1994); Susan R. Braden, *The Architecture of Leisure: The Florida Resort Hotels*

of Henry Flagler and Henry Plant (Gainesville: University Press of Florida, 2002); and Catherine Cocks, *Doing the Town: The Rise of Urban Tourism in the United States, 1850–1915* (Berkeley: University of California Press, 2001).

11. Henry James, *The American Scene* (1907; repr. New York: Horizon Press, 1967), 102–7.

CHAPTER ONE: The Emergence of the American First-Class Hotel, 1820s

Subtitle: *National Intelligencer,* June 18, 1827.

1. Philip Hone, *The Diary of Philip Hone, 1828–1851,* Allan Nevins, ed. (New York: Dodd, Mead and Company, 1927), 2:849.

2. Paul Johnson, *The Birth of the Modern: World Society 1815–1830* (New York: Harper Collins, 1991), xvii.

3. Jefferson Williamson, *The American Hotel: An Anecdotal History* (New York: Alfred A. Knopf, 1930), 3.

4. Kym S. Rice, *Early American Taverns: For the Entertainment of Friends and Strangers* (Chicago: Regnery Gateway, 1983), xv.

5. Quotations from Frances Trollope, *Domestic Manners of the Americans* (London, 1839), 1:276 and James Stuart, *Three Years in North America* (New York: J. & J. Harper, 1833), 1:316, as quoted in Paton Yoder, *Taverns and Travelers: Inns of the Early Midwest* (Bloomington: Indiana University Press, 1969), 146–48.

6. Isaac Weld Jr., *Travels through the States of North America, and the Provinces of Upper and Lower Canada, during the years 1795, 1796, and 1797,* 2 vols. (London: Printed for John Stockdale, Piccadilly, 1800), 1:27–31. See also "The Robert S. and Margaret A. Ames collection of Illustrated Books," available at www.brown.edu/Facilities/University_Library/exhibits/ames/early.html, accessed January 30, 2006.

7. Rice, *Early American Taverns,* 35–38.

8. Richard Bushman, *The Refinement of America: Persons, Houses, Cities* (New York: Vintage Books, 1992), 160–64.

9. Michael Dennis, *Court and Garden: From the French Hotel to the City of Modern Architecture* (Cambridge: MIT Press, 1986), 4.

10. Doris Elizabeth King, "Hotels of the Old South, 1793–1860" (Ph.D. diss., Duke University, 1952), 1–10; and Bushman, *Refinement of America,* 357.

11. Meryle Evans, "Knickerbocker Hotels and Restaurants, 1800–1850," *New York Historical Quarterly* 36 (October 1952): 382; and Williamson, *American Hotel,* 10.

12. Hone, *Diary,* 1:91; and David Longworth, *The American Almanack, New-York Register and City Directory* (New York, 1796), engraving facing title page.

13. *The New Trade Directory, for New-York* (New York, 1800), 173.

14. For example, the January 3, 1794, *New York Daily Gazette* advertises a ball to be held at Corrie's Hotel.

15. Evans, *Knickerbocker Hotels,* 382–83.

16. Doris Elizabeth King, "Early Hotel Entrepreneurs and Promoters, 1793–1860," *Explorations in Entrepreneurial History* 8 (February 1956): 151.

17. Weld, *Travels,* 1:84.

18. A. K. Sandoval-Strausz, *Hotel: An American History* (New Haven: Yale University Press, 2007), 25–30.

19. *The American Traveller,* July 8, 1825. 1.

20. *National Intelligencer,* June 18, 1827.

21. "The New Baltimore Hotel," *Baltimore Gazette and Daily Advertiser,* October 25, 1826.

22. James Stuart, *Three Years in North America* (New York: J. & J. Harper, 1833), 1:249; Tyrone Power, *Impressions of America; During the Years 1833, 1834, and 1835* (Philadelphia: Carey, Lea, and Blanchard, 1836), 124; Charles Dickens, *American Notes: A Journey* (1842; repr. New York: Fromm International Publishing Corporation, 1985), 137; and Godrey T. Vigne, Esq. of Lincoln's Inn, Barrister at Law, *Six Months in America* (Philadelphia: Thomas T. Ash, 1833), 104.

23. King, "Entrepreneurs," 154.

24. Daniel J. Boorstin, *The Americans: The National Experience* (New York: Vintage Books, 1965), 143.

25. "The New Baltimore Hotel," *Baltimore Gazette and Daily Advertiser,* October 25, 1826.

26. King, "Hotels of the Old South," 45.

27. "City Hotel" advertisement, *Baltimore Gazette and Daily Advertiser,* November 4, 1826.

28. "The Indian Queen Hotel," *Baltimore Gazette and Daily Advertiser,* November 9, 1826; and "Indian Queen Hotel and Baltimore House" advertisement, *Baltimore Gazette and Daily Advertiser,* November 9, 1826.

29. Bushman, *Refinement of America,* 25.

30. Stuart, *Three Years in North America,* 1:249.

31. Thomas Hamilton, *Men and Manners in America* (Philadelphia: Carey, Lea, and Blanchard, 1833), 2:5.

32. *National Intelligencer,* June 18, 1827.

33. "The New Baltimore Hotel."

34. *The American Traveller,* July 29, 1825.

35. Marshall Berman, *All That Is Solid Melts into Air,* 2nd ed. (1982; repr. New York: Penguin Books, 1988), 17, 131–71; and David Harvey, *The Condition of Postmodernity* (Oxford: Basil Blackwell, 1989), 9–10.

36. "The Mechanical Age," *American Traveller,* November 20, 1829. See also Thomas Carlyle, "Signs of the Times," in *The Collected Works of Thomas Carlyle* (London: London Chapman and Hall, 1858), 3:99–118.

37. Harvey, *Condition of Postmodernity,* 12.

38. Berman, *All That Is Solid Melts into Air,* 15.

39. Lydia Maria Child, *The American Frugal Housewife, Dedicated to Those Who Are Not Ashamed of Economy,* 12th ed. (1832; repr. Worthington, OH: Worthington Historical Society, 1965), 99.

40. Drew McCoy, *The Elusive Republic: Political Economy in Jeffersonian America* (New York: W. W. Norton, 1980), 95–100; Merritt Roe Smith, "Technology, Industrialization, and the Idea of Progress in America," in *Responsible Science: The Impact of Technology on Society,* ed. Kevin B. Byrne (San Francisco: Harper and Row, 1987), 4; and Leo Marx, "The Idea of 'Technology' and Postmodern Pessimism," in *Does Technology Drive History?* ed. Merritt Roe Smith and Leo Marx (Cambridge: MIT Press, 1994), 240–41.

41. Rush Welter, "The Idea of Progress in America," *Journal of the History of Ideas* 16 (1955): 405.

42. John Sekora, *Luxury: The Concept in Western Thought, Eden to Smollett*

(Baltimore: Johns Hopkins University Press, 1977), 1, 68, 84, 103; Christopher J. Berry, *The Idea of Luxury: A Conceptual and Historical Investigation* (Cambridge: Cambridge University Press, 1994), xii; Charles Sellers, *The Market Revolution: Jacksonian America, 1815–1846* (New York: Oxford University Press, 1991); and Richard E. Ellis, "The Market Revolution and the Transformation of American Politics, 1801–1837," in *The Market Revolution in America*, ed. Melvyn Stokes and Stephen Conway (Charlottesville: University Press of Virginia, 1996), 161.

43. Adam Smith, *An Inquiry Into the Nature and Causes of the Wealth of Nations* (1776; repr. New York: P. F. Collier & Son, 1909), 80; and McCoy, *Elusive Republic,* 20–32.

44. Steven Watts, *The Republic Reborn: War and the Making of Liberal America, 1790–1820* (Baltimore: Johns Hopkins University Press, 1987), 250–51.

45. Lewis Mumford, *Art and Technics* (New York: Twayne Publishers, 1992), 111.

46. Stuart, *Three Years in North America,* 1:249.

47. Frances Trollope, *Domestic Manners of the Americans* (1839; repr. London: Century Publishing, 1984), 20.

48. As a point of comparison, in the early twentieth century professional hotelmen recommended a one-hundred-room hotel or smaller as an appropriately sized building for a small city.

49. Kermit L. Hall, *The Magic Mirror: Law in American History* (New York: Oxford University Press, 1989), 97–98; and Oscar Handlin and Mary Flug Handlin, *Commonwealth: A Study of the Role of Government in the American Economy: Massachusetts, 1774–1861,* 2nd ed. (1947; Cambridge: Belknap Press, 1969), 146–47.

50. Yoder, *Taverns and Travelers,* 49.

51. Alfred D. Chandler Jr., *The Visible Hand: The Managerial Revolution in American Business* (Cambridge: Belknap Press, 1977), 3; and Lisa Pfueller Davidson, "Consumption and Efficiency in the 'City within a City': Commercial Hotel Architecture and the Emergence of Modern American Culture, 1890–1930" (Ph.D. diss., George Washington University, 2003).

52. "Hotels," Warshaw Collection of Business Americana, Archives Center, National Museum of American History, Smithsonian Institution, Washington, D.C. This collection contains thousands of pieces of printed hotel ephemera, such as stationery, envelopes, postcards, and bills. Hereafter referred to as Warshaw Collection.

53. Harvey, *Condition of Postmodernity,* 17, 106.

54. Denis E. Cosgrove, *Social Formation and Symbolic Landscape* (Totowa, NJ: Barnes and Noble Books, 1985); and Karl B. Raitz and John Paul Jones III, "The City Hotel as Landscape Artifact and Community Symbol," *Journal of Cultural Geography* 9 (Fall/Winter 1988): 17–36.

CHAPTER TWO: The Tremont House, Boston, 1829

Subtitle: Samuel Eliot, "Remarks to Bostonian Society," newspaper clipping (no attribution), April 10, 1895. Vertical file, "Tremont Hotel," Bostonian Society.

1. Edward Frank Allen, "The Mysterious Columns in Institute Park," *Worcester Telegram,* May 25, 1957; "Another Look at Worcester," *Worcester Gazette,* August 29, 1951; Richard D. Carreno, "Monumental Mysteries," *Worcester Phoenix,* August 19–25, 1994. I wish to thank the library and volunteer staff of the American Antiquarian Society

who initially pointed out the columns to me during my tenure as a Kate B. and Hall J. Peterson Fellow in early 1994.

2. William Havard Eliot, *A Description of Tremont House* (Boston: Gray and Bowen, 1830), 5; and James Stuart and Nicholas Revett, *The Antiquities of Athens and Other Monuments of Greece,* 3rd ed. (London: Henry G. Bohn, 1858), 18–19. Both Sandoval-Strausz and Levinson Wilk mischaracterize my argument by saying I claim the Tremont is the first hotel. The Tremont is the first American hotel that incorporated all the characteristics of a modern hotel, including professional design as a purpose-built building, advanced technology, commercial architecture, corporate financing, spatial differentiation, and modern management; it unquestionably became a model for all modern luxury hotels that followed. A. K. Sandoval-Strausz, *Hotel: An American History* (New Haven: Yale University Press, 2007), 54; and Daniel Levinson Wilk, "Cliff Dwellers: Modern Service in New York City, 1800–1945" (Ph.D. diss., Duke University, 2005), 52.

3. For an extensive bibliography of published travel diaries that serves as an excellent entry into the literature, see Doris Elizabeth King, "Hotels of the Old South, 1793–1860" (Ph.D. diss., Duke University, 1952).

4. Charles Augustus Murray, *Travels in North America during the Years 1834, 1835, & 1836. Including a Summer Residence with the Pawneed Tribe of Indians, in the Remote Prairies of the Missouri, and a Visit to Cuba and the Azore Islands* (London: Richard Bentley, 1839), 1:102.

5. Thomas Hamilton, *Men and Manners in America* (Philadelphia: Carey, Lea, and Blanchard, 1833), 1:90.

6. Peter Dobkin Hall, *The Organization of American Culture, 1700–1900: Private Institutions, Elites, and the Origins of American Nationality* (New York: New York University Press, 1984), 19, 100.

7. Oscar Handlin and Mary Flug Handlin, *Commonwealth: A Study of the Role of Government in the American Economy: Massachusetts, 1774–1861,* 2nd ed. (1947; Cambridge: Belknap Press, 1969), 61.

8. Hall, *Organization of American Culture,* 89–90.

9. For citations of hotels with the name "Tremont House," see Directory, *Hotel World,* 1878; and *J. Stratton McKay & Co.'s Commercial Directory,* 1876, Warshaw Collection of Business Americana-Hotels, Archives Center, Smithsonian Institution.

10. Justin Winsor, ed., *The Memorial History of Boston including Suffolk County, Massachusetts* (Boston: James R. Osgood and Co., 1880), 4:8. Eliot's father, Samuel, son of Samuel Eliot, the bookseller, founded one of the three largest overseas shipping houses. William H. Eliot's brother, Samuel, was treasurer of Harvard and then mayor of Boston; Samuel's son, Charles W. Eliot, became president of Harvard. William H. Eliot's son, Dr. Samuel Eliot, was president of Trinity College in Hartford and, later, Boston's superintendent of schools.

11. *The American Traveller,* July 5, 1825.

12. Nathaniel Dearborn, *Dearborn's Reminiscences of Boston and Guide Through the City and Environs* (Boston: Nathaniel Dearborn, 1851), 142; and Handlin and Handlin, *Commonwealth,* 119. See also Sandoval-Strausz, *Hotel,* 27–29.

13. Cleveland Amory, *The Proper Bostonians* (New York: E. P. Dutton, 1947), 32.

14. Doris Elizabeth King, "Early Hotel Entrepreneurs and Promoters, 1793–1860," *Explorations in Entrepreneurial History* 8 (February 1956): 151–52.

15. *Shaw's History of Boston* (Boston, 1817), as quoted in James Henry Stark, *Antique Views of Boston* (Boston: Burdette and Co., 1967), 71.

16. Eliot, *Description of Tremont House,* 1.

17. Dearborn, *Dearborn's Reminiscences,* 143.

18. Eliot, *Description of Tremont House,* 1.

19. Handlin and Handlin, *Commonwealth,* 171–73; and Stanley I. Kutler, *Privilege and Creative Destruction, The Charles River Bridge Case* (1971; repr. Baltimore: Johns Hopkins University Press, 1990), 27.

20. *Boston Newsletter and City Record,* April 29, 1826.

21. James Perkins was T. H. Perkins's nephew, the son of Thomas H. Perkins's brother, James.

22. For a contemporary assessment of the "invisible power and presence" of the Boston elite, see *"Our First Men": A Calendar of Wealth, Fashion and Gentility; Containing a List of Those Persons Taxed in the City of Boston Credibly Reported to be Worth One Hundred Thousand Dollars with Biographical Notices of the Principal Persons* (Boston: Published by all the Booksellers, 1846), 3–8.

23. *The Boston Directory Containing Names of the Inhabitants, their Occupations, Places of Business and Dwelling Houses, with lists of the Streets, Lanes and Wharves, the City Offices and Banks, and other Useful Information* (Boston: Hunt and Simpson, 1828).

24. Frederic Cople Jaher, *The Urban Establishment: Upper Strata in Boston, New York, Charleston, Chicago, and Los Angeles* (Urbana: University of Illinois Press, 1982), 21.

25. *Boston Daily Advertiser,* June 15, 1828.

26. Newspaper clipping (no attribution) April 10, 1895. Vertical file, "Tremont Hotel," Bostonian Society.

27. *Boston Daily Advertiser,* June 15, 1828.

28. See *Boston Daily Advertiser;* the *Evening Bulletin;* the *Boston Courier;* and the *Boston Commercial Gazette.*

29. *Boston Daily Advertiser,* June 15, 1828; *Boston Commercial Gazette,* July 10, 1828; Handlin and Handlin, *Commonwealth,* 79–81; and Kutler, *Privilege and Creative Destruction,* 155–60.

30. *Boston Daily Advertiser,* July 1, 1828; and *Boston Commercial Gazette,* July 10, 1828.

31. Kutler, *Privilege and Creative Destruction,* 18–34. The Charles River Bridge controversy concerned the construction of a new, free access bridge over the Charles River that would, in effect, put its predecessor, a hugely successful toll bridge, out of business. Because the state legislature chartered both bridges, the debate centered on whether or not the state had the power to infringe upon the inviolability of property rights on behalf of the general welfare of the greater community.

32. *Boston Commercial Gazette,* July 7, 1828, emphasis in original.

33. Abel Bowen, *Bowen's Picture of Boston or the Citizen's and Stranger's Guide to the Metropolis of Massachusetts, and its Environs to Which is Prefixed the Annals of Boston,* 2nd ed. (Boston: Lilly Wait and Co./Lorenzo H. Bowen, 1833), 213.

34. David E. Nye, *American Technological Sublime* (Cambridge: MIT Press, 1994), 48–49.

35. Moses King, *King's Handbook of Boston,* 4th ed. (Cambridge: Moses King, 1881); *Boston Courier,* July 7, 1828; *Boston Commercial Gazette,* July 7, 1828; and Mary Ryan,

"The American Parade: Representations of the Nineteenth-Century Social Order," in *The New Cultural History,* ed. Lynn Hunt (Berkeley: University of California Press, 1989), 132.

36. Joseph T. Buckingham, *Annals of the Massachusetts Charitable Mechanic Association* (Boston: Press of Crocker and Brewster, 1853), 222.

37. The inscription read: "The cornerstone of Tremont House was laid by Samuel Turrell Armstrong, president of the Massachusetts Charitable Mechanic Association on the fourth day of July A.D. 1828, and the 52nd anniversary of American independence, Levi Lincoln being governor of Massachusetts and Josiah Quincy mayor of Boston. A desire to promote the welfare and to contribute to the embellishment of their native city led the proprietors, Thomas Handasyd Perkins, James Perkins, Andrew Eliot Belknap, William Havard Eliot and Samuel Atkins Eliot, to undertake this work. In its accomplishment they were aided by the liberality of the persons whose names are enrolled on the parchment in the glass case beneath. Isaiah Rogers, Architect." *Boston Commercial Gazette,* July 7, 1828. The silver plate and glass case were recovered "after nearly three months of incessant searching" during the Tremont's demolition. However, the parchment bearing the subscriber's names completely deteriorated over the years and nothing remains of it. The plate was added to the collection of the Bostonian Society. *Boston Globe,* June 15, 1895.

38. *Boston Commercial Gazette,* July 7, 1828.

39. Jaher, *Urban Establishment,* 61.

40. *Massachusetts Charitable Mechanic Association, Its Act of Incorporation—Historical and Statistical Memoranda—Constitution—List of Officers and Members, etc., etc.* (Boston: Warren Richardson, 1879), 70 passim.

41. Robert Rich, "'A Wilderness of Whigs': The Wealthy Men of Boston," *Journal of Social History* 4 (Spring 1971): 274.

42. Boston was originally sited on three large hills, Copp's, Fort, and Beacon. Beacon Hill had three "eminences, distinctly in plain sight from the low grounds of Charlestown, so elevated as to give this hill the appearance of a mountain." These three hills are called Mt. Vernon, Beacon, and Pemberton Hills and are clearly marked on the 1828 Boston map. The city's original name, Trimountain, was changed to Boston on September 17, 1630, by an order of the Court of Assistants, Governor Winthrop presiding. Winsor, *Memorial History of Boston,* 1:116; Caleb Hopkins Snow, *A Geography of Boston, County of Suffolk and Adjacent Town with Historical Notes* (Boston: Carter and Hendee, 1830), 10–11; and *The Boston Directory,* 1828.

43. Bowen, *Picture of Boston,* 16.

44. Samuel Adams Drake, *Old Landmarks and Historic Personages of Boston* (Boston: James P. Osgood, 1873), 291; and "Boston's Idea," newspaper clipping (no attribution), April 10, 1895. Vertical file, "Tremont Hotel," Bostonian Society.

45. This information was gained by looking up the address of each of the investors in the city directory and plotting the location of their residences on an 1828 map. In addition, nearly all their businesses were within three-quarters of a mile of the hotel, although these were concentrated somewhat to the east along State, Kilby, and Broad Streets and on the wharves.

46. Beacon Street travels west from Tremont Street (or Common Street) at the hotel's corner. School Street travels east from the same intersection. Thus, King's Chapel is at the corner of School and Tremont.

47. Walter H. Kilham, *Boston after Bullfinch: An Account of its Architecture, 1800–1900* (Cambridge: Harvard University Press, 1946), 29; and Denys Peter Myers, "Isaiah Rogers," *MacMillan Encyclopedia of Architects*, Adolf Paczed, ed. (New York: Free Press, 1982), 3:599–600.

48. December 13, 1940, Isaiah Rogers Diaries, 1838–1856, 1861, 1867, transcribed from the original notebooks by Denys Peter Myers, Avery Architectural and Fine Arts Library, Columbia University.

49. Oliver W. Larkin, *Art and Life in America* (New York: Rinehart and Company, 1945), 158.

50. Talbot Hamlin, *Greek Revival Architecture in America* (New York: Dover Publications, 1964), 28, 90–118.

51. Kilham, *Boston after Bullfinch*, 21.

52. *Views in Philadelphia and Its Environs, from Original Drawings Taken in 1827–30* (Philadelphia: C. G. Childs, 1830), quoted in Bates Lowry, *Building a National Image: Architectural Drawings for the American Democracy, 1789–1912* (Washington, DC: National Building Museum, 1985), 38.

53. Karl B. Raitz and John Paul Jones III, "The City Hotel as Landscape, Artifact and Community Symbol," *Journal of Cultural Geography* 9 (Fall/Winter 1988): 25.

54. Eliot, *Description of Tremont House*, 8–9 passim.

55. Kilham, *Boston after Bullfinch*, 26–27; Hamlin, *Greek Revival Architecture*, 102–3; Winsor, *Memorial History of Boston*, 4:117–18; Snow, *A Geography of Boston*, 158; Bowen, *Picture of Boston*, 286–88; and Freeman Hunt, *Lives of American Merchants* (New York: Office of *Hunt's Merchants' Magazine*, 1856), 1:75.

56. Kilham, *Boston after Bullfinch*, 29.

57. "Monolithic Stones," 1887, Vertical File, "Tremont House," Bostonian Society.

58. Bowen, *Picture of Boston*, 286–88. Historian Frederick C. Gamst disputes the Granite Railway's "first" designation, contending that the 1805 Beacon Hill Railway deserves that honor. Gamst claims that the Granite Railroad was actually twelfth of at least twenty-five railroads operating by 1830 in the United States and Canada. However, Bostonians perceived their Granite Railway to be the first and judged it to be extremely innovative. Gamst, "The Context and Significance of America's First Railroad on Boston's Beacon Hill," *Technology and Culture* 33 (January 1992): 66.

59. Jefferson Williamson, *The American Hotel: An Anecdotal History* (New York: Alfred A. Knopf, 1930), 22.

60. Bowen, *Picture of Boston*, 218.

61. Samuel Eliot Morison, *One Boy's Boston, 1887–1901* (Cambridge: Riverside Press, 1962), 4.

62. Eliot, *Description of Tremont House*, 13–16.

63. Susan Strasser, *Never Done: A History of American Housework* (New York: Pantheon Books, 1982), 96–97.

64. Richard Bushman, *The Refinement of America: Persons, Houses, Cities* (New York: Vintage Books, 1992), 127.

65. Sally Pierce and Catherina Slautterback, *Boston Lithography, 1825–1880: The Boston Athenaeum Collection* (Boston: Boston Athenaeum, 1991), 57; see also lithographs in the Boston Athenaeum Print Collection.

66. Advertisement for Seth Fuller, *Evening Transcript*, March 24, 1831.

67. Eliot, *Description of Tremont House*, 35–36.

68. Henry Lee, "Boston's Greatest Hotel," *Old Time New England* 55 (Spring 1965): 99; and *Gleason's Pictorial*, vol. 2, no. 16, April 17, 1852, 52.

69. *Boston Daily Advertiser*, October 19, 1829; *Boston Commercial Gazette*, October 19, 22, 1829; *Evening Bulletin*, October 22, 1829; and *American Traveller*, October 20, 1829.

70. *Boston Daily Advertiser*, October 19, 1829, emphasis in original.

71. The MCMA organized as a benevolent organization for tradesmen in 1795. Paul Revere served as its first president. Its goals included financial help for "distressed" workers and their families, the promotion of invention and improvement through the offering of prizes, loans for young mechanics, and the establishment of schools and libraries for apprentices and continuing education. Moses King, *King's Handbook of Boston*, 283–84.

72. *Boston Courier*, October 22, 1829.

73. The *Boston Commercial Gazette* reprinted the letter from the *Daily Advertiser* together with a commentary from the *American Traveller*, October 19, 1829.

74. John F. Kasson, *Rudeness and Civility: Manners in Nineteenth-Century Urban America* (New York: Hill and Wang, 1990), 112–26.

75. Charles Dickens, *American Notes* (1842; repr. New York: Fromm International, 1985), 122–24.

76. *Boston Commercial Gazette*, October 2, 1829.

77. Katherine C. Grier, *Culture and Comfort: People, Parlors, and Upholstery, 1859–1930* (Rochester, NY: Strong Museum, 1988), 29–30.

78. Henry Tudor, *Narrative of a Tour in North America; Comprising Mexico, the Mines of Real Del Monte, the United States, and the British Colonies: With an Excursion to the Island of Cuba* (London: James Duncan, 1834), 358.

79. Bills in Prints and Photographs division of the American Antiquarian Society.

80. Costard Sly, *Sayings and Doings at the Tremont House. In the Year 1832. Extracted from the note book of Costard Sly, Solicitor and Short-Hand Writer, of London*, ed. Zachary Philemon Vangrifter (Boston: Allen and Ticknor, 1833), 1:9–10.

81. Sly, *Sayings and Doings*, 1:70–71, 76, 138.

82. Eliot, *Description of Tremont House*, 10–11; and Tyrone Power, *Impressions of America; During the Years 1833, 1834, and 1835* (Philadelphia: Carey, Lea, and Blanchard, 1836), 1:76.

83. Two dollars in 1830 money converts to nearly $41 in 2005 money. However, most working-class and low-end white collar workers earned about one dollar per day.

84. Power, *Impressions of America*, 1:78–80.

85. E. T. Coke, *A Subaltern's Furlough: Descriptive Scenes in Various Parts of the United States, Upper and Lower Canada, New-Brunswick and Nova Scotia, during the Summer and Autumn of 1832* (New York: J. & J. Harper, 1833), 1:32.

86. James Boardman, *America and the Americans* (London: Longman, Rees, Orme, Brown, Green, and Longman, 1833), 26; Kasson, *Rudeness and Civility*, 186–87; and Harvey Levenstein, *Revolution at the Table: The Transformation of the American Diet* (New York: Oxford University Press, 1988), 3–9.

87. *National Intelligencer*, November 20, 1836, as quoted in Murray, *Travels in North America*, 2:63.

88. Tudor, *Narrative of a Tour in North America*, 37–38.

89. Boardman, *America and the Americans*, 280–81.

90. Godfrey T. Vigne, *Six Months in America* (Philadelphia: Thomas T. Ash, 1833), 104.

91. Tudor, *Narrative of a Tour in North America*, 356.

92. *Gleason's Pictorial Drawing-Room Companion*, vol. 2, February 7, 1852, 88.

93. The 1831 city directory listed fourteen men living at the Tremont House. *Stimpson's Boston Directory; Containing Names of the Inhabitants, their Occupations, Places of Business, and Dwelling Houses, and the City Register, with lists of the Streets, Lanes and Wharves, the City Offices and Banks, and other Useful Information* (Boston: Simpson and Clapp, 1831).

94. *Boston Traveller*, October 16, 1879.

95. Bowen, *Picture of Boston* (1829, 1833, 1837); and "The Tremont House," *Parley's Magazine for Children and Youth*, vol. 1, July 6, 1833, 156.

96. Power, *Impressions of America*, 1:78.

97. "A Fitting Farewell," *Boston Herald*, November 25, 1894.

98. Leslie Dorsey and Janice Devine, *Fare Thee Well* (New York: Crown Publishers, 1964), 190–91.

99. The publication of Eliot's book represented the understated, carefully restrained demonstration of emotion for which Bostonians were known. (For a description of the Boston character, see Power, *Impressions of America*, 1:81.) An editorial in the *American Traveller*, referring to the Tremont House, compared the eternal "*breeze* or *stew*" in which New Yorkers found themselves to the steady deliberations of Bostonians: "The Bostonians build a tremendous great house, eat a dinner in it, and say nothing more about it. Anon, when this is worn threadbare, these Gothamites begin to complain about bad smells, filth, dead cats, &c. No smell like a New York bad smell, no dead cat like a New York dead cat." *American Traveller*, October 30, 1829.

100. John Holloway, *The Mechanicks Quick Step* (Boston: John Ashten, 1835). Collection of the Boston Athenaeum.

101. See, for example, George Russell French, "Palaces Abroad and at Home," *The Builder*, vol. 4 (1846), 459.

102. For examples of Boston salaries, in 1833 the mayor earned $2,500 per year. A messenger in his office, presumably of the middling class, whose job consisted of conveying messages, lighting the rooms, and attending the mayor throughout the day, earned $600 per year. Bowen, *Picture of Boston*, 20. W. H. Eliot's combined personal and real estate was estimated at $53,000 in 1836. *City of Boston: List of Persons, Copartnerships, and Corporations, who were Taxed Twenty Five dollars and upwards, in the city of Boston, in the year 1836* (Boston: John H. Eastburn, 1837).

103. Edward Pessen, "The Egalitarian Myth and the American Social Reality: Wealth, Mobility, and Equality in the 'Era of the Common Man.'" *American Historical Review* 76 (October 1971): 1028.

CHAPTER THREE: The Proliferation of Antebellum Hotels, 1830–1860

Subtitle: "The St. Nicholas Hotel, New York," *Graham's Illustrated Magazine*, vol. 50, no. 2 (February 1857), 168.

1. "The St. Nicholas Hotel, New York," *Graham's Illustrated Magazine*, vol. 50, no. 2 (February 1857), 168.

2. Katherine C. Grier, *Culture and Comfort: Parlor Making and Middle-Class Iden-*

tity, 1850–1930 (Washington, DC: Smithsonian Institution Press, 1988), 22–43; and Carolyn Brucken, "In the Public Eye; Women and the American Luxury Hotel," *Winterthur Portfolio* 31, no. 4 (Winter 1996): 203–20.

3. See Max Page's analysis of Joseph Schumpeter's theory of creative destruction in *The Creative Destruction of Manhattan, 1900–1940* (Chicago: University of Chicago Press, 1999), 2.

4. Philip Hone, *The Diary of Philip Hone, 1828–1851,* ed. Allan Nevins (New York: Dodd, Mead, and Company, 1927), 1:127; and Mary Griffith, "Three Hundred Years Hence" (1836), cited in Carol Farley Kessler, *Daring to Dream: Utopian Stories by United States Women, 1836–1919* (Boston: Pandora Press, 1984), 43–44.

5. Jefferson Williamson, *The American Hotel: An Anecdotal History* (New York: Alfred A. Knopf, 1930), 39.

6. See Hone, *Diary*. These two edited volumes represent about half of the contents of the twenty-eight manuscript volumes, which averaged about four hundred pages each. The originals and photocopies are in the collection of the New-York Historical Society. See also Louis Auchincloss, ed., *The Hone and Strong Diaries of Old Manhattan* (New York: Abbeville Press, 1989).

7. Auchincloss, ed., *Hone and Strong Diaries*, 12.

8. Hone, *Diary*, 1:215, 1:288–89, 2:268, 2:703.

9. Hone, *Diary*, 1:101, 2:612.

10. Hone, *Diary*, 2:520, 2:529, 2:699. The Whig convention was the same that witnessed the first news telegraph message on May 1, 1844, from Annapolis Junction to Washington, D.C., announcing Henry Clay's nomination for president. "What Hath God Wrought," from Washington, D.C., to Baltimore, followed on May 24, 1844.

11. *Gleason's Drawing-Room Companion,* vol. 1, no. 35, December 27, 1851, 553.

12. *Gleason's Drawing-Room Companion,* vol. 1, no. 24, October 11, 1851, 373.

13. *Gleason's Drawing-Room Companion,* vol. 1, no. 3, May 17, 1851, 37.

14. *Gleason's Drawing-Room Companion,* vol. 1, no. 6, June 7, 1851, 86.

15. Nathaniel Hawthorne, "Chiefly about War Matters," *Atlantic Monthly* (July 1862). Accessed in Willard Family Papers 76-45757, container 133, file 2, Library of Congress.

16. Ledger for 1861 (Cash Book), Bill Book No. 3, Willard's Hotel, 6. In Willard Family Papers 76-45757, container 136, Library of Congress.

17. Williamson, *American Hotel,* 280–81.

18. "A Frenchman's Idea of the Astor House," *The New Mirror,* vol. 1, no. 20, August 19, 1843, 311.

19. "My Hotel," *Putnam's Monthly,* vol. 9 (April 1857), 375–83.

20. William Chambers, *Things as They Are in America* (1854; repr. New York: Negro Universities Press, 1968), 182.

21. "New Kind of Hotel Up Town," *New York Weekly Mirror,* December 7, 1844, 131.

22. "The New St. Charles Hotel," *Gleason's Drawing-Room Companion,* vol. 5, no. 3, July 16, 1853, 33; and "Charleston Hotel," *Gleason's Drawing-Room Companion,* vol. 2, no. 17, April 24, 1852, 265.

23. Thomas Walsh, "'The Lindel,' of St. Louis," *The Builder,* vol. 21, February 7, 1863, 92–93.

24. "Astor's Park Hotel, New York," *Atkinson's Casket,* no. 4 (April 1835), 217.

25. John Denis Haeger, *John Jacob Astor: Business and Finance in the Early Republic* (Detroit: Wayne State University Press, 1991), 262–65; Williamson, *American Hotel,* 32–

37; Meryle R. Evans, "Knickerbocker Hotels and Restaurants, 1800–1850," *The New-York Historical Society Quarterly* 36 (October 1952): 387–89; and Vaughn L. Glasgow, "The Hotels of New York City prior to the American Civil War" (Ph.D. diss., Pennsylvania State University, 1970), 70–77.

26. "Astor House," *The New Yorker,* vol. 1, June 4, 1836, 173.

27. *American Traveller,* June 14, 1836.

28. As quoted in Williamson, *American Hotel,* 32.

29. Glasgow, "Hotels of New York City," 75; and Williamson, *American Hotel,* 32.

30. Karen Halttunen, *Confidence Men and Painted Women: A Study of Middle-Class Culture in America, 1830–1870* (New Haven: Yale University Press, 1982), 103–5.

31. "Astor House," *The New Yorker,* vol. 1, June 4, 1836, 173.

32. "A Frenchman's Idea of the Astor House," *The New Mirror,* vol. 1, no. 20, August 19, 1843, 311.

33. "The Astor House, Origin and History of the World-Famous Hotel," *New York Times,* January 31, 1875.

34. "The Large Hotel Question," *Chambers's Journal,* vol. 21 (1854), 153.

35. "Metropolitan Hotel, New York," *Gleason's Pictorial Drawing-Room Companion,* vol. 3, no. 12, September 25, 1852, 201; the *Herald,* as cited in Glasgow, "Hotels of New York City," 97; and "Large Hotel Question," 153.

36. Williamson, *American Hotel,* 53; and Glasgow, "Hotels of New York City," 114.

37. *New York Times,* January 7, 1853.

38. "Phalon's Saloon," *Gleason's Pictorial,* vol. 4, no. 6, February 5, 1853, 112; "St. Nicholas Hotel, New York," *Gleason's Pictorial,* vol. 4, no. 11, March 12, 1853, 161; "The St. Nicholas Hotel," *New York Times,* January 7, 1853; *New York Daily Tribune,* December 3, 1852; and "The St. Nicholas Hotel, New York," *Graham's Illustrated Magazine,* vol. 50, no. 2 (February 1857), 167–68.

39. "St. Nicholas Hotel," typescript of 1856 descriptive booklet published by St. Nicholas Hotel, Quinn Collection, New-York Historical Society, File 1856, in collection of New York Public Library.

40. *New York Daily Tribune,* December 3, 1852.

41. "St. Nicholas Hotel," Quinn Collection.

42. John Henry Vessey, *Mr. Vessey of England, Being the Incidents and Reminiscences of Travel in Twelve Weeks' Tour through the United States and Canada in the Year 1859,* Brian Waters, ed. (New York: G. P. Putnam's and Sons, 1956), 30–31.

43. Elaine Abelson, *When Ladies Go A-Thieving: Middle-Class Shoplifters in the Victorian Department Store* (New York: Oxford University Press, 1989), 13–14; and Halttunen, *Confidence Men and Painted Women,* 64–65.

44. "The St. Nicholas and the Five Points," *Putnam's Monthly,* vol. 1 (May 1853), 510.

45. "St. Nicholas Hotel," *Graham's,* 168.

46. Cecil D. Elliott, *Technics and Architecture: The Development of Materials and Systems for Buildings* (Cambridge: MIT Press, 1992), 330–31.

47. United States Patent Office, Patent no. 32,441, May 28, 1861.

48. "Steam Versus Stairs," *New York Times,* January 23, 1860.

49. "The Fifth Avenue Hotel," *New York Times,* August 23, 1859; and "Fifth Avenue Hotel," *Granite State Monthly* (1875), 317–25, as cited in Glasgow, "Hotels of New York City," 124.

50. *New-York Traveller and United States Hotel Directory,* July 30, 1859; and *New York Times,* August 23, 1859.

51. George A. Sala, "American Hotels and American Food," *Temple Bar Magazine,* vol. 2 (July 1861), 348.

52. *Granite State Monthly,* as cited in Glasgow, "Hotels of New York City," 122.

53. "The Fall Season," *New-York Traveller and United States Hotel Directory,* September 3, 1859.

54. "Girard House, Philadelphia," *Gleason's Pictorial,* vol. 2, no. 8, February 21, 1852, 114; "Niblo's Hotel," *Gleason's Pictorial,* vol. 1, no. 20, September 13, 1851, 306; and "Burnet House, Cincinnati," *Gleason's Drawing Room Companion,* vol. 1, no. 5, May 31, 1851, 66.

55. Anthony Trollope, *North America* (1862; repr. New York: Alfred A. Knopf, 1951), 480.

56. Domingo Faustino Sarmiento, *Travels in the United States in 1847,* trans. Michael Aaron Rockland (Princeton: Princeton University Press, 1970), 141–45.

57. Sala, "American Hotels and American Food," 345.

58. Harvey, *Condition of Postmodernity,* 17; and Page, *Creative Destruction,* 1–5.

59. "American House, Hanover Street, Boston," *Ballou's Pictorial Drawing-Room Companion,* vol. 16, no. 10, March 5, 1859, 157.

CHAPTER FOUR: The Continental Hotel, Philadelphia, 1860

Subtitle: Title page, *The Hotel Folly* (Philadelphia, 1857).

1. *Public Ledger,* March 21, 1857; and *Philadelphia Evening Journal,* April 17 and May 26, 1857.

2. *Philadelphia Evening Journal,* March 6 to May 14, 1857.

3. Julio Rae, *Rae's Philadelphia Pictorial Directory and Panoramic Advertiser, Chestnut Street, from Second to Tenth Streets* (Philadelphia: Julio H. Rae, 1851); George G. Foster, "Philadelphia in Slices," *Pennsylvania Magazine of History and Biography* 93 (1969): 55–60; *Sunday Dispatch,* March 22, 1857; *Sunday Dispatch,* March 29, 1857; and *Philadelphia Evening Journal,* April 27, 1857.

4. Sidney George Fisher, *A Philadelphia Perspective: The Diary of Sidney George Fisher Covering the Years 1834–1871,* Nicholas Wainwright, ed. (Philadelphia: Historical Society of Pennsylvania, 1967), 83, emphasis in original.

5. Fisher, *Philadelphia Perspective,* 22.

6. Frances Trollope, *Domestic Manners of the Americans* (1839; repr. London: Century Publishing, 1984), 221, 231, 297, 311.

7. Charles Dickens, *American Notes: A Journey* (1842; repr. New York: International Publishing Corporation, 1985), 80.

8. Foster, "Philadelphia in Slices," 56.

9. Dickens, *American Notes,* 98.

10. Foster, "Philadelphia in Slices," 28.

11. Foster, "Philadelphia in Slices," 31.

12. Edwin T. Freedley, *Philadelphia and Its Manufactures: A Handbook Exhibiting the Development, Variety, and Statistics of the Manufacturing Industry of Philadelphia in 1857 together with Sketches of Remarkable Manufactories; and a List of Articles Now Made in Philadelphia* (Philadelphia: Edward Young, 1859), 17.

13. Freedley, *Philadelphia and Its Manufactures,* 17, 77, 85.

14. *Philadelphia Evening Journal,* April 17, 25, 1857.

15. J. Thomas Scharf and Thompson Westcott, *History of Philadelphia, 1609–1884* (Philadelphia: L. H. Everts and Co., 1884), 2:734.

16. *The Hotel Folly* (Philadelphia, 1857).

17. Sharf and Westcott, *History of Philadelphia,* 2:762.

18. *Hotel Folly,* 5.

19. *Hotel Folly,* 17.

20. *Hotel Folly,* 12; and "The New Hotel," *Boston Commercial Gazette,* July 7, 1828. A correspondent to the *Gazette* wrote in opposition to the proposed tax abatement: "Suppose that hundreds of [mechanics] were to present themselves to the Mayor, and say 'we are starving, and must be employed.' . . . Would he propose the erection of a new hotel? Would he request the City Council to consider the expediency of building another milldam, or another market-house? Would he even suggest the idea of a Bunker-Hill Monument for the sake of his poor petitioners?"

21. *Sunday Dispatch,* March 29, 1857; Foster, "Philadelphia in Slices," 58–59; and Rae, *Rae's Pictorial.*

22. *Hotel Folly,* 8; and "The New Chestnut Street Hotel," *Philadelphia Evening Journal,* March 10, 1857.

23. *Hotel Folly,* 7; and Scharf and Westcott, *History of Philadelphia,* 2:994.

24. "The Girard House, Philadelphia," *Gleason's Pictorial,* vol. 2, no. 8, February 21, 1852, 114; and "La Pierre House," *Gleason's Pictorial,* vol. 5, no. 15, October 8, 1853, 225.

25. Thompson Westcott, *The Historic Mansions and Buildings of Philadelphia* (Philadelphia: Porter and Coates, 1877), 356; E. Digby Baltzell, *Puritan Boston and Quaker Philadelphia, Two Protestant Ethics and the Spirit of Class Authority and Leadership* (New York: Free Press, 1979), 195; and "Robert Morris," *Gleason's Pictorial,* vol. 5, no. 6, August 6, 1853, 84–85.

26. *Hotel Folly,* 4.

27. *Hotel Folly,* 18.

28. Stanley I. Kutler, *Privilege and Creative Destruction: The Charles River Bridge Case* (1971; repr. Baltimore: Johns Hopkins University Press, 1990), 20–21.

29. Stuart M. Blumin, "Mobility and Change in Antebellum Philadelphia," in *Nineteenth-Century Cities, Essays in the New Urban History,* ed. Stephan Thernstrom and Richard Sennett (New Haven: Yale University Press, 1969), 204.

30. *Hotel Folly,* 8, 10, 17, 25. The act to incorporate a hotel company passed the State Legislature on March 14, 1856, under the name "The Butler House Hotel Company." The proposal under discussion in this chapter represents the third attempt to organize a viable venture. See Butler House Hotel Company, Continental Hotel Company Papers, Historical Society of Pennsylvania, hereafter referred to as CHCP.

31. See Butler House Hotel Company, CHCP, for a complete list.

32. A. McElroy, *McElroy's Philadelphia City Directory for 1861* (Philadelphia: E. C. and J. Biddles and Co., 1861); *Traveler's Sketch* (Philadelphia, 1861), 33–34; and Butler House Hotel Company, CHCP.

33. "The New Hotel Project," *Sunday Dispatch,* March 22, 1857.

34. Scharf and Westcott, *History of Philadelphia,* 1:708.

35. *Hotel Folly,* 21; and "Rice Jobbers and Railroad Subsidyists," *Sunday Dispatch,* April 26, 1857.

36. Scharf and Westcott, *History of Philadelphia*, 1:708.

37. *Hotel Folly*, 23.

38. "The Monster Hotel Folly, Second Series, Letter IV," *Philadelphia Evening Journal*, April 25, 1857.

39. S. K. Hoxsie to Messrs. Hulings, Cowperthwait, Myers, Haddock, and Orne, September 1857, CHCP.

40. S. K. Hoxsie to Messrs. Hulings, Cowperthwait, Myers, Haddock, and Orne, September 1857, CHCP.

41. S. K. Hoxsie to Messrs. Hulings, Cowperthwait, Myers, Haddock, and Orne, September 1857, CHCP; and J. Sergeant Price to S. K. Hoxsie, March 4, 1859, CHCP.

42. Holt's Hotel (New York City, 1833) used a steam-powered hoist to lift baggage as well as food from the kitchen to the dining room. *Atkinson's Casket*, vol. 8 (August 1833).

43. United States Patent Office, *Elevator or Hoisting Apparatus for Hotels, &c.*, Patent no. 25,061, August 9, 1859.

44. Cecil D. Elliott, *Technics and Architecture: The Development of Materials and Systems of Buildings* (Cambridge: MIT Press, 1992), 331; and "Steam Versus Stairs," *New York Times*, January 23, 1860, 2.

45. "Steam Versus Stairs."

46. J. Sergeant Price to Paran Stevens, January 30, 1860, CHCP.

47. Elliot, *Technics and Architecture*, 331–32.

48. J. Sergeant Price to Paran Stevens, January 30, 1860, CHCP.

49. "John McArthur, Jr.," *Architecture and Building*, vol. 12, January 18, 1890, 25; and Henry F. and Elsie R. Withey, "John McArthur, Jr.," *Biographical Dictionary of American Architects (Deceased)* (Los Angeles: New Age Publishing, 1956), 402.

50. Edwin Wolf 2nd, *Philadelphia: Portrait of an American City* (Harrisburg, PA: Stackpole Books, 1975), 188.

51. Foster, "Philadelphia in Slices," 56.

52. Nicholas B. Wainwright, *Philadelphia in the Romantic Age of Lithography* (Philadelphia: Historical Society of Pennsylvania, 1958), 224, 299.

53. *Sunday Dispatch*, March 22, 1857; and *Philadelphia Evening Journal*, April 27, 1857.

54. *Public Ledger*, January 30, 1860.

55. Winston Weisman, "Commercial Palaces of New York: 1845–1875," *Art Bulletin* 36 (December 1954): 285; "New York Daguerreotyped," *Putnam's Monthly*, vol. 1 (1853), 129; and *Frank Leslie's Illustrated Newspaper*, October 6, 1860, 313, both cited in Weisman, "Commercial Palaces," 285.

56. *Traveler's Sketch*, 12.

57. "What an English Sportsman Thinks of Us," *New York Times*, March 6, 1860. See also Molly W. Berger, "A House Divided: Technology, Gender, and Consumption in America's Luxury Hotels, 1825–1860," in *His and Hers: Gender, Consumption, and Technology*, ed. Roger Horowitz and Arwen Mohun (Charlottesville: University Press of Virginia, 1998), 39–65.

58. Architectural Plans, Continental Hotel, Philadelphia, 1871, Historical Society of Pennsylvania.

59. "Palace Homes for the Traveller," *Godey's Lady's Book and Magazine*, vol. 60 (May 1860), 465–66.

60. *Traveler's Sketch*, 3–7; and *Public Ledger*, February 13, 1860.

61. "Palace Homes for the Traveller," 465.

62. Theodore B. White, *Philadelphia Architecture in the Nineteenth Century* (Philadelphia: University of Pennsylvania Press, 1953), 29.

63. "The Fifth Avenue Hotel," *Harper's Weekly*, vol. 3, October 1, 1859, 634.

64. *Public Ledger*, February 13, 1860.

65. "Palace Homes for the Traveller," 466.

66. *Traveler's Sketch*, 16–17.

67. *Traveler's Sketch*, 19–23.

68. *Public Ledger*, January 30, February 13, 1860.

69. *Traveler's Sketch*, 19–21.

70. "The Continental Hotel.—Historical and Descriptive Account of It," newspaper article (no attribution), Philadelphia Prints, Library Company of Philadelphia; and "Palace Homes for the Traveller," 465.

71. *Traveler's Sketch*, 23; and "The Large Hotel Question," *Chambers's Journal*, vol. 1 (January–June 1854), 153.

72. "Mechanical Equipment Boosts the Payroll," *The Hotel Monthly*, vol. 31 (November 1923), 21.

73. Scharf and Westcott, *History of Philadelphia*, 1:733–50, 1:760–62.

74. "Modern Hotels," *Scribner's*, vol. 6 (1873), 487.

75. Architectural Plans, Continental Hotel, Historical Society of Pennsylvania.

CHAPTER FIVE: Production and Consumption in an American Palace, 1850–1875

Subtitle: "American Hotels by a Cosmopolitan," *Putnam's Magazine*, New Series, vol. 5 (January 1870), 29.

1. See David A. Hounshell, *From the American System to Mass Production, 1800–1932: The Development of Manufacturing Technology in the United States* (Baltimore: Johns Hopkins University Press, 1984).

2. "Hotel Life in New York," *New York Times*, November 21, 1865.

3. "First Impressions of America and Its People, Hotel Life," *The Leisure Hour*, vol. 20 (1871), 205.

4. E. Anthony Rotundo, *American Manhood: Transformations in Masculinity from the Revolution to the Modern Era* (New York: Basic Books, 1993), 196–205.

5. Ralph Keeler, "The Great American Hotel," *Lippincott's Magazine of Popular Literature and Science*, vol. 10 (September 1872), 295.

6. "First Impressions," 205.

7. David Kuchta, "The Making of the Self-Made Man: Class, Clothing, and English Masculinity, 1688–1832," in *The Sex of Things*, ed. Victoria de Grazia (Berkeley: University of California Press, 1996), 54–78, quoted in Mary Louise Roberts, "Gender, Consumption, and Commodity Culture," *American Historical Review* 103 (June 1998): 824–25. See also Leora Auslander, "The Gendering of Consumer Practices in Nineteenth-Century France," in *The Sex of Things*, ed. de Grazia, 79–112.

8. For example, Regina Lee Blaszczyk begins her study of American consumerism in 1865, after the Civil War. Regina Lee Blaszczyk, *American Consumer Society, 1865–2005* (Wheeling, IL: Harlan Davidson, 2009).

9. "English Hotels by an American," *Every Saturday,* vol. 5, May 30, 1868, 693; "American Hotels by a Cosmopolitan," 28; *New York Times,* November 11, 1865; "Modern Hotels," *Scribner's Monthly,* vol. 6 (1873), 488; and William Laird MacGregor, *Hotels and Hotel Life at San Francisco, California, in 1876* (San Francisco: S. F. News Company, 1877), 41.

10. G. A. Sala, "American Hotels and American Food," *Temple Bar Magazine,* vol. 2 (July 1861), 345.

11. Anthony Trollope, *North America* (1862; repr. New York: Alfred A. Knopf, 1951), 482–83.

12. "Hotels," *Chambers's Journal,* vol. 39 (1863), 346.

13. "Hotel Reform," *Chambers's Journal,* vol. 42, August 5, 1865, 481–83; "English Hotels by an American," 691–94; and G. A. Sala, "The Philosophy of Grand Hotels," *Belgravia,* vol. 20 (June 1873), 140–41.

14. Albert Smith, *The English Hotel Nuisance,* 2nd ed. (London: Bradbury and Evans, 1858), 25.

15. Sala, "American Hotels and American Food," 345–46.

16. "Hotel Life in New-York," *New York Times,* November 21, 1865; and "First Impressions of America and Its People, Hotel Life," 206.

17. Sala, "American Hotels and American Food," 346.

18. Trollope, *North America,* 486–87.

19. Keeler, "Great American Hotel," 298; "The American Hotel Clerk," *Hotel World,* vol. 6, February 9, 1878, 7; and "English Hotels by an American," 693.

20. See also A. K. Sandoval-Strausz and Daniel Levinson Wilk, "Princes and Maids of the City Hotel: The Cultural Politics of Commercial Hospitality in America," in *The American Hotel, Journal of Decorative and Propaganda Arts,* vol. 5, ed. Molly W. Berger (Cambridge: MIT Press, 2005), 161–67.

21. *Putnam's,* vol. 5, no. 25.

22. Quotation from "Hotels," *Chambers's Journal,* 347; Trollope, *North America,* 489; and "How We Live," *New York Times,* November 22, 1873.

23. Halttunen, *Confidence Men and Painted Women,* 29.

24. Sala, "Philosophy of Grand Hotels," 138–39.

25. William Leach, *Land of Desire: Merchants, Power, and the Rise of a New American Culture* (New York: Pantheon Books, 1993), 6–7; Stuart Ewen, *All Consuming Images: The Politics of Style in Contemporary Culture* (New York: Basic Books, 1988), 58–59; and Richard Bushman, *The Refinement of America: Persons, Houses, Cities* (New York: Vintage Books, 1992), xix.

26. Sala makes reference to the aforementioned early 1850s pamphlet written by "poor" Albert Smith, which drew attention to extortionary tipping practices and provoked weeks' worth of correspondence to the *London Times.* Sala, "Philosophy of Grand Hotels," 140.

27. George B. Post (the second "father of the American hotel"), Arthur Gilman, and George H. Kendall designed the Equitable Life Assurance Building. Winston Weisman, "Commercial Palaces of New York: 1845–1875," *Art Bulletin* 36 (December 1954): 297–98; and Carl W. Condit, *American Building,* 2nd ed. (Chicago: University of Chicago Press, 1982), 115–16.

28. "American Hotels by a Cosmopolitan," 26.

29. Continental Hotel Floor Plans, 1871, Historical Society of Pennsylvania. These

plans were made from actual measurements of the 1860 building as certified by the architect, John McArthur Jr.

30. *New York Times,* November 22, 1873.

31. "Modern Hotels," 487.

32. Sala, "American Hotels and American Food," 349; Anthony Trollope, *North America,* 490; "First Impressions of America and Its People, Hotel Life," 206; and "American Hotels by a Cosmopolitan," 26.

33. "English Hotels by an American," 692.

34. *New York Times,* November 22, 1873.

35. Sala, *Temple Bar,* 350.

36. "American Hotels by a Cosmopolitan," 27. A navvy was a (chiefly) British laborer, usually employed on construction or excavation projects.

37. Anthony Trollope, *North America,* 490.

38. Sala, *Temple Bar,* 352.

39. Gillian Brown, *Domestic Individualism: Imagining Self in Nineteenth-Century America* (Berkeley: University of California Press, 1990), 18.

40. "American Hotels by a Cosmopolitan," 27.

41. Keeler, "Great American Hotel," 296.

42. Sala, *Temple Bar,* 348–52.

43. "Decline and Fall of Hotel Life," *Harper's Weekly,* vol. 1, May 2, 1857, 274.

44. *New York Times,* November 21, 1865; "American Hotels by a Cosmopolitan," 27; and Keeler, "Great American Hotel," 296.

45. "American Hotels by a Cosmopolitan," 25.

46. "American Hotels by a Cosmopolitan," 25.

47. Molly W. Berger, "A House Divided: The Culture of the American Luxury Hotel, 1825–1860," in *His and Hers: Gender, Consumption, and Technology,* ed. Roger Horowitz and Arwen Mohun (Charlottesville: University Press of Virginia, 1998), 48–50.

48. "The Fifth Avenue Hotel, New York," *Harper's Weekly,* vol. 3, October 1, 1859, 634.

49. Halttunen, *Confidence Men and Painted Women,* 114–15.

50. Keeler, "Great American Hotel," 300–301.

51. John Brewer and Roy Porter, "Introduction" and T. H. Breen, "The Meanings of Things: Interpreting the Consumer Economy in the Eighteenth Century," both in *Consumption and the World of Goods,* ed. John Brewer and Roy Porter (London: Routledge, 1993), 5, 249–60; and Halttunen, *Confidence Men and Painted Women,* 64–65, 160–61.

52. Man About Town, "Hotel Morals," *Harper's Weekly,* September 5, 1857.

53. "My Hotel," *Putnam's Monthly, A Magazine of Literature, Science, and Art,* vol. 9, no. 52 (April 1857), 162–74.

54. "Another Glimpse at My Hotel," *Putnam's Monthly,* vol. 10 (August 1875), 162–74.

55. *Gleason's Pictorial Drawing-Room Companion,* vol. 3, no. 13 (September 25, 1852), 199.

56. *New York Times,* November 22, 1873.

57. Frederick William Sharon, "Junior Forensic Essay," Sharon Family Papers, C-B 777, part 1, box 24. Bancroft Library, University of California at Berkeley.

58. "Modern Hotels," 488.

59. "American Hotels by a Cosmopolitan," 28; Keeler, "Great American Hotel," 295; and Sala, "American Hotels and American Food," 355.

60. "Modern Hotels," 487.

61. "Modern Hotels," 486–92.

62. "Modern Hotels," 488.

63. "Modern Hotels," 483.

64. "Modern Hotels," 488.

65. Cecil D. Elliott, *Technics and Architecture: The Development of Materials and Systems for Buildings* (Cambridge: MIT Press, 1992), 332–33.

66. "Modern Hotels," 490.

67. "Modern Hotels," 487.

CHAPTER SIX: The Palace Hotel, San Francisco, 1875

Subtitle: "The Palace Hotel," *Postscript to the San Francisco News-Letter,* vol. 25, October 30, 1875, 1.

1. John S. Hittell, *A History of the City of San Francisco and Incidentally of the State of California* (San Francisco: A. L. Bancroft and Co., 1878), 404.

2. Martyn J. Bowden, "The Dynamics of City Growth: An Historical Geography of the San Francisco Central District, 1850–1931" (Ph.D. diss., University of California, Berkeley, 1967), 114.

3. *The Elite Directory for San Francisco and Oakland* (San Francisco: Argonaut Publishing Co., 1879), 12–13. See also Dorothy Harriet Huggins, "San Francisco Society," *California Historical Society Quarterly* 19 (September 1940): 225–39.

4. See *Gleason's Pictorial Drawing Room Companion,* vol. 1, no. 15, August 23, 1851, 261; vol. 2, no. 11, March 13, 1852, 172; vol. 2, no. 13, March 27, 1852, 196–97; vol. 3, no. 18, October 30, 1852, 276–77.

5. *Elite Directory,* 17.

6. John S. Hittell, *The Resources of California comprising the Society, Climate, Salubrity, Scenery, Commerce, and Industry of the State,* 7th ed. (San Francisco: A. L. Bancroft and Co., 1879), 22. See also B. E. Lloyd, *Lights and Shades in San Francisco* (San Francisco: A. L. Bancroft and Co., 1876), 449; William Laird MacGregor, *Hotels and Hotel Life at San Francisco, California in 1876* (San Francisco: S. F. News and Co., 1877), 36–37; "Caravansaries of San Francisco," *Overland Monthly,* vol. 5 (August 1870), 176, 181; and Oscar Lewis and Carroll D. Hall, *Bonanza Inn: America's First Luxury Hotel* (New York: Alfred A. Knopf, 1939), 10.

7. "Caravansaries of San Francisco," 176.

8. Bowden, "Dynamics of City Growth," 242, 283.

9. *San Francisco Real Estate Circular,* March 7, 1873, 3.

10. *San Francisco Real Estate Circular,* August 8, 1874, 3.

11. Lloyd, *Lights and Shades,* 450; MacGregor, *Hotels and Hotel Life,* 26, 36; and "Caravansaries of San Francisco," 178.

12. "Caravansaries of San Francisco," 181; MacGregor, *Hotels and Hotel Life,* 25; and "Otis Passenger Elevator," Trade Literature Collection, National Museum of American History, Smithsonian Institution, Washington, D.C.

13. John P. Young, *San Francisco: A History of the Pacific Coast Metropolis* (San Francisco: S. J. Clarke Publishing Co., 1912), 570.

14. "Caravansaries of San Francisco," 177–78.

15. *Bancroft's Tourist's Guide, San Francisco and Vicinity* (n.d.), 133. San Francisco Public Library.

16. "The Palace Hotel," *California Spirit of the Times,* December 25, 1875.

17. David Lavender, *Nothing Seemed Impossible: William C. Ralston and Early San Francisco* (Palo Alto, CA: American West Publishing, 1975), 17; and Hittell, *History of San Francisco,* 409.

18. Lavender, *Nothing Seemed Impossible,* 167.

19. Charles Lee Tilden, "William C. Ralston and His Times," unpublished notes on a talk delivered before the California Historical Society, April 12, 1938. C-D 5127, Bancroft Library, University of California at Berkeley.

20. Lewis and Hall, *Bonanza Inn,* 12.

21. See Hittell, *History of San Francisco,* 411–12; Young, *San Francisco,* 505; Lavender, *Nothing Seemed Impossible;* and Lewis and Hall, *Bonanza Inn,* 13–14.

22. Lavender, *Nothing Seemed Impossible,* 13.

23. Hubert H. Bancroft, *Biography of Men Important in the Building of the West* (n.d.), 1. F591-B212x, Bancroft Library, University of California at Berkeley; and "William C. Ralston," *Commercial Herald and Market Review,* September 2, 1875.

24. Lavender, *Nothing Seemed Impossible,* 242–44.

25. *San Francisco Real Estate Circular* (April 1869), 3; and David Pinkney, *Napoleon III and the Rebuilding of Paris* (Princeton: Princeton University Press, 1958), 57–58.

26. "Caravansaries of San Francisco," 177; and *San Francisco Real Estate Circular* (March 1873), 3.

27. *Bancroft's Tourist's Guide,* 133; "Caravansaries of San Francisco," 177; Joseph Armstrong Baird Jr., *Time's Wondrous Changes: San Francisco Architecture, 1776–1915* (San Francisco: California Historical Society, 1962), 27.

28. *San Francisco Real Estate Circular* (January 1873), 3; *San Francisco Real Estate Circular* (February 1873), 1; *San Francisco Real Estate Circular* (December 1873), 1.

29. *The Builder,* vol. 36, September 21, 1878, 988; *Coast Review* (1875), 345–47; "The Palace Hotel," Bancroft Library, University of California at Berkeley; *Sunday Chronicle,* October 17, 1875. This article puts the Palace's ground floor square footage at 96,256, Chicago's Palmer House at 56,350, and the Grand Pacific at 58,140. *Commercial Herald and Market Review,* vol. 8, July 9, 1874, 2.

30. Margot Gayle, *Cast-Iron Architecture in New York* (New York: Dover Publications, 1974), 142–43; Carl Condit, *American Building: Materials and Techniques from the First Colonial Settlements to the Present* (Chicago: University of Chicago Press, 1982), 84; "The Palace Hotel," *Overland Monthly,* vol. 15 (September 1875), 298; "The Palace Hotel," *California Spirit of the Times,* December 25, 1875; "The Great Caravansary of the Western World," *Frank Leslie's Illustrated Newspaper,* vol. 40, no. 1, October 9, 1875, 74.

31. "Great Caravansary," 74; *Coast Review* (1875), 345; *The Builder,* vol. 36, September 21, 1878, 988; "The Palace Hotel," Bancroft Library, University of California at Berkeley, 3.

32. Mrs. Frank Leslie [Miriam Follin Leslie], *A Pleasure Trip from Gotham to the Golden Gate. (April, May, June, 1877)* (New York: G. W. Carleton and Co., 1877), 115; *Sunday Chronicle,* October 17, 1875; *Daily Alta California,* October 17, 1875; "The Palace Hotel," Bancroft Library, University of California at Berkeley, 3; and Harold Kirker, *California's Architectural Frontier: Style and Tradition in the Nineteenth Century* (San Marino, CA: Huntington Library, 1960), 108.

33. Mrs. Frank Leslie, *A Pleasure Trip,* 116.

34. *California Spirit of the Times,* December 25, 1875, 3; "Great Caravansary," 74;

Horatio F. Stoll, "The Palace Hotel Court, Old and New," *The Architect and Engineer of California,* vol. 19 (January 1910), 43; and "The Palace Hotel," *Postscript to the San Francisco News-Letter,* vol. 25, October 30, 1875, 1.

35. *California Spirit of the Times,* December 25, 1875.

36. MacGregor, *Hotels and Hotel Life,* 12; and *Coast Review* (1875), 345.

37. *San Francisco Real Estate Circular* (November 1874), 4.

38. *San Francisco Real Estate Circular* (June 1874), 2; and *California Spirit of the Times,* December 25, 1875. Jefferson Williamson claims that Lewis Leland, one of the third generation of Lelands, was the first proprietor of the Palace. Williamson, *American Hotel,* 155–56.

39. *California Spirit of the Times,* December 25, 1875.

40. *San Francisco News-Letter,* vol. 25, October 30, 1875, 3.

41. "Rules, Regulations, and Prices Per Diem of the Grand Pacific Hotel," August 1, 1874, 20, Chicago History Museum; Harvey Levenstein, *Revolution at the Table: The Transformation of the American Diet* (New York: Oxford University Press, 1988), 16–20; and John F. Kasson, *Rudeness and Civility: Manners in Nineteenth-Century Urban America* (New York: Hill and Wang, 1990), 170.

42. Ned Polsky, *Hustlers, Beats, and Others* (Chicago: University of Chicago Press, 1985), 13–27; and The Brunswick-Balke-Collender Co., "Billiards—The Home Magnet" (Chicago, 1914). Trade Literature Collection, National Museum of American History, Smithsonian Institution.

43. Henry Louis King Manuscript, C-D 327, Bancroft Library, University of California at Berkeley.

44. *San Francisco News-Letter,* vol. 25, October 30, 1875; "Great Caravansary," 74; *California Spirit of the Times,* December 25, 1875; "The Palace Hotel," Bancroft Library, University of California at Berkeley; and *The Builder,* vol. 36, September 21, 1878, 988.

45. "The Palace," *San Francisco Chronicle,* October 17, 1895, 1ff.

46. *The Builder,* vol. 36, September 21, 1878, 988.

47. Michael O'Malley, *Keeping Watch: A History of American Time* (New York: Penguin Books, 1990), 12, 30, 82, 88.

48. John S. Hittell, *The Commerce and Industries of the Pacific Coast of North America,* 2nd ed. (San Francisco: A. L. Bancroft and Co., 1882), 433; *California Spirit of the Times,* December 25, 1875; and "The Ruins of the Lindell Hotel, St. Louis," *Frank Leslie's Illustrated Newspaper,* April 20, 1867, 78.

49. *News-Letter,* vol. 25, October 30, 1875, 2; *California Spirit of the Times,* December 25, 1875; and *Coast Review* (1875), 347.

50. Lewis and Hall, *Bonanza Inn,* 23–27; George D. Lyman, *Ralston's Ring, California Plunders the Comstock Lode* (New York: Charles Scribner's Sons, 1937), 221–25; *News-Letter,* vol. 25, October 30, 1875, 2; and Lavender, *Nothing Seemed Impossible,* 363.

51. Letter, September 12, 1874, from William C. Ralston to Joseph Russell, William C. Ralston Papers (box 6, 11561); letter, August 13, 1874, from John Robertson to William C. Ralston, William C. Ralston Papers (box 5, OBC 18); letter, September 12, 1874, from John Robertson to William C. Ralston, William C. Ralston Papers (box 5, OBC 18); letter, August 4, 1874, from George F. Seward to William C. Ralston, William C. Ralston Papers (box 7, 12055), Bancroft Library, University of California at Berkeley.

52. "Great Caravansary," 74; *News-Letter,* vol. 25, October 30, 1875; *California Spirit of the Times,* December 25, 1875.

53. Gunther Barth, *Instant Cities: Urbanization and the Rise of San Francisco and Denver* (New York: Oxford University Press, 1975), 185–86.

54. Hittell, *Resources of California*, xvii.

55. *Hotel World*, August 17, 1878; *California Spirit of the Times*, July 4, 1876; Mrs. Frank Leslie, *A Pleasure Trip*, 116; *California Spirit of the Times*, December 25, 1875; "The Palace Hotel Court, Old and New," 46; "Great Caravansary," 74; MacGregor, *Hotels and Hotel Life*, 22, 42; "The Grand Court of the Palace Hotel, San Francisco," *Frank Leslie's Illustrated Newspaper*, vol. 46, June 29, 1878, 282–83.

56. Mrs. Frank Leslie, *A Pleasure Trip*, 116; *Daily Alta California*, October 17, 1875; "Warren Leland," *Frank Leslie's Illustrated Newspaper*, October 9, 1875, 69; Williamson, *American Hotel*, 155; *News-Letter*, vol. 25, October 30, 1875; *Hotel World*, vol. 6, January 26, 1878.

57. "Warren Leland," 69; "The New Hotel," *New York Daily Times*, September 2, 1852; "Metropolitan Hotel," *Gleason's Drawing Room Companion*, vol. 3, September 25, 1852, 201; *California Spirit of the Times*, December 25, 1875; menu from Café Leland, 1876. Menu collection at the American Antiquarian Society; *San Francisco Real Estate Circular*, vol. 6 (April 1872), 1; Lloyd, *Lights and Shades*, 51, 56; and letter, August 28, 1874, from W. F. Coolbaugh to W. C. Ralston, William C. Ralston Papers (box 2, 3395), Bancroft Library, University of California at Berkeley.

58. Stephen Boyd Shiring, "American Hotelkeepers and Higher Learning: An Early Era of This Emerging Profession" (Ph.D. diss., University of Pittsburgh, 1995), 61; *Hotel World*, vol. 6, January 12, 1878; letter, Coolbaugh to Ralston.

59. *News-Letter*, vol. 25, October 30, 1875, 2.

60. MacGregor, *Hotels and Hotel Life*, 38–39.

61. *News-Letter*, vol. 25, October 30, 1875, 2; and *Daily Alta California*, October 17, 1875.

62. Douglas Henry Daniels, *Pioneer Urbanites: A Social and Cultural History of Black San Francisco* (Philadelphia: Temple University Press, 1980), 35–40. See also Tunis G. Campbell, *Hotel Keepers, Head Waiters, and Housekeepers' Guide* (Boston, 1848), reprinted as Doris Elizabeth King, ed., *Never Let People Be Kept Waiting* (Raleigh, NC: Graphic Press, 1973). This volume, originally 192 pages long, written by a black New York headwaiter, is a guide to dining room drills, food preparation, and other useful information for proper regimented service.

63. Letter, December 1, 1896, from John Kirkpatrick to F. W. Sharon, Sharon Family Papers, C-B 777, box 8, part 1, folder 1, Bancroft Library, University of California at Berkeley.

64. Lavender, *Nothing Seemed Impossible*, 372–79.

65. Hittell, *History of the City of San Francisco*, 408–9; David Oliver Sr., [A Statement of Arguments in Favor of William C. Ralston], Guide to Manuscripts, 3:345, Bancroft Library, University of California at Berkeley.

66. *Daily Alta California*, October 18, 1875.

67. *Daily Alta California*, October 18, 1875.

68. *American Traveller*, October 30, 1829.

69. "A Visitor Looks at Ralston's Hotel," *San Francisco Chronicle*, April 2, 1877. Reprinted in Joseph Henry Jackson, ed. *The Western Gate, A San Francisco Reader* (New York: Farrar Straus and Young, 1952), 339–41.

70. "Mammoth Caravansaries," *Hotel World*, vol. 7, August 17, 1878, 4.

71. Patricia G. Sikes, "George Roe and California's Centennial of Light," *California History,* vol. 58 (Fall 1979), 241.

72. *Harper's Weekly,* vol. 23, October 25, 1879, 849; "Home Again," *Frank Leslie's Illustrated Newspaper,* October 11, 1879: and Sikes, "George Roe," 242.

73. "Souvenir of the Palace Hotel" San Francisco: H. S. Crocker Company, 1891, San Francisco Public Library; letter, May 1, 1892, from Frederick W. Sharon to Frank S. Newlands, Sharon Family Papers, C-B 777, part 1, box 24, Bancroft Library, University of California at Berkeley; letter, January 13, 1899, from John C. Kirkpatrick to Frederick W. Sharon; letter, January 16, 1900, from John C. Kirkpatrick to Frederick W. Sharon, Sharon Family Papers C-B 777, part 1, box 8, folder 1, Bancroft Library, University of California at Berkeley.

74. Gladys Hansen and Emmet Condon, *Denial of Disaster* (San Francisco: Cameron and Company, 1989), 27.

75. Hansen and Condon, *Denial of Disaster,* 27.

76. Hansen and Condon, *Denial of Disaster,* 65–66.

77. Hansen and Condon, *Denial of Disaster,* 127; Charles Derleth, "The Destructive Extent of the California Earthquake of 1906; Its Effect Upon Structures and Structural Materials, within the Earthquake Belt," in *The California Earthquake of 1906,* ed. David Starr Jordan (San Francisco: A. M. Robertson, 1907), 135–36; Department of the Interior, *The San Francisco Earthquake and Fire of April 18, 1906,* Bulletin no. 324 (Washington, D.C.: Government Printing Office, 1907), 97; and "Facts on the Palace Hotel," California Historical Society, 3.

78. Grand Pacific Hotel (Old) folder, Chicago History Museum; Frank A. Randall, *History of the Development of Building Construction in Chicago* (Chicago: University of Illinois Press, 1949), 172; and Miles Berger, *They Built Chicago* (Chicago: Bonus Books, 1992), 94.

79. Condit, *American Building,* 120–30; "Hotel St. Francis," Hotels, Warshaw Collection of Business Americana, Archives Center, National Museum of American History, Smithsonian Institution, Washington, D.C.; A. C. David, "Three New Hotels," *The Architectural Record,* vol. 17 (March 1905), 167–88; and "The Far-Famed Hotel St. Regis," Hotels, Warshaw Collection.

80. "Palace Hotel," c. 1930. State Library of California; "The Palace Hotel of San Francisco," *The Hotel Monthly,* vol. 18, no. 208 (July 1910), 42; Stoll, "The Palace Hotel Court, Old and New," 45; Palace Hotel Company, "Palace Hotel, San Francisco," c. 1921, State Library of California; and Thomas W. Sweeney, "Legendary Palace Hotel Restored in San Francisco," *Historic Preservation News* (April 1991), 10.

CHAPTER SEVEN: The "New" Modern Hotel, 1880–1920

Subtitle: Jesse Lynch Williams, "A Great Hotel (The Conduct of Great Businesses—Second Paper)," *Scribner's Magazine,* vol. 21 (February 1897), 140.

1. "Chicago Ahead of the World," *Land Owner,* vol. 5, no. 5 (May 1873), 79.

2. John R. Riggleman, "Building Cycles in the United States, 1875–1932," *Journal of the American Statistical Association* 28 (1933): 178, 181.

3. Jefferson Williamson, *The American Hotel, An Anecdotal History* (New York: Alfred A. Knopf, 1930), 261, 264.

4. George Augustus Sala, *America Revisited: From the Bay of New York to the Gulf*

of Mexico, and From Lake Michigan to the Pacific, 2nd ed. (London: Vizetelly and Co., 1882), 1:121–26, 1:138–39, 1:151; on the Continental Hotel, Philadelphia, 1:167–81.

5. See *Daily National Hotel Reporter* (Chicago), the *Hotel World* (Chicago), *J. Stratton McKay & Co.'s Commercial Directory* (Chicago), and the *United States (Official) Hotel Directory for 1886 (Hotel Red Book)* (New York) as examples.

6. Anonymous, *The Horrors of Hotel Life, by a Reformed Landlord* (New York: Connelly and Curtis, 1884). See also Sam B. Harrison, *"Front!" or Ten Years with the Traveling Men* (New York: The American News Co., 1889).

7. Reyner Banham, *The Architecture of the Well-Tempered Environment* (Chicago: University of Chicago Press, 1984), 72. Banham chastises most historical treatments of the skyscraper that limit the necessary components for its development to the steel frame and the elevator.

8. Arthur C. David, "The St. Regis, The Best Type of Metropolitan Hotel," *Architectural Record,* vol. 15 (June 1904), 554. Skyscraper histories routinely ignore hotels, most likely because corporate skyscrapers consistently outpaced them. For example, in Spiro Kostof's *America by Design* (New York: Oxford University Press, 1967), the index entry for "Skyscrapers" reads "See office buildings," as if they are interchangeable terms. The sole exception is Girouard, who, in *Cities and People,* introduces his chapter on skyscrapers with a short history of American luxury hotels as examples of large buildings that antedated the skyscraper. Mark Girouard, *Cities and People: A Social and Architectural History* (New Haven: Yale University Press, 1985), 302–3.

9. Albert Bigelow-Paine, "The Workings of a Modern Hotel," *The World's Work,* vol. 5 (March 1903), 3171, 3184.

10. Bigelow-Paine, "Workings of a Modern Hotel," 3175.

11. Sinclair Lewis, *Babbitt* (1922; repr. New York: NAL Penguin, 1980), 153.

12. Frank Crowninshield, ed., *The Unofficial Palace of New York, A Tribute to the Waldorf-Astoria* (New York: The Waldorf-Astoria Corporation, 1939), x. Crowninshield was editor of *Vanity Fair* for twenty-three years. Another chronicler of the Waldorf-Astoria wrote of New York's earlier hotels, "But to stay at a hotel was not to taste something in the way of living usually reserved only for the rich. Hotel men simply didn't consider themselves called upon to provide anything more than the necessities and the luxuries of life in as attractive—but not as magnificent—a way as possible." James Remington McCarthy, *Peacock Alley, The Romance of the Waldorf-Astoria* (New York: Harper and Brothers, 1931). These two histories, like so many of the others of the Waldorf-Astoria, are antiquarian and anecdotal in nature. Their assessments of nineteenth-century luxury hotels are self-serving.

13. Arthur C. David, "Three New Hotels," *Architectural Record,* vol. 17 (March 1905), 168.

14. Matlack Price, "Great Modern Hotels of America," *Arts and Decoration* 21 (June 1924): 78.

15. Albert Stevens Crockett, *Peacocks on Parade* (1931; repr. New York: Arno Press, 1975), 42–44.

16. Edward Hungerford, *The Story of The Waldorf-Astoria* (New York: G. P. Putnam's Sons, 1925), 129.

17. Montgomery Schuyler, "Henry Janeway Hardenbergh," *Architectural Record,* vol. 6 (January–March 1897), 335–38; "Henry Janeway Hardenbergh," *Dictionary of American Biography* (New York: Charles Scribner's Sons, 1932), 8:240–41; and Henry F. Withey

and Elsie Rathburn Withey, "Henry Janeway Hardenbergh," *Biographical Dictionary of American Architects (Deceased)* (Los Angeles: New Age Publishing, 1956), 263–64. Hardenbergh also designed office buildings, country homes, and city dwellings, particularly New York row houses.

18. Mona Domosh, *Invented Cities: The Creation of Landscape in Nineteenth-Century New York and Boston* (New Haven: Yale University Press, 1996), 73–76.

19. H. J. Hardenbergh, "Hotel," in *A Dictionary of Architecture and Building, Biographical, Historical, and Descriptive,* ed. Russell Sturgis (New York: The Macmillan Co., 1901), 2:410.

20. "Architectural Aberrations, No. 17, The New York Family Hotel," *Architectural Record,* vol. 11 (July 1901), 700.

21. Hardenbergh, "Hotel," 2:410.

22. Bigelow-Paine, "The Workings of a Modern Hotel," 3181.

23. "The Knickerbocker Hotel," *Architectural Record,* vol. 21 (January 1907), 1.

24. David, "Three New Hotels," 185.

25. "The Knickerbocker Hotel," 2. In 1934, the King Cole mural was moved to the St. Regis Hotel, where it continues to reign in the King Cole Bar. See *Hotel St. Regis* (New York: ITT Sheraton, c. 1993).

26. Walter T. Stephenson, "Hotels and Hotel Life in New York," *Pall Mall Magazine,* vol. 31 (September–December 1903), 253.

27. Crockett, *Peacocks on Parade,* 54–55; and McCarthy, *Peacock Alley,* 60–61.

28. David, "Three New Hotels," 175. The best example of a still-chi-chi Palm Room that I know of is the Palm Court of New York's Plaza at 59th St. and Central Park South. The Plaza underwent a restoration and a conversion in part to condominiums, but the Palm Court has been brilliantly restored. While Hardenbergh does not acknowledge this, the idea of the Palm Room draws on the San Francisco Palace's palm-decorated Grand Court, which predated Hardenbergh's design for the Astoria's 34th Street indoor carriage turn-around; Hardenbergh acknowledges Paris's Grand Hotel, which also served as a model for the Palace's Grand Court. McCarthy, *Peacock Alley,* 84; and Hardenbergh, "Hotel," 410–11.

29. See, for example, William Hutchins, "New York Hotels, II," *Architectural Record,* vol. 12 (November 1902), 627; and "A Twentieth-Century Creation," *Harper's Weekly,* vol. 47, January 24, 1903, 156.

30. Arthur E. McFarlane, "Lodgings for the Rich," *Everybody's Magazine,* vol. 21 (September 1909), 336.

31. *New York Daily Tribune,* December 6, 1852.

32. David, "Three New Hotels," 183.

33. Price, "Great Modern Hotels of America," 78.

34. "The Past and Present of Hotels," *Atlantic Monthly,* vol. 70 (September 1892), 426.

35. Jesse Lynch Williams, "A Great Hotel (The Conduct of Great Businesses—Second Paper)," *Scribner's Magazine,* vol. 21 (February 1897), 140.

36. Bigelow-Paine, "The Workings of a Modern Hotel," 3176; McFarlane, "Lodgings for the Rich," 338; and *The St. Regis Hotel* (New York: The St. Regis Hotel Corp., 1905), 24. Warshaw Collection of Business Americana, Archive Center, National Museum of American History, Smithsonian Institution, Washington, D.C.

37. W. Sydney Wagner, "The Hotel Plan," *Architectural Forum,* vol. 39 (November 1923), 218; Hutchins, "New York Hotels, II," 622; and Williams, "A Great Hotel," 143.

38. Williams, "A Great Hotel," 139–43; and Hutchins, "New York Hotels, II," 622.

39. "Electric Service in World's Largest Hotel," *Electrical World*, vol. 73, no. 22, May 31, 1919, 1102; and Frederick G. Colton, "Mechanical and Kitchen Equipment of Hotel Pennsylvania," *Architectural Forum*, vol. 30, no. 4 (April 1919), 97.

40. Fred Warren Parks, "A Study in California Hotel Management, A Look Behind the Scenes," *Overland Monthly*, vol. 29, 2nd Series (April 1897), 403; Williams, "A Great Hotel," 138; and letter, September 27, 1927, from Ernest J. Stevens to Henry Hoppe, Laundry Manager, box 30, Ernest J. Stevens Papers, Chicago History Museum.

41. J. O. Dahl, *Kitchen Management* (New York: Harper and Bros., 1928), xix; and *The Stevens Prospectus*, 20–27, Ernest J. Stevens Papers, box 31, file 10, Chicago History Museum. For an extended discussion of the hotel kitchen, see Molly W. Berger, "The Magic of Fine Dining: Invisible Technology and the Hotel Kitchen," *ICON, Journal of the International Committee for the History of Technology* 1 (1995): 106–19.

42. A 1933 list of booked conventions from the Stevens Hotel, Chicago, suggests the incredibly wide variety of groups that became professional "conventioneers." Some of these groups included the National Canners, the National Food Brokers, General Motors, the Automotive Electric Association, the Mississippi Valley Conference of State Highway Departments, the International Association of Dental Schools, the Toy Manufacturers of the USA, the Radio Manufacturers Association, the American Oil Burner Association, the American Proctologic Society, the Central Conference of American Rabbis, various national fraternities and sororities, the National Education Association, the Society of Industrial Engineers, the Mystic Order of Veiled Prophets of the Enchanted Realm, the International Baby Chick Association, the International Apple Association, the National Society of Denture Prosthetists, the National Shorthand Reporters Association, the American Chemical Society, and the American Gas Association.

43. Waldorf-Astoria Hotel, Entertainment Correspondence Mar.–Dec. 1906, A–Z box 1, New York Public Library.

44. Williams, "A Great Hotel," 146.

45. Williams, "A Great Hotel," 147–50; and "The Workings of a Modern Hotel," 3173.

46. "Kitchen and Bakery, The Blackstone, Chicago," *Hotel Monthly*, vol. 18, no. 207 (June 1910), 48D–F.

47. Williams, "A Great Hotel," 149–50.

48. *Chicago Post*, August 25, 1909. Ernest J. Stevens Papers, Chicago History Museum.

49. Hutchins, "New York Hotels, II," 621.

50. David, "The St. Regis, The Best Type of Metropolitan Hotel," 556.

51. David, "The St. Regis, The Best Type of Metropolitan Hotel," 550, 567–68.

52. "A Twentieth-Century Creation," *Harper's Weekly*, vol. 47 (January 1903), 157.

53. Hardenbergh, "Hotel," 412; Hutchins, "New York Hotels, II," 624; and Charles D. Wetmore, "The Development of the Modern Hotel," *Architectural Review*, vol. 2, new series (April 1913), 38.

54. Wetmore, "Development of the Modern Hotel," 38. The Hotel McAlpin (New York City, 1913), at the time the world's largest hotel, had fifteen hundred rooms and eleven hundred baths. "Hotel McAlpin, New York," *Architecture and Building*, vol. 45 (February 1913), 236.

55. "Growing Attractiveness of the Hotel Business," *Hotel Monthly*, vol. 31 (April 1923), 20.

56. "Mechanical Equipment Boosts the Payroll," *Hotel Monthly*, vol. 31 (November 1923), 21. Standard practice in this time period spelled the word *employee* with a single *e*.

57. Hiram Hitchcock, "The Hotels of America," in *One Hundred Years of American Commerce, 1795–1895,* ed. Chauncey M. Depew (New York, D. O. Haynes and Co., 1895), 155.

58. Williams, "A Great Hotel"; McFarlane, "Lodgings for the Rich," 337–40; and Bigelow-Paine, "Workings of a Modern Hotel," 3179.

59. As quoted in Martha Banta, *Taylored Lives: Narrative Productions in the Age of Taylor, Veblen, and Ford* (Chicago: University of Chicago Press, 1993), 173.

60. Horace Leland Wiggins, "Service and Administration Requirements," *Architectural Forum*, vol. 39, no. 5 (November 1923), 240–42; and "The Coming of the Chain Hotel," *The Literary Digest*, vol. 90, July 3, 1926, 48.

61. Wiggins, "Service and Administration Requirements," 242.

62. W. Sydney Wagner, "The Statler Idea in Hotel Planning and Equipment," *Architectural Forum*, vol. 27 (November 1917), 115.

63. Sinclair Lewis, *Work of Art* (New York: Doubleday, Doran and Co., 1934), 359.

64. *New York World*, April 22, 1928; and *Buffalo Evening Times*, n.d. Both in Ellsworth Statler Collection (3879, box 1), Rare and Manuscript Collections, Carl A. Kroch Library, Cornell University Library. Hereafter cited as Statler Collection.

65. Floyd Miller, *Statler, America's Extraordinary Hotelman* (New York: Statler Foundation, 1968), 22. See also Rufus Jarman, *A Bed for the Night: The Story of the Wheeling Bellboy, E. M. Statler, and His Remarkable Hotels* (New York: Harper and Brothers, 1952), 15. Both of these biographies are celebratory, but Miller's is more historically conscientious.

66. Miller, *Statler*, 33.

67. Jarman, *Bed for the Night*, 21.

68. Miller, *Statler*, 29–60; and Jarman, *Bed for the Night*.

69. Frank Luther Mott reports that the hotel dailies only lasted through the 1880s. *The Hotel Monthly* and the *Hotel Gazette* began publication in 1893, the *Hotel Bulletin* in 1900. Frank Luther Mott, *A History of American Magazines* (Cambridge: Harvard University Press, 1957), 3:136; 4:189. *Hotel Gazette* (August 1901), quoted in Miller, *Statler*, 65.

70. Scrapbook, Statler Collection (box 2); *New York Journal*, 1901; and "The Pan-American," *Hotel World*, August 10, 1901, Statler Collection (box 2).

71. David E. Nye, *American Technological Sublime* (Cambridge: MIT Press, 1996), 149.

72. "Trip to the Big Show," n.d., Statler Collection (box 2).

73. "St. Louis and the World's Fair," *The Hotel Monthly*, vol. 12, no. 134 (May 1904), 21. One thing a wooden hotel compromised on was privacy. In an article entitled "Over the Transom," one writer noted that she had "always doubted the stories she had heard about the brides making love, now [after staying at The Inside Inn] she [was] forced to admit that in some cases it is true." Statler Collection (box 2 scrapbook).

74. Advertisement, "The Inside Inn!" Statler Collection (box 2). This concept predates Disney's first in-park hotels, most notably the 1971 A-frame Contemporary Resort Hotel that straddles the monorail at Walt Disney World, Orlando, Florida, and sold convenience as one of its most attractive characteristics.

75. Miller, *Statler*, 103; Jarman, *Bed for the Night*, 74; "St. Louis and the World's Fair,"

Hotel Monthly, vol. 12, no. 134 (May 1904), 20–24; "Inside and Outside the Inside Inn," *Hotel Monthly,* vol. 12, no. 137 (August 1904), 18–25; and "Friendship Tokens at the Inn," *Hotel Monthly,* vol. 12, no. 140 (November 1904), 28.

76. "Friendship Tokens at the Inn," 28.

77. The Buckingham Hotel (New York, 1876), known for its luxurious appointments, had a bathroom with each bedroom suite. "The Buckingham," Warshaw Collection.

78. *Hotel Monthly* (n.d.), quoted in Miller, *Statler,* 95.

79. Miller, *Statler,* 87–102; "History Re the Acquisition of, and Financial Facts Relative to, Properties Owned by Hotels Statler Company, Inc.," Manuscript, Statler Collection (box 1); "A Statler Hotel for Cleveland," *Hotel Monthly,* vol. 18, no. 213 (December 1910), 57; and "Hotel Statler, Cleveland, Fireproof," *Hotel Monthly,* vol. 21, no. 238 (January 1913), 54.

80. "Hotel Statler, Cleveland, Fireproof," 54.

81. See Lisa Pfueller Davidson, "Consumption and Efficiency in the 'City within a City': Commercial Hotel Architecture and the Emergence of Modern American Culture, 1890–1930" (Ph.D. diss., George Washington University, 2003), 52–55.

82. "George Brown Post," in *Dictionary of American Biography,* ed. Allen Johnson (New York: Charles Scribner's Sons, 1929), 8:116.

83. "Hotel Statler, Cleveland, Fireproof," 54.

84. Miller, *Statler,* 112.

85. "Hotel Statler, Cleveland, Fireproof," 56. See also "What the 'Adam Style' Is," *Statler Salesmanship,* vol. 3, no. 5 (February 1915), 71; David E. Tarn, "New York's Newest Hotel, Notes on the Hotel McAlpin," *The Architectural Record,* vol. 33 (March 1913), 236; and Walter S. Schneider, "The Hotel Biltmore," *The Architectural Record,* vol. 35 (March 1914), 232.

86. Certificate of Incorporation, March 4, 1914, Statler Collection (box 1, folder 4). Papers showed a 1913 net profit of $187,087.66.

87. "Statler Hotels Perennial Source of News," *Hotel Monthly,* vol. 23, no. 272 (November 1915), 88.

88. "For Betterment of Service, The Statler Service Code," *Hotel Monthly,* vol. 18, no. 211 (October 1910), 37; and "Statler Service Codes," *Hotel Monthly,* vol. 23, no. 266 (May 1915), 26–27.

89. "Statler's Talk on Tipping," *Hotel Monthly,* vol. 20, no. 229 (April 1912), 78.

90. "Statler Service Codes," 27.

91. See Philip Scranton, "Manufacturing Diversity: Production Systems, Markets, and an American Consumer Society, 1870–1930," *Technology and Culture* 35 (July 1994): 487.

92. E. M. Statler, "Golden Rules of Hotel Keeping," *American Magazine,* vol. 83 (May 1917), 121.

93. E. M. Statler, "What Salesmanship Is and Means to You," *Statler Salesmanship,* vol. 1, no. 1 (October 1913), 2.

94. E. M. Statler, "How We Practice Business Good Manners," *System, The Magazine of Business,* vol. 31 (April 1917), 371.

95. A complete run of *Statler Salesmanship* can be found uncatalogued in the Manuscript Division of the Olin Library, Cornell University Library.

96. "Special Bulletins for Department Meetings," Statler Collection (box 1).

97. "Hotel Statler, Cleveland, Fireproof," 52–79; "Statler-Detroit, The Complete Hotel," *Hotel Monthly*, vol. 23, no. 264 (March 1915), 44–76; "Fourth Hotel Statler Is Located in St. Louis," *Hotel Monthly*, vol. 26, no. 299 (February 1918), 40–70; "Hotel Statler of Boston, 1300 Rooms," *Hotel Monthly*, vol. 36, no. 418 (January 1928), 30–67; and *Buffalo Courier-Express*, February 5, 1928, Statler Collection (box 2).

98. W. Sydney Wagner, "The Statler Idea in Hotel Planning and Equipment, I. Introduction," *The Architectural Forum*, vol. 27, no. 5 (November 1917), 115–18; "II. The Development of the Typical Floor Plan," *The Architectural Forum*, vol. 27, no. 6 (December 1917), 165–70; and "III. Sample Room Floors and Restaurant Service," *The Architectural Forum*, vol. 28, no. 1 (January 1918), 15–18.

99. Wagner, "II, The Development of the Typical Floor Plan," 167.

100. Lewis, *Babbitt*, 153.

101. "The Coming of the Chain Hotel," *The Literary Digest*, vol. 90, July 3, 1926, 48–49.

102. L. M. Boomer, "How We Fitted Ford's Principles to Our Business," *System, The Magazine of Business*, vol. 44 (October 1923), 421–24, 491–92.

103. W. L. Cook, "Radio for Every Statler Guest," *National Hotel Review*, February 11, 1928, 64, Statler Collection (box 2).

104. Tarn, "New York's Newest Hotel, Notes on the Hotel McAlpin," 241.

105. McCarthy, *Peacock Alley*, 193; and Henry B. Lent, *The Waldorf-Astoria* (New York: Hotel Waldorf-Astoria Corporation, 1934), 20.

CHAPTER EIGHT: The Stevens Hotel, Chicago, 1927

Subtitle: "The Stevens of Chicago; 3000 Rooms," *Hotel Monthly*, vol. 35 (July 1927), 84.

1. William Cronon, *Nature's Metropolis: Chicago and the Great West* (New York: W. W. Norton and Company, 1991), 63; *Chicago Herald*, January 5, 1908, Ernest J. Stevens Papers (box 15 scrapbook), Chicago History Museum. Hereafter cited as Stevens Papers; and Frederic Cople Jaher, *The Urban Establishment, Upper Strata in Boston, New York, Charleston, Chicago, and Los Angeles* (Urbana: University of Illinois Press, 1982), 459, 541.

2. "Historical Sketch of Ernest J. Stevens and the Stevens Family," Finders Guide, Stevens Papers.

3. *Chicago Tribune*, February 9, 1972, Stevens Papers (box 1); "Abstract of Record," *People of the State of Illinois v. Ernest J. Stevens*, Criminal Court, Cook County, 421–23, hereafter cited as "Abstract," Stevens Papers (box 21, file 5).

4. Chicago's Loop is the central business district, its name deriving from the elevated train whose rectangular path defines its boundaries. Carl W. Condit, *Chicago 1910–1929: Building, Planning, and Urban Technology* (Chicago: University of Chicago Press, 1973), 7; and *Chicago Tribune*, January 4, 1908, Stevens Papers (box 15 scrapbook).

5. Condit, *Chicago 1910–1929*, 97, 151–52. See also Col. W. A. Starrett, *Skyscrapers and the Men Who Build Them* (New York: Charles Scribner's Sons, 1928), 32.

6. The Chicago fire of 1871 destroyed the entire business district, so all buildings in the Loop were "post-fire." *Chicago News*, May 1, 1908; and *Chicago News*, September 11, 1908, Stevens Papers (box 15 scrapbook).

7. "Chicago's New Hotel LaSalle," *Hotel Monthly*, vol. 18, no. 204 (March 1910), 40; and "Abstract," 422.

8. "Chicago's New Hotel LaSalle," 54.

9. Letter, George Gazley to W. S. Quimby and Co., Stevens Papers (box 9, file 3); and letter, George Gazley to Froehling and Heppe, Stevens Papers (box 9, file 3).

10. "Abstract," 422–23.

11. "Abstract," 452.

12. Stevens Papers (box 9, file 6).

13. "Abstract," 424.

14. "Abstract," 424; and E. J. Stevens, "Radio Speech," *Daily National Hotel Reporter,* March 17, 1925, Stevens Papers (box 42, file 9).

15. W. Sydney Wagner, "The Statler Idea in Hotel Planning and Equipment, I. Introduction," *The Architectural Forum,* vol. 27 (November 1917), 116.

16. Robert Bruegmann, *Holabird and Roche, Holabird and Root: An Illustrated Catalog of Works, 1880–1940* (New York: Garland Publishing Co., in cooperation with the Chicago History Museum, 1991), 2:241.

17. "Abstract," 425; Stevens, "Radio Speech"; telegram, May 15, 1922, from E. J. Stevens to J. S. Stevens, Stevens Papers (box 18, file 9); and Bruegmann, *Holabird and Roche, Holabird and Root,* 2:244.

18. Eugene S. Taylor, "Chicago and the Lake Front," *Hotel World,* vol. 104, no. 19, May 7, 1927, 20. For more on the Chicago Plan, see Daniel Bluestone, *Constructing Chicago* (New Haven: Yale University Press, 1991), 183–204; and Condit, *Chicago, 1910–1929,* 59–85.

19. The Auditorium was a multiuse building containing the hotel, an opera house, and an office building. For more on the Auditorium Building, see George A. Larson and Jay Pridmore, *Chicago Architecture and Design* (New York: Harry N. Abrams, 1993), 62–67 and passim; and Joseph M. Siry, *The Chicago Auditorium Building, Adler and Sullivan's Architecture and the City* (Chicago: University of Chicago Press, 2002).

20. "Abstract," 426–28.

21. On page 428 of the "Abstract," E. J. Stevens reported the purchase price as $2.5 million. The $4.5 million figure comes from the *Economist,* March 4, 1922, as published in Bruegmann, *Holabird and Roche, Holabird and Root,* 2:242. The valuation of the land was $6 million.

22. "Appraisal Report, Stevens Hotel, Chicago, Illinois," Stevens Papers (box 70, file 11).

23. "Prospectus, The Stevens Hotel," September 9, 1924, Stevens Papers (box 31, file 10).

24. The recent restorations of this genre of hotels have resulted in hotels with less than half as many rooms as when built. Bedrooms were typically so small that two needed to be combined to make one normal late-twentieth-century room. Examples include the Cleveland Renaissance (formerly the Hotel Cleveland) and the Chicago Hilton and Towers (formerly the Stevens). When this author stayed at the Chicago Hilton and Towers for the annual meeting of the American Historical Association, she had a room with two bathrooms, the result of two bedrooms, each with bath, being combined. This plan continues to accommodate conventioneers, who often room with unrelated, although not usually unacquainted, persons.

25. "The Stevens of Chicago; 3000 Rooms," *Hotel Monthly,* vol. 35 (July 1927), 45.

26. Starrett, *Skyscrapers and the Men Who Build Them,* 32–33. Starrett's excellent memoir details the history of skyscraper construction from the perspective of his family's various construction engineering firms. He credited Fuller, who died in 1900, with upgrading the contractor from a "boss carpenter," creating "both an industry and a profession, visualizing the building problem in its entirety—promotion, finance, en-

gineering, labor and materials; [while] the architect reverted to his original function of design." The increasing complexity of tall building construction required an equally complex construction business organization to execute the growing demands of architects and their clients.

27. "Abstract," 445.

28. "The Forest of Steel for World's Largest Hotel," *Hotel Monthly*, vol. 34, no. 396 (March 1926), 42.

29. "Steel Trusses Carry 22 Stories in Chicago Hotel," *Engineering News-Record*, vol. 96, April 22, 1926, 638–41; "When Engineering Design Problems Were Difficult," *Engineering and Contracting*, vol. 66 (June 1927), 253–61; and *Western Society of Engineers* (December 1926), as quoted in Bruegmann, *Holabird and Roche, Holabird and Root*, 245.

30. "Forest of Steel," 43–44.

31. E. J. Stevens, "Speech to Steel Men," Stevens Papers (box 31, file 15).

32. David E. Nye, *American Technological Sublime* (Cambridge: MIT Press, 1994), xiii, 4; and "The Flag on the Big Hotel," *Chicago Tribune*, as reprinted in *Hotel Monthly*, vol. 34, no. 399 (June 1926), 31.

33. "Abstract," 446–47.

34. Earle Ludgin, "Creating the Perfect Hotel," *Hotel World*, vol. 104, no. 19, May 7, 1927, 5, 10.

35. *Stepping at the Stevens*, vol. 5, no. 113, September 28, 1929, 6; and *Hotel World*, 104, no. 19, May 7, 1927, 42.

36. B. K. Gibson, "The Stevens Architecturally," *Hotel World*, vol. 104, no. 19, May 7, 1927, 23; "The Stevens, World's Largest Hotel," *Hotel Bulletin* (June 1927), 3; and Henry J. B. Haskins, "The Stevens Hotel, Chicago," *The Architectural Forum*, vol. 47, no. 2 (August 1927), 97.

37. "The Stevens of Chicago," 36.

38. Ludgin, "Creating the Perfect Hotel," 6–8.

39. Richard Bushman, *The Refinement of America: Persons, Houses, Cities* (New York: Vintage Books, 1992), 118–20; and Ludgin, "Creating the Perfect Hotel," 8–9.

40. Ludgin, "Creating the Perfect Hotel," 14.

41. Earle Ludgin, "The Kitchens of the Stevens Establish New Precedents," *Hotel World*, vol. 104, no. 19, May 7, 1927, 39; "The Stevens, World's Largest Hotel," 14; and Ludgin, "Creating the Perfect Hotel," 14–15.

42. "The Stevens of Chicago; 3000 Rooms," 84.

43. Gibson, "The Stevens Architecturally," 31; "The Stevens, World's Largest Hotel," 24; and "The Stevens of Chicago; 3000 Rooms," 53.

44. See Thomas P. Hughes, *Networks of Power: Electrification in Western Society, 1880–1930* (Baltimore: Johns Hopkins University Press, 1983), 201–26; Harold L. Platt, *The Electric City* (Chicago: University of Chicago Press, 1991). On the Stevens's electrical plant, see Thomas G. Thurston, "Mechanical and Electrical Equipment of the World's Largest Hotel," *National Engineer*, vol. 30, no. 7 (July 1926), 285–89; "World's Greatest Hotel to Have Own Power Plant," *Power*, vol. 64, no. 1, July 6, 1926, 32–33; J. A. Sutherland, "Demand Factors and Heat Balances in Plant of World's Largest Hotel—The Stevens," *Power*, vol. 65, no. 23, June 7, 1927, 879–81; and Egbert Douglas and Charles A. Crytser, "The Mighty Power Plant and Mechanical Department," *Hotel World*, vol. 104, no. 19, May 7, 1927, 45–47.

45. Douglas and Crytser, "Mighty Power Plant," 48–49.

46. "Forest of Steel," 42–46; "The Stevens of Chicago; 3000 Rooms," 82; Bruegmann, *Holabird and Roche, Holabird and Root*, 240–41; and "The Stevens, World's Largest Hotel," 28.

47. "Two Great Hotels Soon to be Finished," *Chicago Commerce*, vol. 22, November 13, 1926, 9.

48. Douglas and Cryster, "Mighty Power Plant," 44.

49. "The Stevens, World's Largest Hotel," 24; and Ludgin, "Creating the Perfect Hotel," 1.

50. Thurston, "Mechanical and Electrical Equipment," 288.

51. Douglas and Cryster, "Mighty Power Plant," 50–51.

52. "The Stevens of Chicago; 3000 Rooms," 36.

53. "First Floor Plan," *The Stevens, The World's Greatest Hotel* (Chicago: Stevens Hotel Company, n.d.). Collections of the Hagley Museum and Library.

54. Haskins, "The Stevens Hotel, Chicago," 102.

55. Ludgin, "Creating the Perfect Hotel," 16.

56. *Stopping at the Stevens*, September 24, 1927.

57. Letter, Irene C. Murphy to E. J. Stevens, October 18, 1929, Stevens Papers (box 45, file 8).

58. *The Stevens, The World's Greatest Hotel.* See Stevens Papers for clippings about other hotels (box 7, file 5); letters for advice on Hotel LaSalle (box 8, file 2); file on Morrison Hotel, Chicago (box 9, file 5); expense reports from Statler Hotels (box 9, file 6); the Savoy Hotel (box 9, file 8); and letters about sharing brochures (box 31, file 1).

59. *Stopping at the Stevens*, September 28, 1929; *The Stevens, The World's Greatest Hotel*; and Ludgin, "Creating the Perfect Hotel," 17–18.

60. Ernest J. Stevens, "The Hotel and the Field of Institution Economics," *Hotel Monthly* (September 1923), 48–50; and Ernest J. Stevens, "I am glad to meet," speech manuscript, Stevens Papers (box 18, file 11).

61. J. O. Dahl, *Kitchen Management* (New York: Harper and Bros., 1928), 101, 125, 183.

62. Letter, September 27, 1927, from E. J. Stevens to Mr. Henry Hoppe, Stevens Papers (box 30, file 12). Mr. Hoppe probably earned about $400 per month. Other salaries per month included washmen, $120; wringermen, $120; head washer and wringermen, $160; loaders, $115. Women laundry workers earned considerably less: ironers, $65; starchers, $70; bosom press operators, $80; lady clothes ironers, $75; folders/feeders, $65.

63. "The Stevens Personnel," *Hotel World*, vol. 104, no. 19, May 7, 1927, 54; letter, May 19, 1927, from E. J. Stevens to Ferdinand Karcher, Chef, Stevens Papers (box 36, file 6); and letter, September 2, 1927, from E. J. Stevens to Ferdinand Karcher, Stevens Papers (box 36, file 6).

64. Gibson, "The Stevens Architecturally," 24. Nicholas Pevsner states that Moscow's Hotel Rosslya (1964–67) relieved the Stevens of its title with its 3,128 rooms and suites. Nicholas Pevsner, *A History of Building Types* (Princeton: Princeton University Press, 1976); and "Be There—When the Whistle Blows!" Stevens Papers (box 27, file 9). Malaysia's First World Hotel reputedly has a cozy 6,118 rooms. Seventeen of the world's twenty largest hotels are located in Las Vegas. The smallest of the twenty has 2,916 rooms. See "20 Largest Hotels in the World," available at www.insidervlv.com/hotelslargestworld.html, accessed July 9, 2010.

65. "Revenue/Expense Statements, Stevens Hotel Company," Stevens Papers (box

39, file 4); "Abstract," 430; and "Statements of Registrations, House Counts and Averages for Years 1929–1928–1927," Stevens Papers (box 75, file 2).

66. "Statements of Registrations, House Counts and Averages for Years 1932–1931–1930," Stevens Papers (box 75, file 5).

67. "Statements of Registrations, House Counts and Averages for Years 1932–1931–1930"; and Stevens Hotel Company Report on Accounts Year Ended December 31, 1932, Stevens Papers (box 75, file 5).

68. *Daily National Hotel Reporter*, June 7, 1932, Stevens Papers (box 45, file 4).

69. See clippings in Stevens Papers (box 7, file 5; box 9, files 4, 5, 8; box 45, file 4).

70. "Abstract," 464.

71. *State of Illinois v. James W. Stevens, Raymond W. Stevens, Ernest J. Stevens*, no. 68027, Stevens Papers (box 45, file 7).

72. Unsigned note on Raymond Stevens's death, Stevens Papers (box 52, file 4).

73. See Stevens Papers (box 19, files 1–5).

74. *Chicago Herald*, February 11, 1934.

75. *Chicago Herald*, October 27, 1934.

76. *Insurance Index* (July 1942), Stevens Papers (box 42, file 1); *Insurance Index* (October 1933), Stevens Papers (box 42, file 2). See also list of jurors, Stevens Papers (box 59, file 12). The jurors included a musician, a teamster, a florist, a manager of an A & P store, an unemployed electrical worker, a freight claims adjustor, a steel worker, a tinplate foreman, an auto parts salesman, an engineer, and a printer.

77. *The Chicago American*, June 4, 1932.

78. Letter, April 10, 1933, from E. J. Stevens to Elizabeth Street Stevens, Stevens Papers (box 31, file 12).

79. "Appraisal Report," 9–10.

80. *Chicago Herald*, October 15, 1935; *Chicago Herald*, June 13, 1936; letter, June 3, 1936, from E. J. Stevens to James W. Stevens II and Raymond W. Stevens Jr., Stevens Papers (box 46, file 1); and Finder's Guide, 2, Stevens Papers.

81. Letter, June 11, 1932, from Burridge D. Butler to Ernest J. Stevens, Stevens Papers (box 19, file 3).

Conclusion

1. Sinclair Lewis, *Babbitt* (1922; repr. New York: NAL Penguin, 1980), 170, 14–15, 138, 151.

2. Frederic Cople Jaher, *The Urban Establishment: Upper Strata in Boston, New York, Charleston, Chicago, and Los Angeles* (Urbana: University of Illinois Press, 1982), 275–81; See also Molly W. Berger, "The Rich Man's City: Hotels and Mansions of Gilded Age New York," in *The American Hotel, The Journal of Decorative and Propaganda Arts*, vol. 25, ed. Molly W. Berger (Cambridge: MIT Press, 2005): 68–69.

3. "New Kind of Hotel Up Town," *New York Weekly Mirror*, December 7, 1944, 131.

4. Miriam Levin makes a similar point about the 1889 Eiffel Tower construction, that it served as an active agent of community building during the early Third Republic. Miriam Levin, *When the Eiffel Tower Was New* (South Hadley, MA: Mount Holyoke College Art Museum, 1989), 22–23.

5. Marianne Finch, *An Englishwoman's Experience in America* (London: Richard Bentley, 1853), 336.

6. *Gleason's Pictorial Drawing Room Companion,* vol. 1, no. 5, May 31, 1851, 1.

7. *Ohio Cultivator,* November 1, 1850.

8. As cited in, "Burnet House Relics Disappear but Long Search Reveals Them," *Times-Star,* April 6, 1927, Cincinnati Historical Society, Hotels and Motels H-832 vol. 1, A–B (clippings scrapbook); J. Stacy Hill, *A History of the Inns and Hotels of Cincinnati, 1793–1923* (Cincinnati: Hotel Gibson, 1922); and *Gleason's Pictorial Drawing Room Companion,* vol. 1, no. 5, May 31, 1851, 1.

9. "Order for Apartments for Lord Renfrew and Suite, Burnet House, Friday, September 28th, 1860." Cincinnati Historical Society, Mss q347 RM.

10. "Movements of the Prince," *New York Times,* October 1, 1860.

11. "The City of Red Wing," *Goodhue County Republican,* January 15, 1874.

12. "Red Wing," *St. Paul Daily Press,* August 6, 1874.

13. *Argus,* November 6, 1874.

14. *St. Paul Daily Pioneer Press,* November 27, 1875.

15. Eric Johannesen, *Cleveland Architecture, 1876–1976* (Cleveland: Western Reserve Historical Society, 1979), 131–33.

16. "A Twentieth-Century Creation," *Harper's Weekly* 47, January 24, 1903, 131.

17. "Van Sweringen, Oris Paxton and Mantis James," *Encyclopedia of Cleveland History,* available at http://ech.case.edu/ech-cgi/article.pl?id=VSOP, accessed March 21, 2010.

18. Ian S. Haberman, *The Van Sweringens of Cleveland: The Biography of an Empire* (Cleveland: Western Reserve Historical Society, 1979), 35.

19. Burnham's famous Plan of Chicago followed in 1909.

20. John J. Grabowski and Walter C. Leedy Jr., *The Terminal Tower, Tower City Center: A Historical Perspective* (Cleveland Western Reserve Historical Society, 1990), 26–27.

21. "Hotel Cleveland and Terminals Section," *Cleveland Plain Dealer,* December 22, 1918.

22. "Hotel Cleveland and Terminals Section," *Cleveland Plain Dealer,* December 22, 1918, 1, 3.

23. "Hotel Life of the Future," *Hotel Monthly,* vol. 28, no. 328 (July 1920), 30–31.

24. "The Stevens is Opened, A Hotel of Superlatives," *Chicago Daily Tribune,* May 3, 1927.

25. Robert Twombly, *Power and Style: A Critique of Twentieth-Century Architecture in the United States* (New York: Hill and Wang, 1995), 28.

26. "FutureHotel's Hotel Room of the Future," *Hotels, The Magazine of the Worldwide Hotel Industry,* August 13, 2009, available at www.hotelsmag.com/article/CA6676594 .html?industryid=47564, accessed August 15, 2009. *The Jetsons* was an early 1960s animated cartoon sitcom set in 2062. The middle-class Jetson family lived in a futuristic technological utopia that featured Space Age inventions. See Molly W. Berger, "Popular Culture and Technology in the Twentieth Century," in *A Companion to American Technology,* ed. Carroll Pursell (Oxford, UK: Blackwell Publishing, 2005), 395–96.

27. See Taj Hotels website, available at www.tajhotels.com, accessed August 15, 2009; see also Charles Allen, "The Taj Mahal Hotel Will, as Before, Survive the Threat of Destruction," *The Guardian,* December 3, 2008, 35, available at www.guardian.co.uk/ commentisfree/2008/dec/03/taj-mahal-hotel-mumbai, accessed August 15, 2009.

ESSAY ON SOURCES

Primary Sources

The endnotes document the primary sources used in the preparation of this book. It is useful, though, to discuss the range of sources and how they changed as the long nineteenth century advanced. For the early chapters, I made extensive use of published travel narratives, newspapers (including articles, want ads, social columns, and advertisements), periodicals, sheet music, city directories, handbooks, stranger's guides, pictorial histories, centennial histories, architectural descriptions and plans, diaries, maps, and many kinds of ephemera, such as hotel bills, menus, and brochures. These were located mostly at research, university, and public libraries and local historical societies. Several electronic databases now streamline the process of finding sources, especially for early newspapers and periodicals, with the inevitable drawback of "hits" numbering in the thousands.

For the middle third of the nineteenth century, I relied more heavily on American and British newspapers and mass circulation periodicals, particularly *Gleason's Pictorial Drawing-Room Companion, Ballou's Pictorial Drawing-Room Companion, Frank Leslie's Illustrated Newspaper, Harper's Weekly,* and *Putnam's.* Published observations about cities were also useful, such as Matthew Hale Smith, *Sunshine and Shadow in New York* (Harford: J. B. Burr and Co., 1869); George G. Foster's 1848 series of essays, "Philadelphia in Slices," reprinted in *Pennsylvania Magazine of History and Biography* 93 (1969): 55–60; and Junius Henri Browne, *The Great Metropolis: A Mirror of New York* (1869; repr. New York: Arno Press, 1975), to name just a representative few. By this time, there were more published hotel booklets and brochures, such as the Continental's *Traveller's Sketch.* The New-York Historical Society's Quinn Collection has a good collection for New York hotels. The best source for ephemera is the Warshaw Collection of Business Americana—Hotels located in the Archives Center, National Museum of American History, Smithsonian Institution. Hotel booklets and other publications can often be found in local research libraries or historical societies as well. The NMAH's massive Trade Literature Collection is useful for investigating various construction materials and systems, such as plumbing fittings or any of the thousands of physical components that go into a building. There are several published diaries that I found useful for social history, as have many other historians, such as Philip Hone, *The Diary of Philip Hone, 1828–1851,* ed. Allan Nevins, 2 vols. (New York: Dodd, Mead and Company, 1927); Louis Auchincloss, ed., *The Hone and Strong Diaries of Old Manhattan* (New York: Abbeville Press, 1989); and Sidney George Fisher, *A Philadelphia Perspective, The Diary of Sidney George Fisher Covering the Years 1834–1871,* Nicholas Wainwright, ed. (Philadelphia: Historical Society of Pennsylvania, 1967). In addition, the Isaiah Rogers diaries and their transcriptions by Denys Peter Myers are located at the Avery Architectural Library at Columbia University in New York City. These last would make an excellent source for someone writing a social history of nineteenth-century architects.

Beginning in the 1870s, the industry began to publish local trade periodicals such as *Daily Hotel Gazette,* the *New-York Traveler and United States Hotel Directory,* and the *Boston Daily Hotel Reporter.* Many of the larger American cities had similar publications. These recorded the comings and goings of both travelers and hoteliers. Toward the end of the nineteenth century, national trade journals such as *Hotel World, Hotel Bulletin,* and *Hotel Monthly* began publication. Architectural journals, such as *The Builder, Architectural Forum,* and *Architectural Record* had frequent articles and picture essays on hotels. I also found engineering journals useful, including *Engineering and Building, Architecture and Building,* and *Engineering News-Record.* Each of the engineering sub-specialties had its own journal as well. The shift in the types of sources is consonant with the growth of the professions and the national structures created to ensure professional standards and certification. The seminal source on this process is Robert Wiebe, *The Search For Order, 1877–1920* (New York: Hill and Wang, 1967).

Finally, several hotel archives exist that supported research on the hotels in this study. The Barnum City Hotel Papers are housed at the Maryland Historical Society. The Historical Society of Pennsylvania has a small archive on the Continental Hotel Company. The Bancroft Library at University of California at Berkeley holds the William C. Ralston Papers. The New York Public Library has manuscripts from the old Waldorf-Astoria. The Ernest J. Stevens Papers, which are often used by researchers working on Supreme Court Justice John Paul Stevens, are at the Chicago History Museum. The Rare and Manuscript Collections at the Carl A. Kroch Library at Cornell University houses the Ellsworth Statler Collection. While the short passage on the St. James Hotel in Red Wing, Minnesota, does not represent the extensive research I conducted there, the St. James Hotel graciously gave me access to their historic archive, and I also spent time profitably at the Goodhue County Historical Society and the Minnesota Historical Society.

Hotels

The literature on hotels is still thin, although interest in them as a topic for academic inquiry is developing. Paul Groth's excellent *Living Downtown: The History of Residential Hotels in the United States* (Berkeley: University of California Press, 1994) has an introductory chapter on luxury hotels but focuses on San Francisco's single-room occupancy hotels. My own edited volume, *The American Hotel, The Journal of Decorative and Propaganda Arts,* vol. 25 (Cambridge: MIT Press, published by the Wolfsonian-Florida International University, 2005) focuses on the 1880–1940 period and has a range of excellent essays that can serve as jumping off points for further study. A companion volume to the Wolfsonian's 2005 exhibit "The Architecture of Schultze and Weaver" is also valuable. See Marianne Lamonaca and Jonathan Mogul, eds., *Grand Hotels of the Jazz Age* (New York: Princeton Architectural Press, 2005). Andrew Sandoval-Strausz's beautifully illustrated *Hotel: An American History* (New Haven: Yale University Press, 2008) is the most recent scholarly book on hotels and particularly adds to the literature on hotels and law and civil rights. Annabel Jane Wharton's *Building the Cold War: Hilton International Hotels and Modern Architecture* (Chicago: University of Chicago Press, 2001) covers post–World War II Hilton International Hotels and the proliferation of American corporatism. Catherine Cocks, *Doing the Town: The Rise of Urban Tour-*

ism in the United States, 1850–1915 (Berkeley: University of California Press, 2001) has a long and useful chapter on urban hotels. No list of books on hotels would be complete without Jefferson Williamson's *The American Hotel, An Anecdotal History* (New York: Alfred A. Knopf, 1930). Published as it was at the end of the long nineteenth century, Williamson's book is, at this point, a primary source for a storehouse of minutiae and anecdotes that will lead scholars to a myriad of sources.

A number of fine dissertations are worth reading. Chief among these are Daniel Levinson Wilk, "Cliff Dwellers: Modern Service in New York City, 1800–1945" (Ph.D. diss., Duke University, 2005); Carolyn Brucken, "Consuming Luxury: Hotels and the Rise of Middle-Class Public Space, 1825–1860" (Ph.D. diss., George Washington University, 1997); and Lisa Pfueller Davidson, "Consumption and Efficiency in the 'City within a City': Commercial Hotel Architecture and the Emergence of Modern American Culture, 1890–1930" (Ph.D. diss., George Washington University, 2003). Older, but still valuable, are Doris Elizabeth King, "Hotels of the Old South, 1793–1860" (Ph.D. diss., Duke University, 1952); and Martin Bowden, "The Dynamics of City Growth: An Historical Geography of the San Francisco Central District, 1850–1931" (Ph.D. diss. University of California at Berkeley, 1967).

There are several books on tourist or resort hotels and tourism. These include Susan R. Braden, *The Architecture of Leisure: The Florida Resort Hotels of Henry Flagler and Henry Plant* (Gainesville: University Press of Florida, 2002); Jon Sterngass, *First Resorts: Pursuing Pleasure at Saratoga Springs, Newport, and Coney Island* (Baltimore: Johns Hopkins University Press, 2001); and Dona Brown, *Inventing New England: Regional Tourism in the Nineteenth Century* (Washington, D.C.: Smithsonian Institution Press, 1995). Additionally, sources on taverns include David W. Conroy, *In Public Houses: Drink and the Revolution of Authority in Colonial Massachusetts* (Chapel Hill: University of North Carolina Press, 1995); Kym S. Rice, *Early American Taverns: For the Entertainment of Friends and Strangers* (Chicago: Regnery Gateway, 1983); Peter Thompson, *Rum Punch and Revolution: Taverngoing and Public Life in the Eighteenth Century* (Philadelphia: University of Pennsylvania Press, 1999); and Sharon V. Salinger, *Taverns and Drinking in Early America* (Baltimore: Johns Hopkins University Press, 2002).

Urban Architecture and Buildings

Because this book focused on the material aspect of the hotels, I consulted many books on city architecture and building systems. Books that are particularly instructive for learning how building systems work are Carl W. Condit's, *Chicago 1910–1929: Building, Planning, and Urban Technology* (Chicago: University of Chicago Press, 1973) and *American Building*, 2nd ed. (Chicago: University of Chicago Press, 1982). Also valuable are Cecil D. Elliott, *Technics and Architecture: The Development of Materials and Systems for Buildings* (Cambridge: MIT Press, 1992); Reyner Banham, *The Architecture of the Well-Tempered Environment* (Chicago: University of Chicago Press, 1984); and Stewart Brand, *How Buildings Learn: What Happens after They're Built* (New York: Penguin Books, 1995). Books that focus on American architecture include Spiro Kostof, *America by Design* (New York: Oxford University Press, 1967); Carter Wiseman, *Shaping a Nation: Twentieth-Century American Architecture and Its Makers* (New York: W. W. Norton, 1998); Oliver Larkin, *Art and Life in America* (New York: Rinehart and

Co., 1945); Bates Lowry, *Building a National Image: Architectural Drawings for the American Democracy, 1789–1912* (Washington, DC: National Building Museum, 1985); and Robert Twombly, *Power and Style: A Critique of Twentieth-Century Architecture in the United States* (New York: Hill and Wang, 1995).

Many of the books already mentioned contained useful analyses of the rise of the American skyscraper. Additional sources include Paul Goldberger, *The Skyscraper* (New York: Alfred A. Knopf, 1981); Sarah Bradford Landau and Carl W. Condit, *The Rise of the New York Skyscraper, 1865–1913* (New Haven: Yale University Press, 1996); and Roberta Moudry, *The American Skyscraper: Cultural Histories* (Cambridge: Cambridge University Press, 2005). Because Chicago was central to the development of the American skyscraper, books that focus on Chicago architecture are particularly important. They include Miles Berger, *They Built Chicago* (Chicago: Bonus Books, 1992); Daniel Bluestone, *Constructing Chicago* (New Haven: Yale University Press, 1991); Robert Bruegmann, *Holabird and Roche, Holabird and Root: An Illustrated Catalog of Works, 1880–1940*, 3 vols. (New York: Garland Publishing, 1991); and George A. Larson and Jay Pridmore, *Chicago Architecture and Design* (New York: Harry N. Abrams). Other sources include Larry R. Ford, *Cities and Buildings, Skyscrapers, Skid Rows, and Suburbs* (Baltimore: Johns Hopkins University Press, 1994); Margot Gayle, *Cast-Iron Architecture in New York* (New York: Dover Publications, 1974); Mark Girouard, *Cities and People: A Social and Architectural History* (New Haven: Yale University Press, 1985); and Nikolaus Pevsner's classic *A History of Building Types* (Princeton: Princeton University Press, 1976). Pevsner's chapter on hotels puts their development in a global context and traces their origins back to early modern European and British inns. Drawing on Jefferson Williamson, Pevsner regards Barnum's, the Tremont House, and the Astor House as the first monumental hotels.

Urban History

Urban histories provide the context for understanding how hotels function in the city. Max Page's *The Creative Destruction of Manhattan, 1900–1940* (Chicago: Chicago University Press, 1999) employs Joseph Schumpeter's theory of creative destruction to explain the continual rebuilding of New York City. Keith Revell's *Building Gotham: Civic Culture and Public Policy in New York City, 1898–1938* (Baltimore: Johns Hopkins University Press, 2003) examines the relationship between local government and efforts to control building development. In *Domesticating the Street: The Reform of Public Space in Hartford, 1850–1930* (Columbus: Ohio State University Press, 1999), Peter C. Baldwin uses Hartford, Connecticut, to study how Progressive reformers tamed the chaotic streets of the early twentieth century. William Cronon's *Nature's Metropolis, Chicago and the Great West* (New York: W. W. Norton, 1991) offers an environmental and economic history of Chicago's development in the nineteenth century. Other books that proved valuable in understanding the growth of nineteenth-century cities include Gunther Barth, *Instant Cities: Urbanization and the Rise of San Francisco and Denver* (New York: Oxford University Press, 1975); Thomas Bender, *Toward an Urban Vision: Ideas and Institutions in Nineteenth-Century America* (Baltimore: Johns Hopkins University Press, 1975); Steven Conn and Max Page, eds., *Building the Nation: Americans Write about Their Architecture, Their Cities, and Their Landscape* (Philadelphia: University of Pennsylvania Press, 2003); Mona Domosh, *Invented Cities, The Creation of*

Landscape in Nineteenth-Century New York and Boston (New Haven: Yale University Press, 1996); Sam Bass Warner, *The Private City: Philadelphia in Three Periods of Its Growth* (Philadelphia: University of Pennsylvania Press, 1968) and *Streetcar Suburbs: The Process of Growth in Boston, 1870–1900* (Cambridge: Harvard University Press, 1978); and James Gilbert, *Perfect Cities: Chicago's Utopias of 1893* (Chicago: Chicago University Press, 1991). Books that give important insight on New York City, including its underworld, include Timothy J. Gilfoyle, *City of Eros: New York City, Prostitution, and the Commercialization of Sex, 1790–1920* (New York: W. W. Norton, 1992); and Edward K. Spann, *The New Metropolis: New York City, 1840–1857* (New York: Columbia University Press, 1981).

Landscape and Space

I also turned to cultural geographers to understand the role played by the hotels in my book in the emerging and shifting built environment of their cities and the cultural ramifications of urban development. Books that provided theoretical structure for this study include Dennis Cosgrove, *Social Formation and Symbolic Landscape* (Totowa, NJ: Barnes and Noble Books, 1985); David Ward and Olivier Zunz, eds., *The Landscape of Modernity: Essays on New York City, 1900–1940* (New York: Russell Sage Foundation, 1992); Jürgen Habermas, *The Structural Transformation of the Public Sphere*, trans. Thomas Burger with the assistance of Frederick Lawrence (Cambridge: MIT Press, 1989); David Harvey, *Social Justice and the City* (Baltimore: Johns Hopkins University Press, 1973); John Brinckerhoff Jackson, *Discovering the Vernacular Landscape* (New Haven: Yale University Press, 1984) and *A Sense of Place, A Sense of Time* (New Haven: Yale University Press, 1994); Henri Lefebvre, *Writings on Cities* (Oxford: Blackwell Publishers, 1996); Karl B. Raitz and John Paul Jones III, "The City Hotel as Landscape, Artifact and Community Symbol," *Journal of Cultural Geography* 9 (Fall/Winter 1988): 17–36; John R. Stilgoe, *Common Landscape of America, 1580–1845* (New Haven: Yale University Press, 1982) and *Metropolitan Corridor* (New Haven: Yale University Press, 1983); Stephen Toulmin, *Cosmopolis* (Chicago: University of Chicago Press, 1992); and David M. Scobey, *Empire City: The Making and Meaning of the New York City Landscape* (Philadelphia: Temple University Press, 2002).

Modernity and Progress

The two books that underpinned my framework for contextualizing the development of luxury hotels within theories of modernity and ideas about progress are Marshall Berman's *All That Is Solid Melts into Air: The Experience of Modernity*, 2nd ed. (1982; New York: Penguin Books, 1988); and David Harvey's *The Condition of Postmodernity: An Enquiry into the Origins of Cultural Change* (Oxford: Basil Blackwell, 1989). Other important books were George Cotkin, *Reluctant Modernism* (New York: Twayne Publishers, 1992); Alphonse Ekirch Jr., *The Idea of Progress in America, 1815–1860* (New York: Columbia University Press, 1941); Anthony Giddens, *The Consequences of Modernity* (Stanford: Stanford University Press, 1990); Paul Johnson, *The Birth of the Modern: World Society 1815–1830* (New York: Harper Collins, 1991); Stephen Kern, *The Culture of Time and Space* (Cambridge: Harvard University Press, 1983); and T. J. Jackson Lears, *No Place of Grace: Antimodernism and the Transformation of American Culture, 1880–1920*

(Chicago: University of Chicago Press, 1983). Lears's book also plays into my analyses of luxury and masculinity.

Political Economy

The evolution of the luxury hotel took place in the years when the United States was sorting out the workings of a new political system and its relationship to emerging forms of capitalism. The earlier chapters were heavily influenced by the literature on republicanism and market and industrial capitalism. Important books include Joyce Appleby, *Capitalism and a New Social Order: The Republican Vision of the 1790s* (New York: New York University Press, 1984); Christopher Clark, *The Roots of Rural Capitalism: Western Massachusetts, 1780–1860* (Ithaca, NY: Cornell University Press, 1990); Alfred D. Chandler Jr., *The Visible Hand: The Managerial Revolution in American Business* (Cambridge, MA: Belknap Press, 1977); Oscar Handlin and Mary Flug Handlin, *Commonwealth: A Study of the Role of Government in the American Economy: Massachusetts, 1774–1861* (Cambridge: Belknap Press, 1969); Stanley I. Kutler, *Privilege and Creative Destruction: The Charles River Bridge Case* (1971; repr. Baltimore: Johns Hopkins University Press, 1990); Stanley Elkins and Eric McKitrick, *The Age of Federalism: The Early American Republic, 1788–1800* (New York: Oxford University Press, 1993); Drew McCoy, *The Elusive Republic, Political Economy in Jeffersonian America* (New York: W. W. Norton, 1980); Charles Sellers, *The Market Revolution: Jacksonian America, 1815–1846* (New York: Oxford University Press, 1991); Carl Siracusa, *A Mechanical People: Perceptions of the Industrial Order in Massachusetts, 1815–1880* (Middletown, CT: Wesleyan University Press, 1979); Steven Watts, *The Republic Reborn: War and the Making of Liberal America, 1790–1820* (Baltimore: Johns Hopkins University Press, 1987); Gordon S. Wood, *The Creation of the American Republic, 1776–1787* (Chapel Hill: University of North Carolina Press, 1969); and Olivier Zunz, *Making America Corporate, 1870–1920* (Chicago: University of Chicago Press, 1990).

Technology

Hotel Dreams is grounded in the literature of the history of technology. George Basalla's *The Evolution of Technology* (New York: Cambridge University Press, 1988) early on helped me to theorize evolutionary ideas about hotel development and the centrality of technology. David A. Hounshell's *From the American System to Mass Production, 1800–1932: The Development of Manufacturing Technology in the United States* (Baltimore: Johns Hopkins University Press, 1984) is one of the key books that traces the emergence of mass production techniques. Phil Scranton's *Proprietary Capitalism, The Textile Manufacture at Philadelphia, 1800–1885* (Philadelphia: Temple University Press, 1983) and *Endless Novelty* (Princeton: Princeton University Press, 1997) are necessary counterpoints to the hegemony of the mass production discourse. Two books that are foundational for those seeking to think about technology through the lens of American Studies are John F. Kasson, *Civilizing the Machine: Technology and Republican Values in America, 1776–1900* (New York: Penguin Books, 1988); and Leo Marx, *The Machine in the Garden: Technology and the Pastoral Ideal in America* (London: Oxford University Press, 1964). Kasson's treatment of the relationship between technology and republicanism is an important addition to the aforementioned works on political economy. Other key works in American Studies that focus on technology are Cecelia Tichi, *Shifting Gears: Technology, Literature, Culture in Modernist America* (Chapel Hill: University of North Carolina Press, 1987); and Martha Banta, *Taylored Lives: Narrative Productions in the Age of Taylor, Veblen, and Ford* (Chicago: University of Chicago Press, 1993). All of David Nye's books are important models for writing about technology, again through the lens of American Studies, foremost among them *American Technological Sublime* (Cambridge: MIT Press, 1994) and *Electrifying America: Social Meanings of a New Technology, 1880–1940* (Cambridge: MIT Press, 1992). All of Lewis Mumford's works are important contributions to thinking about cities, technology, and architecture; *Art and Technics* (New York: Columbia University Press, 1952), a series of lectures, proved especially valuable for this study. Finally, the essays in Merritt Roe Smith and Leo Marx, eds., *Does Technology Drive History?* (Cambridge: MIT Press, 1994) are insightful for understanding the way technology functions culturally.

Luxury

There are only a few books that address ideas about luxury. Of them, John Sekora's *Luxury: The Concept in Western Thought, Eden to Smollett* (Baltimore: Johns Hopkins University Press, 1977) is the most comprehensive and theoretical. Others include Christopher J. Berry, *The Idea of Luxury: A Conceptual and Historical Investigation* (Cambridge: Cambridge University Press, 1994); and Maxine Berg and Helen Clifford, eds., *Consumers and Luxury: Consumer Culture in Europe, 1650–1850* (Manchester: Manchester University Press, 1999).

Elites

Similarly, with histories of elite classes being out of fashion, recent years have seen few such works. The best sources are Digby E. Baltzell, *Philadelphia Gentlemen: The Making of a National Upper Class* (1958; repr. Philadelphia: University of Pennsylvania Press, 1979); Digby E. Baltzell, *Puritan Boston and Quaker Philadelphia: Two Protestant Ethics and the Spirit of Class Authority and Leadership* (New York: Free Press, 1979); Peter Dobkin Hall, *The Organization of American Culture, 1700–1900: Private Institutions, Elites, and the Origins of American Nationality* (New York: New York University Press, 1984); Frederic Cople Jaher, *The Urban Establishment: Upper Strata in Boston, New York, Charleston, Chicago, and Los Angeles* (Urbana: University of Illinois Press, 1982); Ronald Story, *The Forging of an Aristocracy: Harvard and the Boston Upper Class, 1800–1879* (Middletown, CT: Wesleyan University Press, 1980); Tamara Plakins Thornton, *Cultivating Gentlemen: The Meaning of Country Life among the Boston Elite, 1785–1860* (New Haven: Yale University Press, 1989); Robert F. Dalzell Jr., *Enterprising Elites: The Boston Associates and the World They Made* (Cambridge: Harvard University Press, 1987); and the newest contribution, Sven Beckert, *The Monied Metropolis: New York City and the Consolidation of the American Bourgeoisie, 1850–1896* (New York: Cambridge University Press, 2001).

The Middle Class

One of the key features of the long nineteenth century is the emergence of the middle class. Many of the excellent monographs written about this were instrumental in helping to understand the social and cultural issues that defined the controversies surrounding hotel life. Among the best and most important are Elaine S. Abelson, *When Ladies Go A-Thieving: Middle-Class Shoplifters in the Victorian Department Store* (New York: Oxford University Press, 1989); Stuart M. Blumin, *The Emergence of the Middle Class: Social Experience in the American City, 1760–1900* (Cambridge: Cambridge University Press, 1989); Ann Fabian, *Card Sharps, Dream Books, and Bucket Shops* (Ithaca: Cornell University Press, 1990); Karen Halttunen, *Confidence Men and Painted Women: A Study of Middle-Class Culture in America, 1830–1870* (New Haven: Yale University Press, 1982); John F. Kasson, *Rudeness and Civility: Manners in Nineteenth-Century Urban America* (New York: Hill and Wang, 1990); Harvey Levenstein, *Revolution at the Table: The Transformation of the American Diet* (New York: Oxford University Press, 1988); and Mary Ryan, *Cradle of the Middle Class: The Family in Oneida County, New York, 1790–1865* (New York: Cambridge University Press, 1983).

Cultural History

As much as *Hotel Dreams* is grounded in the history of technology, it is equally a cultural history. Important books that provide theoretical insight to cultural history include James W. Cook, Lawrence B. Glickman, Michael O'Malley, eds., *The Cultural Turn in U. S. History: Past, Present, Future* (Chicago: University of Chicago Press, 2008); Neil Harris, *Cultural Excursions* (Chicago: University of Chicago Press, 1990); Warren Susman, *Culture as History: The Transformation of American Society in the Twentieth Century* (New York: Pantheon Books, 1984); Miles Orvell, *The Real Thing, Imitation and Authenticity in American Culture, 1880–1940* (Chapel Hill: University of North Carolina Press, 1989); Alan Trachtenberg, *The Incorporation of America: Culture and Society in the Gilded Age* (New York: Hill and Wang, 1982); and Richard Wightman Fox and T. J. Jackson Lears, *The Power of Culture: Critical Essays in American History* (Chicago: University of Chicago Press, 1993).

Consumer Culture

Books on consumer culture lie at the intersection of cultural, economic, class, and gender history. Excellent examples include Susan Porter Benson, *Counter Cultures: Saleswomen, Managers, and Customers in American Department Stores, 1890–1940* (Urbana: University of Illinois Press, 1988); John Brewer and Roy Porter, eds., *Consumption and the World of Goods* (London: Routledge, 1993); Simon J. Bronner, *Consuming Visions: Accumulation and Display of Goods in America, 1880–1920* (New York: W. W. Norton, 1989); Pierre Bourdieu, *Distinction: A Social Critique of the Judgement of Taste* (1979; trans. Cambridge: Harvard University Press, 1984); T. H. Breen, *Marketplace of Revolution: How Consumer Politics Shaped American Independence* (New York: Oxford University Press, 2004); Victoria de Grazia and Ellen Furlough, eds., *The Sex of Things: Gender and Consumption in Historical Perspective* (Berkeley: University of California Press, 1996); Stuart Ewen, *All Consuming Images: The Politics of Style in*

Contemporary Culture (New York: Basic Books, 1988); William Leach, *Land of Desire: Merchants, Power, and the Rise of a New American Culture* (New York: Pantheon Books, 1993); Jackson Lears, *Fables of Abundance: A Cultural History of Advertising in America* (New York: Basic Books, 1994); Roland Marchand, *Advertising the American Dream: Making Way for Modernity 1920–1940* (Berkeley: University of California Press, 1985); and Susan Strasser, *Satisfaction Guaranteed: The Making of the American Mass Market* (Washington, DC: Smithsonian Institution Press, 1989).

Gender

The books listed below focus on gender analysis, but any of them could easily be included in some of the categories above, such as political economy, consumer culture, architecture, and cultural geography. Seminal books on the women's sphere include Carroll Smith-Rosenberg, *Disorderly Conduct: Visions of Gender in Victorian America* (New York: Oxford University Press, 1985); Nancy F. Cott, *The Bonds of Womanhood: "Woman's Sphere" in New England, 1780–1835* (New Haven: Yale University Press, 1977); and Gillian Brown, *Domestic Individualism: Imagining Self in Nineteenth-Century America* (Berkeley: University of California Press, 1990). Key books that look at gender and the built environment include Delores Hayden, *Grand Domestic Revolution: A History of Feminist Designs for American Homes, Neighborhoods, and Cities* (Cambridge: MIT Press, 1982); Gillian Rose, *Feminism and Geography: The Limits of Geographical Knowledge* (Minneapolis: University of Minnesota Press, 1993); Daphne Spain, *Gendered Spaces* (Chapel Hill: University of North Carolina Press, 1992); Leslie Kanes Weisman, *Discrimination by Design: A Feminist Critique of the Man-Made Environment* (Urbana: University of Illinois Press, 1993); and Diana Agrest, Patricia Conway, and Leslie Kanes Weisman, *The Sex of Architecture* (New York: Harry N. Abrams, 1996). Works on masculinity are equally important for understanding the luxury hotel. These include Gail Bederman, *Manliness and Civilization: A Cultural History of Gender and Race in the United States, 1880–1917* (Chicago: University of Chicago Press, 1995); Mark C. Carnes, *Secret Ritual and Manhood in Victorian America* (New Haven: Yale University Press, 1989); Mark C. Carnes and Clyde Griffen, *Meanings for Manhood: Constructions of Masculinity in Victorian America* (Chicago: Chicago University Press, 1990); and E. Anthony Rotundo, *American Manhood: Transformations in Masculinity from the Revolution to the Modern Era* (New York: Basic Books, 1993).

INDEX

Page numbers in boldface indicate figures.